Theory and Management of Tropical Fisheries

Theory and Management of Tropical Fisheries

Proceedings of the ICLARM/CSIRO Workshop on the Theory and Management of Tropical Multispecies Stocks, 12-21 January 1981, Cronulla, Australia

Edited by

D. Pauly and G.I. Murphy

1982

INTERNATIONAL CENTER FOR LIVING AQUATIC RESOURCES MANAGEMENT
COMMONWEALTH SCIENTIFIC AND INDUSTRIAL RESEARCH ORGANISATION

Theory and Management of Tropical Fisheries

Proceedings of the ICLARM/CSIRO Workshop on the Theory and Management of Tropical Multispecies Stocks, 12-21 January 1981, Cronulla, Australia

EDITED BY

D. PAULY
G.I. MURPHY

1982

Printed in Manila, Philippines

Published by the International Center for Living Aquatic Resources Management, MCC P.O. Box 1501, Makati, Metro Manila, Philippines and Division of Fisheries Research, Commonwealth Scientific and Industrial Research Organisation, Cronulla, Australia

Pauly, D. and G.I. Murphy, Editors. 1982. Theory and management of tropical fisheries. ICLARM Conference Proceedings 9, 360 p. International Center for Living Aquatic Resources Management, Manila, Philippines and Division of Fisheries Research, Commonwealth Scientific and Industrial Research Organisation, Cronulla, Australia.

ISSN 0115-4389
ISBN 971-04-0021-5 cloth
ISBN 971-04-0022-3 paper

Cover: Trawlers laid up, central Thailand. Photo by R.S.V. Pullin.

ICLARM Contribution No. 105

Preface

The workshop of which we present the proceedings was, we believe, the first conference devoted explicitly to the theory and management of tropical fish stocks. The time was ripe for such a conference as fishery development in the 1960s and 1970s generated a number of questions which scientists now have to answer for the continued exploitation of these resources to be put on a rational basis.

The conference and the subject matter of the invited papers were organized in such a way that all aspects of the problems could be covered, hopefully resulting in proceedings that represent the state of the art. However, as readers will notice, we failed to reach a consensus on a number of important matters. The main reason for this is probably that fishery biology is not a "mature" science (in the Kuhnian sense), as evidenced by the fact that we are still wrestling with issues raised at the beginning of this century.

In the tropics, this situation is worsened by factors which in a number of instances have prevented even the right questions being asked – not to mention the lack of data from which answers could be derived.

In this book, some of the right questions are asked and some tentative answers are given. The areas where we have disagreed among ourselves might suggest fruitful research topics.

The initiatives to hold this workshop began in 1978 with John C. Marr, then Director General of ICLARM, who approached the Australian Government to sponsor a workshop on the subject. In 1980, a grant was awarded to ICLARM by the Australian Development Assistance Bureau (ADAB) to hold, in conjunction with the Commonwealth Scientific and Industrial Research Organisation (CSIRO), a workshop on the "Theory and Management of Tropical Multispecies Stocks."

The workshop was held on 12-21 January 1981 at Cronulla, Australia. We take this opportunity to express our appreciation to ADAB.

D. Pauly
ICLARM

G.I. Murphy
CSIRO

October 1982

Contents

Introduction

P. A. LARKIN
(Workshop Chairman)
University of British Columbia
Vancouver, Canada

In the management of fisheries, the assessment of the effects of harvesting of the fish stocks is a matter of prime importance. Accordingly, there is a large scientific literature on the techniques of assessment. For the most part, these techniques have been developed for single species. The possible interactions with other species are commonly set to one side, to be involved as contributing factors if simpler explanations will not suffice.

To some extent this is a practical approach. Where the species of interest are ecologically dominant and where a fishery is only moderately intense, the interactions with other species may well have a relatively minor influence on the validity of stock assessments. However, in fisheries in many parts of the world, sharp changes in the relative abundance of the species of fish in a community have been associated with intensive fishing. The first few times this happened there were extensive debates about whether fishing or some changes in environmental factors (or both) were involved. To be sure, there are natural fluctuations in abundance of all species, but the association of heavy rates of harvesting and marked changes in species composition is now too well documented to be viewed as coincidence. Stock assessments of the future must be more cognizant of species interactions, and must be augmented by assessments of the harvested community.

These developments in the theory of fisheries science are highly relevant to the future prospects for the multispecies fisheries of the tropics. The great variety of species is well documented. A single trawl haul may contain as many as 100 species. Coral reefs are spectacular for their mix of colorful species living side by side. In the circumstances of rapidly expanding fisheries, there should be strong expectations for dramatic fluctuations in the species composition in such associations; indeed, there are already signs that they are occurring. The demersal fisheries of the Gulf of Thailand are a case in point and there are many more that, though less well documented, are no less indicative of change.

The theory concerned with explaining these situations is not well developed, perhaps being better described as "artless" than as at a "state of the art." The literature of theoretical ecology has advanced considerably in the past two decades, prompted in part by almost world-wide concern in the fields of agriculture, forestry and disease control. Despite the wide applicability of the theory, it is still characterized as being far richer in ideas than veracity. The newer perceptions of stability, diversity, resilience, transfer efficiency, competitive coexistence, predator-prey relations, optimal strategies, etc. are still far from routine application to the diagnosis of what has gone on, and most to the point, what could or should go on.

The fisheries literature is no exception to these generalities. A variety of attempts has been made to model associations of two or more species, but the first such effort, by Volterra in the 1920s, is still respectable as an approach today. The multiplicity of possible associations among species, coupled with the difficulty (and expense) of measuring all of the relevant parameters, has consistently prevented the usual progress of science through replacement of a body of theory with a newer and more powerful paradigm.

In these several circumstances it is timely to direct attention to greater efforts to build understanding of multispecies relationships and to try to hasten application of this knowledge to management of fisheries in all parts of the world. To this end several workshops have been held in the past few years, and it is to this pattern of activity that the ICLARM/CSIRO workshop was aimed as a contribution.

The objectives of the workshop were designed to place emphasis on applications. It was perceived that the fisheries are at present so intense and the available information so limited, that a special effort should be made to develop from present understandings a set of tools for rough and ready action, now. Awaiting the development of much more sophisticated understanding before making commitments to particular approaches was not seen as a prudent course of action.

Specifically, the objectives of the workshop were:

- To review models that are presently in use, or that have been proposed for stock assessment in the tropics.
- To define the data requirements of various stock assessment and management methods.
- To review current research on stock assessment in tropical waters.
- To identify the major constraints for stock assessment and management in tropical waters, and to formulate a feasible plan of action to overcome these constraints.

The target area for recommendations was Southeast Asia, although much of what was discussed has world-wide applicability. The tunas and the related pelagic community were specifically excluded from the discussions because they pose a rather special set of questions concerning migration which is atypical. Otherwise, the workshop was not limited either by species or habitat. Attention was given to freshwater, estuarial and marine fisheries.

The papers that were prepared in advance of the meeting collectively comprise a valuable review of the dimensions of the problems that are posed in scientific approaches to multispecies fisheries management and a review of current research. The first paper by Simpson reviews the nature of the data base in Southeast Asia, indicating the lack in some areas of statistics for use in even the simplest techniques of stock assessment. The paper of Pauly indicates the range of techniques available for studying single-species dynamics in the multispecies context, and is followed by that of Munro which indicates some of the practical problems of estimating biological and fishery parameters in coral reef fisheries.

Kirkwood reviews the simpler models for multispecies assessment. James and Stark present an example of a complex model of an inshore marine ecosystem. Larkin and Gazey question the present utility of complex models and present, as an alternative, an example of a simple model that might be useful for testing the central question of the relative importance of biological interactions in the observed changes in the Gulf of Thailand. Murphy indicates that present data on stock and recruitment relations of tropical fishes may be best interpreted as reflecting *intra* rather than *inter* specific interactions, but the data are characterized by great variability, some of which reflect environmental factors. Majkowski reviews the important role of sensitivity analysis in directing attention to those parts of a system that are most critical to its behavior, an important technical consideration in determining priorities for research and sampling programs.

The ecological basis for tropical demersal fisheries is reviewed by Sainsbury, who concludes that there is no adequate model of community dynamics and no adequate method of estimating the parameters of the models that are available. Jones extracts several broad conclusions from the marine and freshwater literature on ecosystems, food chains and fish yields, and presents their implications for management. The structure and dynamics of coral reef fish communities are described by Sale, who also draws several implications for management. Marten and Polovina compare fish yields from a variety of tropical ecosystems and place strong emphasis on habitat considerations in management.

The measures that are taken to manage tropical multispecies fisheries in different parts of the world are reviewed by Gulland, who contrasts the pattern of management that is supposed to occur in advanced countries with what happens in practice in developing countries. Marr describes the realities of contemporary fisheries management in Southeast Asia, which must be considered in any plans for action.

Taken together, these papers strongly indicate the difficulty of the scientific problems, the paucity of information, and the urgency for rapid advances in understanding coupled with immediate development of simple and effective tools to guide corrective measures. It is to these themes that the discussions of the working groups converged, leading to the recommendations which conclude this report.

It was appreciated in these discussions that, as Marr (this vol.) points out, fisheries is only one element in a much broader social and economic context. The solution of what may be seen as problems in fisheries may well lie in other spheres of governmental action. However, the aim of this meeting and the focus of these discussions was what could be done within the fisheries sector to further the causes of understanding and rational use of fisheries resources. That was enough for one workshop.

A Review of the Database on Tropical Multispecies Stocks in the Southeast Asian Region

ARTHUR C. SIMPSON
106 Fambridge Road
Maldon, Essex CM9 6BG, England

SIMPSON, A.C. 1982. A review of the database on tropical multispecies stocks in the Southeast Asian region, p. 5-32. *In* Pauly, D. and G.I. Murphy (eds.) Theory and management of tropical fisheries. ICLARM Conference Proceedings 9, 360 p. International Center for Living Aquatic Resources Management, Manila, Philippines and Division of Fisheries Research, Commonwealth Scientific and Industrial Research Organisation, Cronulla, Australia.

Abstract

This review covers data available in Indonesia, Malaysia, the Philippines and Thailand. These countries have had national fisheries statistics for many years and while those for Thailand and Peninsular Malaysia are now fairly reliable and complete, the extensive artisanal fishing in the Philippines and Indonesian archipelagos has made data collection in these countries very difficult and some of it is less reliable. Only Thailand has good effort data. Indonesia, Malaysia and Thailand have continuing programs of demersal fish surveys which are producing a great deal of useful data. Information on the biology of exploited species is however very limited. All Malaysian and Philippine reports are in English, while some reports published in Indonesia and most in Thailand are not. A number of these reports would merit translation. There is much unpublished data, especially in the Philippines.

Introduction

The study of the interactions between the elements of multispecies stocks, both in temperate and tropical waters resulted from the increasing need for precision in forecasting the wider effects of management measures. To obtain such precision, however, it is necessary to take into account the effects of management measures on one fishery on the abundance and population structure of the fishes in other stocks of an area.

This is particularly true in tropical fisheries where, among other features, the boundaries between demersal and pelagic species are blurred and where large quantities of pelagic species (e.g., chub mackerels and barracuda) are caught by the trawl fishery.

Coral reef areas and other rough, untrawlable grounds are widespread in the tropics. The number of species inhabiting them is very great, there is no clear boundary between demersal and pelagic species, there is probably a great deal of interspecific competition and many of the larger species are predators on the smaller ones. A multispecies approach is essential in these areas.

The fisheries for the many smaller pelagic species in Southeast Asia (e.g., *Rastrelliger* spp., *Decapterus* spp., *Selar* spp., *Sardinella* spp., *Stolephorus* spp., etc.) rarely exploit only one species at a time and the target species vary seasonally and between countries, due to differing market preferences.

Thus, there are few fisheries in the area that can be treated as single species fisheries without taking the risk of making serious errors.

This review is limited to the data from the four major fishing countries of Southeast Asia—Indonesia, Malaysia (including Sabah and Sarawak), the Philippines and Thailand. These four countries exploit what is probably the richest and most extensive area of tropical fisheries in the world. Currently the total landings of marine fish and shellfish from these four countries exceed 5,000,000 t and are continuing to rise (Fig. 1).

Vietnam and Kampuchea fish in the Gulf of Thailand and on the Sunda Shelf. Vietnam is now redeveloping its previous substantial fishing industry; however, few data are available on fisheries or current research. Fishing vessels from Korea, Taiwan (China) and Japan also fish these waters and their catches are recorded in their national statistics, but are not included here.

The data used for stock assessment are of three types:

—fisheries statistics (catch and effort data)
—research surveys data
—biological studies

These three types of data are discussed separately further below.

Fish has been a very important source of protein to the coastal populations of all the countries of Southeast Asia for many centuries. These highly populated countries have had fisheries for very many generations and the

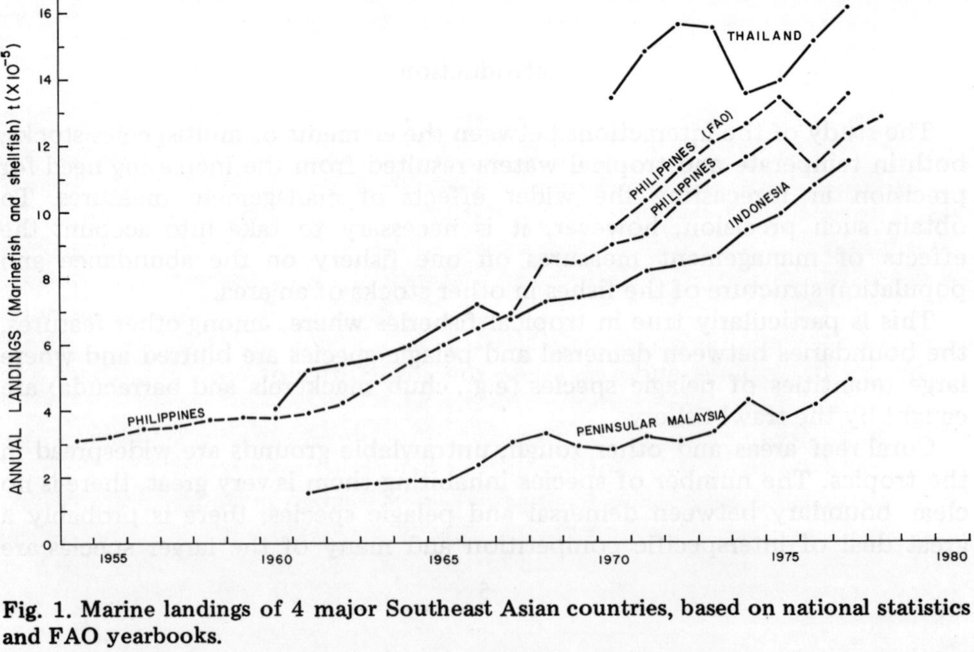

Fig. 1. Marine landings of 4 major Southeast Asian countries, based on national statistics and FAO yearbooks.

numbers of people involved in artisanal fishing and small-scale commercial fishing have been, and still are, large. There is little doubt therefore that fish stocks that come within sight of the coast and considerably further in some areas, have been exploited on a substantial scale long before fisheries statistics or quantitative fisheries research started. Thus, it is doubtful whether any stocks in the area should be considered to have been "virgin" when modern methods of fishing were introduced and scientists began to study the resources.

Highly skilled artisanal fishing has been particularly important in the archipelagos of Indonesia and the Philippines where there are thousands of coastal villages each with a number of fishing boats and also many fishermen without boats using tidal and demersal traps and barriers. As described in greater detail further below, the collection of catch and effort statistics from these fisheries is a difficult task which has been tackled in recent years only.

All four countries covered here are subject to the monsoon winds which make substantial seasonal changes in fishing activities and also in the movements of the fish. In addition the monsoons strongly influence the direction and strength of currents, the upwelling and the surface water temperatures (Wyrtki 1961; Weber 1976).

The extent of these effects varies greatly from area to area, depending on latitude and the configuration of the land masses. On the Pacific coast of the Philippines, for example, fishing is very greatly reduced and stopped in some areas during the five months November through March, due to onshore winds and heavy seas. In the Java Sea, off the north Java coast the trawl catch rates following the southeast monsoon are only about half those following the northwest monsoon (Dwiponggo and Badrudin 1979a). Such seasonal variations must be considered by the users of any of the fishery discussed in this review.

Fisheries Statistics (Catch and Landings)

National and Other Statistics

All countries around the South China Sea collect data on the commercial landings of fish and on the fishing boats and publish national annual reports. Also, they submit data to FAO Headquarters for publication in the annual Yearbooks of Fisheries Statistics. However, the number of years during which these data have been collected and published, the species groups adopted, data collected on fishing effort etc., vary very much from country to country, as do the completeness and accuracy of the data.

In the Philippines and Indonesia the difficulties of collecting these data are immense due to these countries being archipelagos and more than 50% of the fishing being artisanal and done from many hundreds of villages on the large numbers of islands. Both countries have been re-organizing their collection of statistics since 1976 to improve the information on the artisanal fishing. Both have met many problems, especially the appointment and training of adequate staff and the regular flow of sufficient money to carry out their plans on a long-term basis. In consequence, statistics from these two countries have fluctuated in completeness and accuracy during recent years.

Below are the known, current, annual publications which include catch and effort data of the commercial fishing vessels in the area.

Indonesia
"Statistik perikanan–Fisheries Statistics of Indonesia" (Published by the Directorate General of Fisheries, Jakarta, Indonesia). First published as a yearbook in 1972, but catch data have been available since the data for 1940 and 1950-53 were published in 1954. The yearbook was enlarged to include more data from 1978. All titles, captions, etc. are in Indonesian and English.

Malaysia, peninsular
"Perangkaan Tanunan Perikanan–Annual Fisheries Statistics–Malaysia" (Published by the Ministry of Agriculture and Rural Development, Fisheries Division, Kuala Lumpur, Malaysia). Includes summary tables for Sabah and Sarawak. First published in the early 60s in present form. All titles, captions, etc. are in Malay and English.

Malaysia, Sabah
"Laporan Tahunan bagi Jabatan Perikanan Sabah–Annual Report of the Department of Fisheries, Sabah" (Published by the Department of Fisheries, Kota Kinabalu, Sabah) First published in the early 60s. Report, captions, etc. are in English.

Malaysia, Sarawak
"Annual report of the Department of Fisheries, Sarawak." (Published by the Department of Fisheries, Kuching, Sarawak) Started in the late 60s. In English.

Philippines
"Fisheries Statistics of the Philippines" (Published by the Bureau of Fisheries and Aquatic Resources, Manila, Philippines) Started before 1954. In English.

Thailand
—"The Marine Fisheries Statistics, based on the sample survey" First published in 1969. In Thai and English.
—"Fisheries Record of Thailand" First published in 1968. In Thai.
—"Fishing vessel statistics" First published in 1971/72. (Published by the Department of Fisheries, Ministry of Agriculture and Co-operatives, Bangkok, Thailand).
There may be other annual publications of fisheries statistics in Thai, not known to the author.

Southeast Asian Fisheries Development Center (SEAFDEC)
"Fisheries Statistical Bulletin for the South China Sea Area" (Published by Southeast Asian Fisheries Development Center, Phrapradaeng, Samutprakarn, Thailand) First published in 1978 for the year 1976.

Food and Agriculture Organization of the United Nations (FAO)
"Yearbook of Fishery Statistics—catches and landings" (Published by FAO, Rome) First published in 1948 for the year 1947.

With the improvements in the collection and processing of fisheries statistics in each country, the national publications have changed to different degrees since they started publication. Major changes occurred in the statistics of Thailand with the introduction of catch sampling in the mid-70s and in the Philippines with the start of the collection of detailed data from the artisanal fisheries in 1976.

In most countries the FAO statistics of landings have closely mirrored the catch data in national annual statistical tables. However, there are two exceptions to note:

—The Indonesian total catch figures are the same in the national fisheries statistics and the FAO yearbook, but the species breakdown is much more detailed in the FAO yearbook (40 groups of marine fishes compared with only 17 in the national publication).

—In the Philippines before 1976 information on species composition of catches were only obtained for the "commercial" fishery (vessels over 3 t) which contributed about 40% of the marine catch. A species breakdown of this 40% into some 67 categories was given in the national statistics. An estimated total for the "municipal" catch (from vessels under 3 t) was given, but no species breakdown, as none existed for this 60% of the total catch. However, while the FAO tables give total national catches closely similar to the total (commercial + municipal) figure given in the national publication, the FAO species breakdown adds up to the same total, indicating that the species composition of the municipal catch had been estimated and added to the known commercial catches. Therefore these figures of species composition of the Philippines in the FAO yearbook prior to 1975 can not be taken as very reliable. No check has been made on the more recent figures.

The main differences in coverage between the FAO, SEAFDEC and national tables are:

FAO yearbook	:	National totals by species groups and FAO major fishing areas.
SEAFDEC	:	National totals for Hongkong, Thailand, Malaysia (Peninsular, Sarawak, Sabah), Indonesia, Philippines, Brunei. Breakdowns to species and species groups and with some countries, to sub-national areas. Some data on fishing effort. Further breakdown is under discussion.
National	:	Very variable from country to country. Usually some further breakdown than SEAFDEC in fishing areas, by fishing gears and by months. Some data on fishing effort.

Species and Species Group Reporting in National and FAO Statistics

The majority of commercially exploited fish species are present throughout the area covered by this review, though they vary in relative abundance and in relative commercial importance among the countries. The reporting of the landings by species, species groups or by families therefore varies greatly among countries and there are also surprisingly great differences between the groupings used in national statistics compared with those in the FAO yearbooks.

Taking the overall breakdown first, the number of categories is usually higher in the national statistics than in the FAO yearbook, with the exception of Indonesia, where the national statistics give a very limited breakdown (see Table 1).

There are extremely few species that are reported from most countries. Sabah even does not report any species separately. Table 2 records all the single species that are reported separately in national or FAO yearbooks.

Of the 50 species reported by one country or another or FAO, only one, *Megalaspis cordyla* (Carangidae), the hard-tail scad is reported separately by the three adjacent countries Thailand, Peninsular Malaysia and Indonesia (see Table 2).

With the large number of species being caught (especially with trawls) and the large number of mixed small species and juveniles of larger species, there is a difficult problem in the collection of statistics and this problem is solved differently in different countries. In the Philippines nearly all fishes are classified into one or another of some 65 named categories so that in 1974, for example, only 1.4% were left for a "miscellaneous" group in the national statistics, while figures submitted to FAO had no miscellaneous group. In other countries the named categories are less comprehensive and there are large "miscellaneous groups", reaching 61% of the total for Thailand (Table 3).

All countries collect their fisheries statistics through administrative divisions and all but Sabah publish landings by species and species groups for each administrative unit. The Philippines also gives the fishing grounds (51 areas) on which the fish were caught and the landing places (231) grouped by regions (13).

The numbers of areas covered in each country are summarized in Table 4.

Thai landings on both east and west coasts include substantial catches made by large trawlers fishing distant waters. Malaysian catches are largely from grounds close to the landings port; however, some boats based on west coast ports fish grounds off different states from those at which the fish are landed. Some Philippine data on catches from fishing grounds are inaccurate due to the difficulties in determining the origin of fish landed by carrier boats

Table 1. Comparison of the numbers of species and species groups reported separately in national and FAO annual statistics for 1977.

	Marine fish		Cephalopods		Other molluscs		Crustaceans	
	Nat.	FAO	Nat.	FAO	Nat.	FAO	Nat.	FAO
Indonesia	17	47	1	3	0	5	1	4
Malaysia								
Peninsular	51	44	1	1	2	2	3	2
Sabah	10	10	—	—	1	1	1	1
Sarawak	49	34	1	1	5	1	6	4
Philippines	67	60	1	2	0	10	3	8
Thailand	32	32	3	3	7	6	10	11

that have collected fish at sea in the south and landed them at Manila or another large city. Catches of shrimp in Sabah are given by 20 x 20 mile statistical squares and are the only landings in these countries reported in this fashion.

Data on Fishing Effort

A large amount of information is collected by regional fisheries staff in each country on the numbers of fishing boats, their sizes and the types of gear used. However not all the information collected is analyzed and published. Increasingly, these countries are improving the information collected on fishing boats and their gear, and they are starting to collect data from each boat on its landings and the amount of fishing that has produced it. Thailand is very much more advanced in this respect than any of the other countries in the area.

In the Philippine and Indonesian archipelagos, the thousands of small vessels scattered around the islands make the collection of effort data a major problem. With the large number of species, fishing boats and landing places, the amount of data collected is so great and the analysis so laborious that computerization is becoming essential. This is being started, but is running into serious difficulties with funding and training.

A summary of the information related to fishing effort published in each country follows, but it may be slightly out of date as the most recent tables have not been seen by the author.

Indonesia: Since 1973 the information on effort has been greatly increased and there are now annual data on catches in weight by species/species groups by gears (12) for each province (26). The data on gears are the numbers of vessels using different gears landing in each province.

Malaysia, peninsular: For many years the catches by species/species groups (51) have been given for (11) different gears for each state (12). Also the number of fishing gears licensed and in operation in each state is tabulated separately. There are considerable incompatibilities between the numbers in these two tables and there is a good deal of discussion on the usefulness of these figures.

Sabah: The number of licenses for different gears (8) is given for each district (13), but the only catch data corresponding to these measures of effort are estimated total landings of fish and shellfish in each district.

Sarawak: Data are given on total catch by gears (11) and separately on total catch divided into species/species groups.

Philippines: Since 1977 there has been additional information collected on fishing effort in general and for the first time on the "municipal" (artisanal) boats (under 3 t), but this has not yet entered regularly into the annual statistics. The regular data, relating only to vessels of 3 t and over, are as follows: The catches by species/species groups are given for different gears (9) for the country as a whole and the number of fishing vessels, divided into 7 tonnage groups, are given with the gear they use, for the country as a whole. In addition the catch made on each fishing ground (51) by each kind of fishing gear is tabulated.

Thailand: Since the introduction of the sample survey in 1969, the coverage of catch and effort data has greatly increased. For the major fisheries (otter

Table 2. Showing where catches of separate species are recorded in national statistics (o) and/or in FAO yearbooks (x).

ISSCAAP group	Species	Indonesia Nat.	Indonesia FAO	Malaysia Peninsular Nat.	Malaysia Peninsular FAO	Malaysia Sarawak Nat.	Malaysia Sarawak FAO	Philippines Nat.	Philippines FAO	Thailand Nat.	Thailand FAO
24	*Chanos chanos*		x					o	x		
	Anodontostoma chacunda			o	x	o	x		x		
	Hilsa toli		x								
	Pellona ditchela				x		x				
25	*Lates calcarifer*		x								
31	*Psettodes erumei*		x								x
33	*Harpadon nehereus*		x				x				
	Arius leiotetocephalus					o					
	Muraenesox cinereus				x		x				x
	Lactarius lactarius			o	x	o	x		x	o	x
	Caesio erythrogaster			o							
	Lutjanus argentimaculatus			o	x						
	Pomadasys hasta				x						
	Plectorhynchus pictus			o							
	Pristipomoides typus			o							
	Priacanthus tayenus									o	x
	Drepane punctata			o	x	o	x		x		
	Mene maculata							o	x		
	Platax orbicularis								x		
34	*Elagatis bipinnulatus*		x	o				o	x		
	Formio niger		x	o		o	x			o	x
	Pampus argenteus		x	o		o	x			o	x
	Pampus chinensis			o		o					
	Megalaspis cordyla		x	o	x			o	x	o	x
	Decapterus russelli			o	x	o	x		x		
	Alectis indicus					o					
	Selaroides leptolepis				x						
	Selar crumenophthalmus							o	x		
	Rachycentron canadum			o					x		
	Scomberoides commersonianus							o			
	Coryphaena hippurus								x		
	Polynemus sextarius					o					
35	*Sardinella longiceps*	o	x								
	Sardinella fimbriata		x								
	Hilsa macrura			o		o					
	Dussumieria acuta		x						x		
	Setipinna taty					o					
	Chirocentrus dorab			o		o				o	x
	Megalops cyprinoides							o	x		
36	*Scomberomorus commerson*		x								
	Scomberomorus guttatus		x								
	Euthynnus affinis		x		x		x		x		
	Katsuwonus pelamis		x						x		
	Thunnus albacares							o	x		
	Auxis thazard							o	x		
	Makaira nigricans				x						
	Xiphias gladius								x		
37	*Trichiurus lepturus*				x		x			o	x
	Scomber japonicus								x		
	Rastrelliger brachysoma								x		
	Rastrelliger kanagurta					o			x	o	x
	Rastrelliger neglectus									o	x

trawl, pair trawl, beam trawl, Thai seine, Chinese purse seine, anchovy purse seine, mackerel gill-net and bamboo stake net) data are given on catch by species/ species groups and fishing effort for each of 10 regions. The fishing effort is given in number of trips, days fishing and number of hauls for all gear and the number of hours fishing is also given for otter trawling. The data on the trawl fisheries are given separately for vessels of different size groups, three groups for pair trawling and five groups for otter trawling.

The one major shortcoming at present is that the landings of the large trawlers fishing outside Thai waters are included in the data for the region in which they land their catch.

Table 3. Proportion of the catch of marine fish classified as "miscellaneous" (from the FAO Yearbook for 1977).

Country	Total marine fish (t)	Miscellaneous marine fish (t)	% miscellaneous
Indonesia	1,021,350	146,357	14
Pen. Malaysia	378,164	181,568	48
Sabah	27,829	1,462	5
Sarawak	71,116	42,172	59
Philippines	1,325,231	0	0
Thailand	1,213,879	736,996	61

Survey Data

The review in this section of the data from surveys is largely concerned with trawl surveys, from which there is now a great quantity of data relevant to the understanding of tropical multispecies stocks. Surveys of the smaller pelagic species by acoustic methods have now been undertaken for a number of years with FAO/SCSP assistance by all the Southeast Asian countries. However, this research has experienced many difficulties, partly from the problems of interpretation of the results and partly from technical problems, including the training of electronics technicians to operate and maintain the equipment. Although a number of surveys have been made since they started in 1975, few results have yet been published. However, some preliminary results are available in the "Working Papers" and "Workshop Reports" series of publications of the FAO/South China Sea Programme (e.g., SCS 1976a, b, 1978a; Buzeta et al. 1979).

Early Exploratory Trawling

Some trawl surveys were carried out between the beginning of the century and the 1950s by the colonial powers. Most often, these were exploratory research trips to find out what species lived in the areas studied, rather than to start quantitative assessment of the resources. These research trips included trawling by the "Tongkol" in 1926/27 in Malaysian waters (Birtwistle and Green 1927, 1928) and by the "Manahine" off Peninsular Malaysia and Eastern Malaysia (Sabah and Sarawak) in 1955/56 (Ommanney 1961). Exploratory trawling was also done with American assistance in the Philippines (Warfel and Manacop 1950) and by Dutch biologists in Indonesian waters (van Rosendaal and van Kampen 1909).

Systematic Trawl Surveys

Systematic trawl surveys to assess the resources were started as a result of a German aid program in Thailand. The first survey was made in 1961 in the Gulf. Subsequent annual surveys, from 1963 onwards were made with the research vessel "Pramong II" and these have continued to the present day.

In 1967 a joint Malaysian/Thai/German trawl survey using the "Pramong II" was made along the east coast of Peninsular Malaysia and this was followed by a series of trawl surveys off the east and west coasts of Peninsular Malaysia and off Sarawak and Sabah by the Malaysian research vessel "Peneyelidek I" which was subsequently renamed the "K.K. Jenahak".

Table 4. Area breakdown in national catch statistics.

Country	Area units	No. of units	Remarks
Indonesia	Provinces	26	Species breakdown by provinces since 1973.
Pen. Malaysia	States on west coast	8	
	States on east coast	4	
Sabah	Districts	13	Estimated total fish landings.
Sarawak	Divisions	5	
Philippines	Fishing grounds	51	Data from commercial fishing vessels only until 1977.
	Landing places	231	
	Regions	13	
Thailand	Regions on west coast	3	Includes fish caught by trawlers fishing in distant waters.
	Regions on east coast	5	

Again arising from a German aid program, trawl surveys were started in Indonesian waters in 1974 and are continuing.

The Philippines has not yet initiated a program of systematic trawl surveys.

The areas covered by one or more of the standard trawl surveys by Thailand, Malaysia and Indonesia are shown in Fig. 2.

Table 5 gives an overview of systematic trawl surveys conducted in Southeast Asian waters since the early 60s.

In Thailand, Malaysia and Indonesia, the surveys have been made with the German "Engels trawl" and standard hauls of 1 hour have been adopted. A variety of station patterns have been used, most of which have not been planned to a definite environmental stratification.

The catches have been sorted into families or genera, or to species where these were of particular commercial value, very readily identified or the only species in the family. In most cases the separated groups have been counted and weighed and samples of selected species groups measured for length composition. Some other biological observations were also made.

Details of the fishing gear and vessels used in these surveys are given in Table 6.

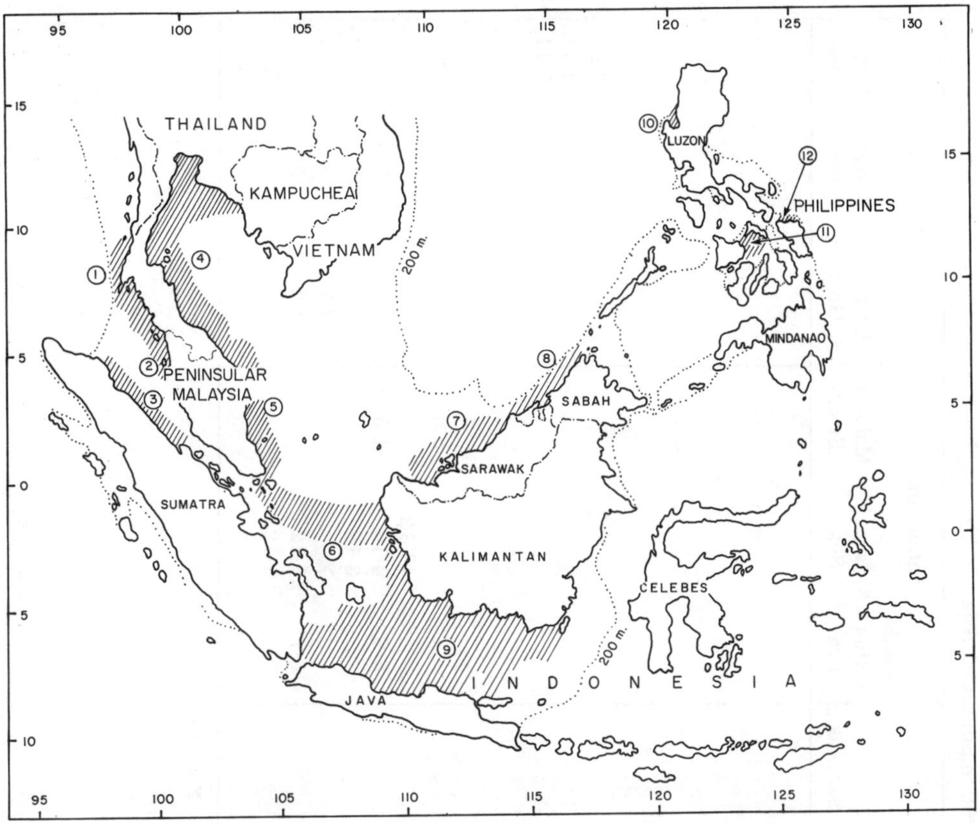

Fig. 2. Areas of Southeast Asia where trawling surveys have been conducted.

Notes on the Data Obtained by National Trawl Surveys

Indonesia. Comprehensive surveys for the assessment of resources have been made each year since November 1974. The areas of trawlable continental shelf (see Fig. 2) are large:

Malacca Strait	54,900 km^2
South China Sea	247,300 km^2
Java Sea	470,300 km^2
Total	772,500 km^2

The surveys, conducted from November 1974 to June 1976, were devoted to covering the whole western Indonesian waters area once. During the remainder of 1976, 1977 and 1978, the South China Sea and the Java Sea were again covered by comprehensive surveys. In 1979 the surveys covered the coastal waters all around the Java Sea. Also test fishing was carried out to assess commercial catch rates.

All trawl hauls in the surveys were of about 1 hour and were made only during daylight.

At each station the catch was sorted and weighed and identified into species (118) and species groups (30), invertebrates (8) and trash fish. Trash fish averaged 3 to 4% by weight of the catch. Length measurements were

Table 5. Overview of major demersal trawl surveys conducted since 1960 in Southeast Asia.[a]

	Thailand			Malaysia				Indonesia			Philippines		
Survey year	West Coast F.D.[b]	East Coast E.U.[b]	Gulf of Thailand	Pen. West Coast	Pen. East Coast	Sabah	Sarawak	South China Sea	Java Sea	Malacca Strait	Lingayen Gulf	Visayan Sea	Samar Sea
1961			1-12										
1962													
1963			1-12										
1964			1-12										
1965		X	1-12										
1966	X	X	1-12										
1967	X	X	1-12		3-4								12
1968	X	X	1-12										1-11
1969	X	X	1-12										7-11
1970	X	X	1-12	12	8-10								
1971	X	X	1-12	1, 11-12	3-5								
1972		X	1-12		8-9	4-5	3-4	12					
1973		X	1-12	4			8-9			7-9			
1974		X	1-12	11-12	7-8				11-12				
1975		X	1-12				7-8	8-9	4-12	1-3		X	
1976		X	1-12						3-12			X	
1977		X	1-12						X		2-12	X	
1978		X	1-12					6-7	1-12		1-2	X	
1979		X	1-12						1-12				X
1980			1-12										X

[a]Numbers are used to indicate the survey months, where these are not known, X is used.
[b]F.D. = Fisheries Department; E.U. = Exploratory Unit.

made on many species and observations on maturity made and otoliths taken on some of the economically more important species.

Most of the published reports (see Table 7) give full details of all stations, the weights of each species or species group caught at each station, and also include brief summaries of the catch rates in different areas and preliminary comments on the results.

Some of the data from the surveys were summarized in the reports of workshops held in Oct/Nov. 1977 and in Dec. 1978 (Buzeta 1978; SCS 1978a; Buzeta et al. 1979).

Since December 1979 a joint, 3-year Indonesian/German/Australian project called JETINDOFISH has been making surveys of the demersal and pelagic fish resources along the Indian Ocean coasts of the Indonesian islands Sumatra to Timor, using acoustic methods and trawling. One survey was made in December 1979 and four in April, May and June 1980.

The main vessels used for trawling were "R/V Jurong" and "R/V Bawal Putih 2". The "Jurong" is of 360 GRT with a towing speed of 3.5 knots and used an Engel high opening bottom trawl with a 40-mm cod-end. The trawling is stratified at 0/15, 50/100, 100/150, 150/300 and over 200-m depths.

The survey data are being put on a computer in Jakarta and it is expected that the first reports will be published soon by the German Agency for Technical Cooperation (GTZ).

Table 6. Details of the research vessels and gear used in selected trawl surveys. All trawls are "Engels trawls".

Country and vessel	Gross tonnage	HP	Towing speed knots	Headline (m)	Bridles (m)	Cod-end mesh (mm)
Indonesia						
"R/V Mutiara IV"	102	266	2.8	36.7	18	40
Malaysia						
"R/V Jenahak"	85	325	2.8	39.3	18	40
Thailand						
"R/V Pramong II"	76	320	2.5	36	13	40

Table 7 gives a detailed overview of reports on trawl surveys conducted in Indonesia.

Malaysia. The fisheries laboratory at Penang, Peninsular Malaysia, has been responsible for the trawl surveys made off Peninsular Malaysia, Sarawak and Sabah. Therefore, methods used have been the same for all these areas. The results of the surveys are published in the "Fisheries Bulletins" of the laboratory. The areas, years and months of the surveys are shown in Fig. 2 and Table 8.

The surveys comprised a large number of 1 hour hauls distributed more or less uniformly over the area, but not on a set grid. Details of station positions are not given in tables but their positions are shown on maps. The weights of some 12 species, 24 species groups, 5 invertebrate groups and "trash" fish occurring in each haul are tabulated in the report of each survey. No information appears to have been kept on the numbers (to give average weights) or size composition of the species. The trash fish group includes commercial

Table 7. Overview of published results of surveys conducted in Indonesian waters.

Time of survey year(s)	months	Areas covered	Author(s) of report	Emphasis of report (besides catch per effort data)	LPPL[a] report no.
1972	12	South China Sea	Anon. (1972)	(In Indonesian)	—
1973	6-12	Malacca Strait	Martosubroto (1973)	(In Indonesian)	—
1974-75	11/74-12/75	Java Sea, Malacca Strait	Saeger et al. (1976)	Standing stocks and potential yield estimates, some environmental data, selection experiments.	1
1974-75	11/74-11/75	Western Indonesia, South China Sea	Widodo (1976)	No catch data, but checklist of fish caught.	1
1974-76	11/74-7/76	Java Sea, South China Sea	Martosubroto and Pauly (1976)	Length-frequency data, oceanographic observations.	2
1973-75	6-12/73-2/75	Malacca Strait	Sujastani et al. (1976)	Catch and effort from the fishery, stock assessment.	—
1974-76	11/74-7/76	Java Sea, South China Sea	Dwiponggo (1977)	Atlast with distributional maps of major resources.	—
1976	6-12	Java Sea	Losse and Dwiponggo (1977)	Deck operating procedures.	3
1976-77	7-12/77, 1/78	Java Sea	Beck and Sudrajat (1978)	Growth parameters of 3 fish species.	4
1978	6-7	South China Sea	Sudrajat and Beck (1978)	Trawling prospects	4
1977	5-9	Throughout Indonesia	Expl. Fish. Div. (1978)	Oceanography, use of non-trawl gears.	—
1977	1-12	Java Sea	Dwiponggo and Badrudin (1978a)	Summary of c/f data.	5
1977	1-12	Java Sea	Dwiponggo and Badrudin (1978b)	c/f and L/F data, per haul.	5A
1978	6-7	South China Sea	Sudrajat and Beck (1978)	Standing stocks and potential yield estimates.	4
1978	1-12	Java Sea	Dwiponggo and Badruddin (1979a)	Seasonality of c/f data.	6
1978	1-12	Java Sea	Dwiponggo and Badrudin (1979b)	c/f and L/F data, per haul.	6A
1979	1-12	Java Sea	Dwiponggo and Badrudin (1980a)	Summary of c/f data.	7
1979	1-12	Java Sea	Dwiponggo and Badrudin (1980b)	c/f data, per haul.	7A
1979	1-12	Java Sea	Dwiponggo and Badrudin (1980c)	L/F data, per haul.	7B
—	—	Java Sea	Buzeta et al. (1979)	Review of earlier work, stock assessment.	—
—	—	Java Sea	Dwiponggo (1979)	Review, including fishery data stock assessment.	—
—	—	Java Sea	Buzeta et al. (1979)	Review of earlier work, stock assessment.	—
—	—	Java Sea	Dwiponggo (1979)	Review, including fishery data stock assessment.	—

fish of unsaleable size and in the 1970/71 survey, for example, trash fish averaged 37% of the catches.

In June/July 1972 a separate survey was made of the shrimps along the west coast of Peninsular Malaysia using German prawn trawls with 13.3-m and 16.3-m headropes and 30-mm and 35-mm cod-end meshes, respectively. A total of 101 hauls was made. The report (Ong and Weber 1977) groups catch data by areas (4) and depth ranges (5), and gives the average weights (kg/hr) and numbers per kg of each species or species group (17), total food fish and total trash fish. Full data on the weights of species/species groups of shrimps and of food fish (19) for individual stations are given for only 12 stations in the most important shrimp area.

Philippines. The 7,000 islands of the Philippine archipelago are largely surrounded by deep seas. Shelf areas large enough to support fisheries for demersal species and smooth enough for trawling are limited and are broken up into several isolated grounds. The Visayan Sea supports the largest demersal stocks and yields over half of the recorded trawler catch. Manila Bay, Bohol Strait and the San Miguel Bay area are the only other substantial trawlable shelf areas, each yielding over 5,000 t per year (SCS 1976c, 1977, 1978b; Simpson 1979).

Exploratory trawling has been carried out by research vessels of the Bureau of Fisheries and Aquatic Resources (BFAR) in many of the trawlable areas of the Philippines over many years in a rather unorganized way and few of the results have gone further than internal reports; some of the published data are in brief reports of the meetings of the Indo-Pacific Fisheries Council. The most extensive sets of data are probably those in the reports of the two international research and development projects led by Tiews and Magnusson (see Table 9 for an overview of published reports of Philippine trawl surveys).

Table 8. Overview of published results of demersal trawl surveys conducted by Malaysia.

Peninsular Malaysia East coast	Peninsular Malaysia West coast	Sarawak	Sabah		Bull. no.
1926	1926	—	—	Birtwistle and Green (1927)	—
1927	1927	—	—	Birtwistle and Green (1928)	—
—	—	1955-56	—	Ommanney (1961)	—
—	1965	—	—	Pathansali et al. (1966)	—
1967	—	—	—	Anon. (1967)	—
Aug-Oct 70	—	—	—	Pathansali et al. (1974)	1
—	Dec 70-Jan 71	—	—	Latiff et al. (1974)	3
Mar-May 71	—	—	—	Jothy et al. (1975)	4
—	Nov 71-Jan 72	—	—	Lam et al. (1975)	7
—	—	Mar 72-May 72	—	Latiff et al. (1976c)	8
—	Jun-Jul 72	—	—	Ong and Weber (1977)	18
Aug-Sep 72	—	—	—	Lam et al. (1975)	9
—	Apr 73	—	—	Latiff et al. (1976b)	10
—	—	Aug-Sep 73	—	Latiff et al. (1976a)	11
Jul-Aug 74	—	—	—	Lamp and Latiff (1976)	12
—	Nov-Dec 74	—	—	Latiff and Chai (1976)	13
—	—	Jul-Aug 75	—	Lui et al. (1976)	14
all areas (review)				Latiff (1976)	—

In view of the considerable amount of exploratory trawling that has been done in the Philippines, the results of which have not been fully worked up and published, a member of the BFAR research team is currently working with ICLARM (Manila) to assemble, standardize, analyze and publish these data.

In the last four or five years, several research surveys have been made by the College of Fisheries of the University of the Philippines (Manila), largely for training. Although these surveys are not planned for the continuing monitoring of the resources, they often provide useful data (see e.g., Saeger 1981). The vessel used is "R/V Albacore", a stern trawler of 32-m LOA and 600 hp with a bottom trawl with a 27-m headrope. Both 40-mm and 20-mm cod-end meshes are used. Few of the results of these or other fishing surveys have yet been published; it is expected that some reports will be available shortly.

Table 9. Overview of reports on Philippine demersal stocks, as based mainly on research surveys.

Manila Bay	San Miguel Bay	Visayan Sea	Samar Sea	Other areas
Warfel & Manacop (1950)	Warfel & Manacop (1950)	Warfel & Manacop (1950)	Warfel & Manacop (1950)	Warfel & Manacop (1950)
Tiews & Caces-Borja (1965)	Tiews & Caces-Borja (1965)	Rasalan (1957)	Magnusson (1972) Armada & Silvestre (1981)	Encina (1972b)
Magnusson (1970)	Pauly and Mines (1982)	Encina (1972a)	Villoso (1981)	Strom et al. (1972)
Caces-Borja (1972) Caces-Borja (1967) Caces-Borja (1976)		Aprieto (1978)	Saeger (1981)	Caces-Borja & Llorca (1975)
Ronquillo et al. (1972)				Aprieto & Patolot (1976)
De Jesus (1976)				

Thailand. The data from the important series of surveys of the Gulf of Thailand started in 1961 have been reported (in Thai) in the Annual Reports of the Demersal Fisheries Investigation Unit of the Marine Fisheries Laboratory, Bangkok and summaries and analyses given from time to time by Ritragsa and others in Thai and in English (see Table 10).

It should be noted that the design of the surveys was altered after the first survey (1961) and after 1966 and again in about 1977 to make a better stratification with depth.

Thailand has carried out two other less well known series of trawl surveys. These were, and one is continuing, off the west (Indian Ocean) coast, and have been carried out by the Department of Fisheries. One covered the years 1966 to 1971 inclusive, used the research vessel "Pramong III" (250 hp stern trawler) and the results were reported (in Thai) from the Department's laboratory at Phuket. Similar observations were made as on the "Pramong II" for the Gulf of Thailand surveys.

The second series has been carried out since 1965 by the Exploratory Fishing Unit of the Department of Fisheries using the "Researcher II" (1,000 hp stern trawler). The results have been regularly published in the annual reports of that unit (in Thai) though with limited analyses.

Some data from these surveys were included in Part II of the report of the workshop on the resources of the Malacca Strait (SCS 1976b) but there is probably a considerable amount of data obtained during these surveys that has not been fully used.

Table 10 gives an overview of reports on Thai trawl surveys and of papers based on the data given in these reports.

Table 10. Overview of published data and analyses of demersal trawl data conducted in Thai waters.

Gulf of Thailand		Andaman Sea
Reports containing mainly c/f data	Reports containing more or less detailed analysis of c/f and/or other data	(Thai waters) (c/f data and stock assessment)
Ritragsa (1976) Ritragsa et al. (1968, 1969, 1970)	Isarankura (1966) Tiews et al. (1967) Wetchagarun et al. (1967) Boonyubol & Hongskul (1978) FAO (1978) Tiews (1965)	Expl. Fish. Div. (1965 onward) Phuket Mar. Fish Stn. (1965-1971) Isarankura (1971)
Vadhanakul et al. (1975)	Buzeta (1978) SCS (1978a) Pope (1979) Pauly (1979, 1980b, 1982)	SCS (1976a, 1976b)

Future Surveys

Thailand, Malaysia and Indonesia now have continuing programs to survey at intervals their stocks of demersal fish to provide information for evaluation and management and there is every reason to believe that these surveys will continue along the lines already described in this section. This means that the whole of the Sunda Shelf from the Gulf of Thailand to the Java Sea and from the Andaman Sea to the coast of Sabah, with the one exception of the deeper water east of Peninsular Malaysia, is being covered by comparable trawl surveys. The Gulf of Thailand survey is annual, the coastal waters of the Java Sea have been covered in most recent years and the Thai west coast survey appears to be annual, but the other areas are likely to be covered every two to four years.

With regard to international aid projects, there are two in the area in which FAO is involved. One is concerned with testing the viability of pole and line fishing in the Gulf of Thailand, while the other (Project GCP/INS/056/CAN) is to evaluate the possibilities of developing fisheries around the Natuna and Sanihe Island groups in Indonesia. This latter project (34 months in 1980-82) will, among other things, obtain catch rates by various fishing methods and should, if suitably planned, provide some data of value for studies of the available resources. Both areas are rocky and untrawlable, the first in the southern South China Sea and the second in the Celebes Sea.

The current Indonesian/German/Australian "JETINDOFISH" project, (see above) although primarily concerned with acoustic surveys of pelagic fish, will continue its trawling program, which covers the resources of the Indian Ocean/Timor Sea coasts of southern Indonesia to North Australia.

OTHER TRAWL SURVEYS

In addition to data from national surveys by the South China Sea countries, there are also some data from other sources.

Between about 1960 and 1974 the South East Asian Fisheries Development Center (SEAFDEC) maintained a research laboratory at Changi on the island of Singapore for training fisheries scientists and for biological research.

This laboratory had a substantial research program and ran a 1,000 hp stern trawler research vessel "Changi".

The research was varied and included trawl surveys both in the Malacca Strait and in the South China Sea, though these were not regular surveys to study changes in the resources with time. The results of this research were reported in the annual reports and in the reports of technical seminars of SEAFDEC. Some results have been published in Japanese fisheries journals. There is a very considerable amount of information from the SEAFDEC investigations that has not been published, though much of this may be held by the individual scientists who worked at Changi.

As part of a series of acoustic and trawl surveys being carried out in various parts of the world by the Norwegian research vessel "Fridtjof Nansen" (under FAO/NORAD auspices), surveys have been made in the following areas during 1980,

off the west coast of Peninsular Malaysia — 36 trawl stations
off the east coast of Peninsular Malaysia — 41 trawl stations
off the west coast of Thailand — 64 trawl stations

It is understood that preliminary cruise reports have been issued by the "Fridtjof Nansen" Project covering these surveys.

Trawl surveys on the Sunda Shelf have also been carried out by research vessels and commercial vessels from Japan, Korea and Taiwan; most of the reports are in languages other than English. References to such reports are: Chang (1968), Liu (1973), Liu and Lai (1978), Hooi (1973), Senta et al. (1973), Tseng (1962), Ushioku (1935), Wanatabe and Tan (1978).

Studies on the Bionomics of Demersal Species

During the majority of trawl surveys and exploratory trawling, size composition data and observations on a variety of biological parameters have been made on a number of the commoner species or species groups. There is undoubtedly a considerable quantity of data on length composition, maturity stages, stomach contents and parasites of demersal and pelagic fishes in existence in all countries of Southeast Asia. Some of these data have formed the basis for papers published in languages other than English, particularly Indonesian and Thai. However the majority of these observations have remained in various stages of compilation and analysis and very few have been published in raw data lists, summaries or in fully analyzed form. Notable exceptions to this are trawl survey reports issued by the Marine Fisheries Research Institute in Jakarta some of which include detailed length-frequency and catch data on per-species and per-haul basis, using the condensed format of Martosubroto and Pauly (1976) (see Table 7).

Table 11 gives an overview of published papers on the biology and population dynamics of demersal fishes from the South China Sea area. The table is not meant to be exhaustive, and more references could be obtained, particularly from bibliographies listing all marine biology work conducted in a given country (e.g, Vicente 1980; Gomez 1980) or from more specialized bibliographies (e.g, Pauly and Wade-Pauly 1981).

Several demersal species present in Southeast Asia, some of which have distributions ranging throughout the Indo-Pacific, have been studied in other countries and the results published outside Southeast Asia, particularly in India. No attempt has been made to locate these publications.

Discussion

The continuing series of trawl surveys by Thailand, Malaysia and Indonesia should produce invaluable data for developing and testing hypotheses on population dynamics and the management of fisheries. However it will take a number of years until there are series of data relating to the fish stocks in the waters of Malaysia and Indonesia comparable with the 20-year series now available for the Gulf of Thailand. These newer surveys will produce good data on catch rates and species composition and changes in size composition of a number of species, though the sampling for size composition may need to be more critically planned to give maximum useful information.

There is a need for similar series of surveys to be undertaken with traps, lines and other gear on rough, untrawlable grounds, as very few observations have been made on the often substantial stocks inhabiting these areas (but see Senta et al. 1973; Watanabe and Tan 1978).

Information on the biology of fish species is limited and there is a clear need for much more on many important species, particularly to identify stocks, determine prey/predator relationships and growth and mortality rates. Obtaining material for the determination of growth, especially by using length composition data, will need planning and execution apart from the surveys, if reliable conclusions are to be drawn (see Pauly this vol. and Munro this vol.). Tagging and other studies are necessary to obtain information to define stocks and to determine migrations and mixing rates.

The collection of statistics of catch and effort of adequate coverage and accuracy is a major problem for most developing countries, particularly if the country is an archipelago. It is therefore most important to know what data are essential and where simplifications can be made without loss of information. With the large numbers of species in tropical waters, there may be little lost by extensive groupings, provided that selected, representative species are kept separate.

Data on fishing effort are inadequate in most countries due to difficulties of collection or insufficient clarity on what the essential data are that need to be collected so that there is a link-up with catch data. To permit the standardization of fishing effort in developing fisheries, there is a need for

Table 11. Selected references on the biology and population dynamics of demersal fish stocks of the South China Sea and adjacent waters.

Family	Species	Catch-per-effort, abundance	Food and feeding habits	Reproduction	Growth	Mortality and/or population dynamics
Leiognathidae	*Leiognathus splendens*	Pauly (1977)	Tiews et al. (1972a)	Tiews & Caces-Borja (1965)	Beck & Sudrajat (1978)	Pauly (1979, 1980a)
Leiognathidae	several spp.	Tiews & Caces-Borja (1965)	Tham (1950), Tiews et al. (1972a)	Tiews & Caces-Borja (1965)	Pauly (1978)	Tiews & Caces-Borja (1965)
Lutjanidae	*Lutjanus sanguineus*	Badrudin (1977), Kungvankij (1973)	—	Kungvankij (1973)	Lai & Liu (1974), Kungvankij (1973)	Pauly (1979)
Mullidae	*Upeneus sulphureus*	Beck & Sudrajat (1978)	—	—	Beck & Sudrajat (1978)	—
Mullidae	*Upeneus moluccensis*	Lee (1974)	Lee (1973)	Lee (1974)	Lee (1974)	Lee (1974)
Nemipteridae	*Nemipterus virgatus*	Eggleston (1970)	Eggleston (1970)	Eggleston (1970)	Eggleston (1970), Kao & Liu (1979)	Eggleston (1970)
Nemipteridae	*Nemipterus japonicus*	Lee (1975)	Eggleston (1972)	Eggleston (1972)	Lee (1975)	Lee (1975)
Nemipteridae	*Nemipterus marginatus*	Pauly & Martosubroto (1980)	—	—	Pauly & Martosubroto (1980)	Pauly & Martosubroto (1980)
Nemipteridae	several spp.	Weber & Jothy (1977)	Eggleston (1972)	Eggleston (1972)	Pauly (1980b)	Pauly (1980b)
Psettodidae	*Psettodes erumei*	Ritragsa (1976)	—	—	Kühlmorgen-Hille (1976)	Pauly (1980b)
Siganidae	several spp.	Alcala (1979)	Von Westernhagen (1973)	Von Westernhagen & Rosenthal (1975)	Hortsman (1975), Carumbana and Luchavez (1979)	—
Synodontidae	*Saurida tumbil*	Tiews et al. (1972b)	Tiews et al. (1972b)	Tiews et al. (1972b)	Yeh et al. (1977)	—
Synodontidae	*Saurida undosquamis*	—	—	Sinoda & Intong (1978)	Sinoda & Intong (1978)	—

more information on the changing technology of capture and the changing strategy of fishing. It may not be necessary to collect data on effort for all artisanal fishing methods if the total catch is known and the catch and effort of other gears exploiting the same stocks are well documented. Also, it is essential that the fisheries scientist is as familiar with the fishery as he or she is with the fish, and that close liaison between the research work and the collection of statistics is maintained.

A major problem with fishery research in Southeast Asia is the fact that results remain generally unpublished or are published (often in languages other than English) in reports the distribution of which is very limited. This has created a large backlog of (largely inaccessible) "grey literature" which will be extremely difficult to reduce. This is particularly unfortunate in the light of the fact that the nature of the resources (e.g., large numbers of single-species stocks) and institutional problems (e.g., lack of well-trained scientists) render each study, even each bit of information, particularly valuable.

A partial solution to this problem has been provided by the proceedings of various symposia and workshops (e.g., Tiews 1969, 1976; Marr 1970), by the series of reports by the South China Sea Development and Coordinating Programme of FAO, in which a large amount of the available information on Southeast Asian stock and fisheries is incorporated, as well as by various ICLARM publications in which the relevant "grey literature" is reviewed (e.g., Pauly 1979; Smith 1979; Smith et al. 1980).

However, there are in the long run no alternatives to formal publication by Southeast Asian scientists of their results in internationally-accessible, national journals or in international journals, and thoughts should be devoted particularly by donor and/or development agencies, on how this process could be accelerated.

Acknowledgments

I wish to thank Dr. D. Pauly (ICLARM) for supplying me with a number of references on biological and stock-assessment studies of Southeast Asian fishes, and for adding to the section on demersal surveys.

References

Alcala, A. 1979. Ecological notes on rabbitfishes (Family Siganidae) and certain economically important animals in Southeastern Negros and environs. Philippines. Silliman J. 26(2-3): 115-133.

Anon. 1967. Results of the joint Thai-Malaysian-German trawling survey off the east coast of the Malay peninsula. Mar. Fish. Lab., Bangkok and Fish. Inst., Malaysia. 64 p.

Anon. 1972. Menguti operasi pair trawl di perairan selat Karimata & Laut Cina Selatan (11-22 Desember 1972). Mar. Fish. Res. Inst., Jakarta. 52 p. (mimeo)

Aoyama, T. 1973. The demersal stocks and fisheries of the South China Sea. SCS/DEV/73/3. 80 p. FAO, Rome.

Aprieto, V.L. 1978. Hydrobiological Studies of the Visayan sea demersal fisheries. Institute of Fisheries Development and Research, Diliman, Quezon City, Philippines. 78 p. (mimeo)

Aprieto, V.L. and J.N. Patolot. 1976. Investigations on potential deep-sea trawl fishing grounds. Fish. J. Philipp. 1(1): 57-85.

Armada, N.B. and G.T. Silvestre. 1981. Demersal fish resource survey in Samar Sea and Carigara Bay. Prepared for UP-NSDB Project 7811.1c Ag. College of Fisheries, University of the Philippines in the Visayas. 56 p. (mimeo)

Badrudin, M. 1977. Lutjanidae, p. 18-19. *In* A. Dwiponggo (ed.) Distribution maps of demersal fisheries resources of the Java and South China Seas. Mar. Fish. Res. Inst., Jakarta. 35 p. (In Indonesian and English)

Beck, U. and A. Sudrajat. 1978. Variation in size and composition of demersal trawl catches from the North Coast of Java with estimated growth parameters for three important food-fish species. Mar. Fish. Res. Rep. (Spec. Rep.)/Contrib. Demersal Fish. Proj., Jakarta 4: 1-80.

Birtwistle, W. and C.F. Green. 1927. Report of the working of the S.T. Tongkol for the period 28 May to 31 December 1926. Gov. Print. Off., Singapore.

Birtwistle, W. and C.F. Green. 1928. Report of the working of the S.T. Tongkol for the year 1927. Gov. Print. Off., Singapore.

Boonyubol, M. and V. Hongskul. 1978. Demersal resources and exploitation in the Gulf of Thailand, 1960-1975, p. 56-70. *In* Report of the workshop on the demersal resources of the Sunda Shelf, 31 October-6 November 1977. Penang, Malaysia. Part II. SCS/GEN/77/13. South China Sea Fisheries Development and Coordinating Programme, Manila.

Buzeta, R.B., Editor. 1978. Report of the workshop on the demersal resources of the Sunda Shelf, 31 October-6 November 1977, Penang, Malaysia. Part I. SCS/GEN/77/12. 51 p. South China Sea Fisheries Development and Coordinating Programme, Manila.

Buzeta, R.B., A. Dwiponggo and T. Sujastani. 1979. Report of the workshop on demersal and pelagic resources of the Java Sea. 5-9 December 1978, Semarang, Indonesia. SCS/GEN/79/20. 60 p. South China Sea Fisheries Development and Coordinating Programme, Manila.

Caces-Borja, P. 1967. Status of trawling data in the Philippines. Proc. Indo-Pac. Fish. Counc. 12(2): 172-180.

Caces-Borja, P. 1972. On the ability of otter trawls to catch pelagic fish in Manila Bay. Philipp. J. Fish. 10(1/2): 39-53.

Caces-Borja, P., R. Bustillo and S. Ganaden. 1972. Further observation on commercial trawl fishery of Manila Bay (1960-62). Proc. Indo-Pac. Fish. Counc. 13(3): 631-637.

Caces-Borja, P. and M.N. Llorca. 1975. Preliminary report on test fishing programme covering the period Nov. 1973 to May 1974. Bureau of Fisheries and Aquatic Resources, Manila. 10 p. (mimeo).

Carumbana, E. and J. Luchavez. 1979. A comparative study of growth rates of *Siganus canaliculatus, S. spinus* and *S. guttatus* under laboratory and semi-natural conditions in Southern Negros Oriental, Philippines. Silliman J. 26: 187-209.

Chang, J.H. 1968. The results of experimental trawl fishing in South China Sea: report of catch per area. Fish. Res. Dev. Agency, Pusan, South Korea. (In Korean)

De Jesus, A.S. 1976. A report on the design, construction and operation of German (Hermann Engel Type) trawl net in Manila Bay. Philipp. J. Fish. 14(1): 117-144.

Dwiponggo, A., Editor. 1977. Distribution maps of demersal fisheries resources of the Java and South China Seas. Mar. Fish. Res. Inst., Jakarta. (In Indonesian and English)

Dwiponggo, A. 1979. Review of the demersal resources and fisheries in the Java Sea. Mar. Fish. Res. Inst., Jakarta. 36 p. (mimeo)

Dwiponggo, A. and M. Badrudin. 1978a. Demersal resources survey in coastal areas of the Java Sea, 1977. Mar. Fish. Res. Rep. (Spec. Rep.)/Contrib. Demersal Fish. Proj., Jakarta. 5. 14 p.

Dwiponggo, A. and M. Badrudin. 1978b. R/V Mutiara IV survey data in 1977. Mar. Fish. Res. Rep. (Spec. Rep.)/Contrib. Demersal Fish. Proj., Jakarta. 5A. 133 p.

Dwiponggo, A. and M. Badrudin. 1979a. Variations in catch rates and species composition of trawl surveys in the Java Sea subareas. Mar. Fish. Res. Rep. (Spec. Rep.)/Contrib. Demersal Fish. Proj., Jakarta. 6. 19 p.

Dwiponggo, A. and M. Badrudin. 1979b. Data of trawl surveys by R/V Mutiara IV in the Java Sea subareas in 1978. Mar. Fish. Res. Rep. (Spec. Rep.)/Contrib. Demersal Fish. Proj., Jakarta. 6A. 128 p.

Dwiponggo, A. and M. Badrudin. 1980a. Results of the Java Sea inshore monitoring surveys, 1979. Mar. Fish. Res. Rep. (Spec. Rep.)/Contrib. Demersal Fish. Proj., Jakarta. 7. 31 p.

Dwiponggo, A. and M. Badrudin. 1980b. Data of the Java Sea inshore monitoring surveys by R/V Mutiara IV in 1979. Mar. Fish. Res. Rep. (Spec. Rep.)/Contrib. Demersal Fish. Proj., Jakarta. 7A. 88 p.

Dwiponggo, A. and M. Badrudin. 1980c. Length-frequency measurements. Mar. Fish. Res. Rep. (Spec. Rep.)/Contrib. Demersal Fish. Proj., Jakarta. 7B. 94 p.

Eggleston, D. 1970. Biology of *Nemipterus virgatus* in the northern part of the South China Sea, p. 417-424. *In* J.C. Marr (ed.) The Kuroshio: a symposium on the Japan current. East-West Center Press, Honolulu.

Eggleston, D. 1972. Patterns of biology in the Nemipteridae. J. Mar. Biol. Assoc. India 14(1): 357-364.

Encina, V.B. 1972a. A preliminary report on the use of plastic bobbins and oval-shaped otter boards for trawling operations in Western Visayan Sea with the aid of echo Sounder. Philipp. J. Fish. 10(1/2): 85-118.

Encina, V.B. 1972b. A report on two-boat bottom trawl fishing in the Philippines. Proc. Indo-Pac. Fish. Counc. 13(3): 653-654.

Exploratory Fishing Division. 1965-1981. Annual reports. Ministry of Agriculture and Cooperatives, Department of Fisheries, Bangkok, Thailand. (In Thai)

Exploratory Fishing Division. 1978. The first Indonesian-Thai fisheries expedition, May-September 1977. Ministry of Agriculture and Cooperatives, Dept. of Fish., Bangkok, Thailand. (1), 183 p. (mimeo)

FAO. 1978. Some scientific problems of multispecies fisheries. Report of the Expert Consultation on Management of Multispecies Fisheries, Rome, Italy, 20-23 September 1977. FAO Fish. Tech. Pap. 181. 42 p.

Gomez, E.D. 1980. Bibliography of Philippine marine science, 1978. Filipinas Foundation, Inc., Makati, Metro Manila.

Hooi, K.K. 1973. A study on the catch data on the Jurong in the South China Sea in 1971 and 1972. SEAFDEC/SCS 73: S-19.

Hortsmann, U. 1975. Some aspects of the mariculture of different siganid species in the Philippines. Philipp. Sci. 12: 5-20.

Isarankura, A.P. 1966. Size selection of some commercially important species of Gulf of Thailand demersal fish by otter trawl cod-end mesh size of 4 cm. Rep. Mar. Fish. Lab., Bangkok.

Isarankura, A.P. 1971. Assessment of stocks of demersal fish off the west coasts of Thailand and Malaysia. IOFC/DEV/71/20. 20 p. Indian Ocean Fish. Comm., FAO, Rome.

Jothy, A.A., G. Rauck, M.S.S.A. Latiff, K.S. Ong, P.T. Liong and J.L. Carvalho. 1975. Demersal fish resources in Malaysian waters. 3. Second trawl survey of the coastal waters off the east coast of Peninsular Malaysia (March-May, 1971). Fish. Bull. (Malays.) No. 4. 36 p.

Kao, C.L. and H.C. Liu. 1979. Age and growth of golden thread, *Nemipterus virgatus* (Houttuyn), from the East and the South China Seas. Acta Oceanogr. Taiwanica 9: 97-110.

Kühlmorgen-Hille, G. 1976. Preliminary study on the life-history of the flatfish *Psettodes erumei*, p. 261-268. *In* K. Tiews (ed.) Proceedings of the international seminar on fisheries resources and their management in Southeast Asia, West Berlin, 19 Nov.-6 Dec. 1974. West Kruez, Druckerei, Berlin.

Kungvankij, P. 1973. Biological study of red snapper *Lutjanus sanguineus* (Cuvier et Valenciennes). Thai Fish. Gaz. 25(3): 349-368.

Lai, H.L. and H.C. Liu. 1974. Age determination and growth of *Lutjanus sanguineus* (C and V) in the South China Sea. J. Fish. Soc. Taiwan 3(1): 39-57.

Lam, W.C., W. Weber, A.K. Lee, K.S. Ong and P.C. Liong. 1975. Demersal fish resources in Malaysian waters. 7. Third east coast trawl survey off the east coast of Peninsular Malaysia, (14 August-20th September 1972). Fish. Bull. (Malays.) No. 9. 19 p.

Lamp, F. and Latiff, M.S.S.A. 1976. Demersal fish resources in Malaysian waters. 10. Fourth trawl survey of the coastal waters off the east coast of Peninsular Malaysia, (13th July-12th August 1974). Fish. Bull. (Malays.) No. 12. 25 p.

Latiff, M.S.S.A., G. Rauck, K.S. Ong and S.P. Tan. 1974. Demersal fish resources in

Malaysian waters. 2. Trawl survey of the coastal waters off the west coast of Peninsular Malaysia (12th December 1970-22nd January 1971). Fish. Bull. (Malays.) No. 3. 41 p.

Latiff, M.S.S.A. 1976. Demersal fish resource surveys and problems of fisheries resource management, p. 224-247. *In* K. Tiews (ed.) Proceedings of the international seminar on fisheries resources and their management in Southeast Asia, West Berlin, 19 Nov.-6 Dec. 1974. West Kruez, Druckerei, Berlin.

Latiff, M.S.S.A. and H.L. Chai. 1976. Demersal fish resources in Malayan waters. 11. Third trawl survey off the west coast of Peninsular Malaysia. Fish. Bull. (Malays.) No. 13. 17 p.

Latiff, M.S.S.A., W. Weber and A.K. Lee. 1976a. Demersal fish resources in Malaysian waters. 9. Second trawl survey off the coast of Sarawak. Fish. Bull. (Malays.) No. 11. 28 p.

Latiff, M.S.S.A., W. Weber and P.C. Liong. 1976b. Demersal fish resources in Malaysian waters. 8. Trawl survey off the west coast of Peninsular Malaysia (southern part of the Malacca Straits) (4th April-30th April 1973). Fish. Bull. (Malays.) No. 10. 21 p.

Latiff, M.S.S.A., W. Weber, A.K. Lee and W.C. Lam. 1976c. Demersal fish resources in Malaysian waters. 6. First trawl survey off the coasts of Sarawak, Brunei and west coast of Sabah. Fish. Bull. (Malays.) No. 8. 64 p.

Lee, C.K.C. 1973. The feeding of *Upeneus molluccensis* (Bleeker) on fishing grounds near Hong Kong. Hong Kong Fish. Bull. 3: 47-53.

Lee, C.K.C. 1974. The reproduction, growth and survival of *Upeneus molluccensis* (Bleeker) in relation to the commercial fishery of Hong Kong. Hong Kong Fish. Bull. 4: 17-32.

Lee, C.K.C. 1975. The exploitation of *Nemipterus japonicus* (Bloch) by Hong Kong vessels in 1972-73, p. 48-52. *In* B. Morton (ed.) The Pacific Science Association Spec. Symp. Mar. Sci., 7-16 Dec. 1973, Hong Kong.

Liu, H.C. 1973. The demersal resources of the South China Sea. SEAFDEC/SCS 73: s-26. 23 p.

Liu, H.C. and H.L. Lai. 1978. Demersal fish resources of the Sunda Shelf, p. 71-92. *In* Report of the workshop on the demersal resources of the Sunda Shelf, 31 October-6 November 1977. Penang, Malaysia. Part II. SCS/GEN/77/13. South China Sea Fisheries Development and Coordinating Programme, Manila.

Losse, G.F. and A. Dwiponggo. 1977. Report on the Java Sea Southeast monsoon trawl survey, June-December 1976. Mar. Fish. Res. Rep. (Spec. Rep.)/Contrib. Demersal Fish. Proj., Jakarta. 3. 119 p.

Lui, Y.P., M.S.S.A. Latiff, I.T.B.M. Yusoff and F. Lamp. 1976. Demersal fish resources of Malaysian waters. 12. Third trawl survey off the coast of Sarawak, (16th July-8 August 1975). Fish. Bull. (Malays.) No. 14. 35 p.

Magnusson, J. 1970. Deep sea fishing development. The Philippine marine fisheries biology. FAO-UNDP report to the Philippine Government. FI:SF/PHIL II. 84 p.

Magnusson, J. 1972. On the trawl fishery of the Philippines. Proc. Indo-Pac. Fish. Counc. 13(3): 597-604.

Marr, J.C., Editor. 1970. The Kuroshio: a symposium on the Japan current. East-West Center Press, Honolulu.

Martosubroto, P. 1973. Investigations on demersal stocks in the Malaka Strait and Jambi Bay. Mar. Fish. Res. Rep., Jakarta. 1: 1-33. (In Indonesian)

Martosubroto, P. and D. Pauly. 1976. R/V Mutiara IV survey data, November 1974 to July 1976. Mar. Fish. Res. Rep. (Spec. Rep.)/Contrib. Demersal Fish. Proj., Jakarta. 2. 136 p.

Ommanney, F.D. 1961. Malaysian offshore trawling grounds. The experimental and exploratory fishing cruises of the F.R.V. Manihine in Malayan and Borneo waters, 1955-1956, with a note on temperatures and salinities in the Singapore Strait. Colon. Off. Fish. Publ. 18. H.M.S.O., London.

Ong, K.S. and W. Weber. 1977. First prawn trawling survey off the west coast of Peninsular Malaysia. Fish. Bull. (Malays.) No. 18. 28 p.

Pathansali, D. 1976. Some observations on catch and effort of commercial and research trawlers with estimates of yield, p. 248-255. *In* K. Tiews (ed.) Proceedings of the international seminar on fisheries resources and their management in Southeast Asia, West Berlin, 19 Nov.-6 Dec. 1974. West Kruez, Druckerei, Berlin.

Pathansali, D., K.S. Ong, M.S.S.A. Latiff and J. Carvalho. 1966. Preliminary results of trawling investigations off Penang. Proc. Indo-Pac. Fish. Counc. 12(2): 181-201.

Pathansali, D., G. Rauck, A.A. Jothy, M.S.S.A. Latiff and T. Curtin. 1974. Demersal fish resources in Malaysian waters. Trawl survey of the coastal waters off the east coast of West Malaysia. Fish. Bull. (Malays.) 1: 1-46.

Pauly, D. 1977. The Leiognathidae (Teleostei); their species, stocks and fisheries in Indonesia, with notes on the biology of *Leiognathus splendens* (Cuvier). Mar. Res. Indones. 19: 73-93.

Pauly, D. 1978. A preliminary compilation of fish length growth parameters. Ber. Inst. Meereskunde, Kiel 55. 200 p.

Pauly, D. 1979. Theory and management of tropical multispecies stocks: a review, with emphasis on the Southeast Asian demersal fisheries. ICLARM Studies and Reviews No. 1. 35 p. International Center for Living Aquatic Resources Management, Manila.

Pauly, D. 1980a. The use of a pseudo catch curve for the estimation of mortality rates in *Leiognathus splendens* (Pisces, Leiognathidae). Meeresforschung 28(1): 56-60.

Pauly, D. 1980b. A new methodology for rapidly acquiring basic information on tropical fish stocks: growth, mortality and stock recruitment relationships, p. 154-172. *In* S.B. Saila and P.M. Roedel (eds.) Stock assessment for tropical small-scale fisheries. International Center for Marine Resource Development, Kingston, Rhode Island.

Pauly, D. 1982. A method to estimate the stock-recruitment relationships of shrimps. Trans. Amer. Fish. Soc. 111: 13-20.

Pauly, D. and P. Martosubroto. 1980. The population dynamics of *Nemipterus marginatus* off Western Kalimantan, South China Sea. J. Fish Biol. 17 : 263-273.

Pauly, D. and S. Wade-Pauly. 1981. An annotated bibliography of the slipmouths (Pisces: Leiognathidae). ICLARM Bibliographies 2, 62 p. International Center for Living Aquatic Resources Management, Manila.

Pauly, D. and A. Mines, Editors. 1982. Small-scale fisheries of San Miguel Bay, Philippines: biology and stock assessment. ICLARM Technical Reports 7. Institute of Fisheries Development and Research, College of Fisheries, University of the Philippines in the Visayas, Quezon City, Philippines; International Center for Living Aquatic Resources Management, Manila, Philippines; and the United Nations University, Tokyo, Japan.

Phuket Marine Fisheries Station. 1965-1971. Annual reports. Department of Fisheries, Thailand. (In Thai)

Pope, J.G. 1979. Stock assessment in multispecies fisheries, with special reference to the trawl fishery in the Gulf of Thailand. SCS/DEV/79/19. 106 p. South China Sea Fisheries Development and Coordinating Programme, Manila.

Rasalan, S.B. 1957. Marine fisheries of Central Visayas. Philipp. J. Fish. 5(1): 53-89.

Ritragsa, S. 1976. Results of the studies on the status of demersal fish resources in the Gulf of Thailand from trawling survey 1963-1972, p. 198-223. *In* K. Tiews (ed.) Proceedings of the international seminar on fisheries resources and their management in Southeast Asia. West Kruez, Druckerei, Berlin.

Ritragsa, S., D. Dhamniyom and S. Sitthichaikasem. 1968. An analysis of demersal fish catches taken from the otterboard trawling survey in the Gulf of Thailand, 1966. Contrib. Mar. Fish. Lab., Bangkok. 11. 104 p. (In Thai)

Ritragsa, S., D. Dhamniyom and S. Sitthichaikasem. 1969. An analysis of demersal fish catches taken from the otterboard trawling survey in the Gulf of Thailand, 1967. Contr. Mar. Fish. Lab., Bangkok. 15. 70 p. (In Thai)

Ritragsa, S. and S. Pramokchutima. 1970. The analysis of demersal fish catches taken from the otterboard trawling survey in the Gulf of Thailand, 1968. Contr. Mar. Fish. Lab., Bangkok. 16. 61 p. (In Thai)

Ronquillo, I.A., P. Caces-Borja and A.N. Mines. 1972. Preliminary observation on the otter trawl fishery of Manila Bay, Proc. Indo-Pac. Fish. Counc. 13(3): 638-648.

Saeger, J. 1981. Do trawling bans work in tropical waters? ICLARM Newsl. 4(1): 3-4.

Saeger, J., P. Martosubroto and D. Pauly. 1976. First report of the Indonesian-German demersal fisheries project. Results of a trawling survey in the Sunda Shelf area. Mar. Fish. Res. Rep. (Spec. Rep.)/Contrib. Demersal Fish. Proj., Jakarta. 1: 1-46.

SCS. 1976a. Report of the workshop on the fishery resources of the Malacca Strait, 29 March-2 April 1976, Jakarta. Part I. SCS/GEN/76/2. 85 p. South China Sea

Fisheries Development and Coordinating Programme, Manila.

SCS. 1976b. Report of the workshop on the fishery resources of the Malacca Strait, 29 March-2 April 1976, Jakarta. Part II. SCS/GEN/76/6. 85 p. South China Sea Fisheries Development and Coordinating Programme, Manila.

SCS. 1976c. Report on the BFAR/SCSP workshop on the fishery resources of the Visayan Sea areas, 18-22 October 1976, Tigbauan, Iloilo, Philippines. SCS/GEN/76/7. 26 p. South China Sea Fisheries Development and Coordinating Programme, Manila.

SCS. 1977. Report on the BFAR/SCSP workshop of the fishery resources of the Sulu Sea, Bohol Sea and Moro Gulf areas, 25-29 April 1977, Cagayan de Oro, Philippines. SCS/GEN/77/11. 58 p. South China Sea Fisheries Development and Coordinating Programme, Manila.

SCS. 1978a. Report of the workshop on the demersal resources of the Sunda Shelf, 31 October-6 November 1977, Penang, Malaysia. Part II. SCS/GEN/77/13. 120 p. South China Sea Fisheries Development and Coordinating Programme, Manila.

SCS. 1978b. Report of the BFAR/SCSP workshop on the fishery resources of the Pacific coast of the Philippines, 18-22 September 1978. SCS/GEN/78/19. 48 p. South China Sea Fisheries Development and Coordinating Programme, Manila.

Senta, T., S.M. Tan and P.Y. Lim. 1973. Results of the experimental trawl fishing in the South China Sea by R/V Changi in the years 1970 to 1972. SEAFDEC/SCS/73: S-10.

Shindo, S. 1973. General review of the trawl fishery and the demersal fish stocks of the South China Sea. FAO Fish. Tech. Pap. 120. 49 p.

Simpson, A.C. 1979. Report of the BFAR/SCSP workshop on the fishery resources of the north and western coasts of Luzon, 18-20 April 1979, Manila, Philippines. SCS/GEN/79/22. 57 p. South China Sea Fisheries Development and Coordinating Programme, Manila.

Sinoda, M. and S. Intong. 1978. Size-frequency distribution of lizard fish *Saurida undosquamis* in the Inner Gulf of Thailand. Bull. Jap. Soc. Sci. Fish. 44(1): 1-6.

Smith, I.R. 1979. A research framework for traditional fisheries. ICLARM Studies and Reviews No. 2. 45 p. International Center for Living Aquatic Resources Management, Manila.

Smith, I.R., M.Y. Puzon and C.N. Vidal-Libunao. 1980. Philippine municipal fisheries: a review of resources, technology and socioeconomics. ICLARM Studies and Reviews No. 4. 87 p. International Center for Living Aquatic Resources Management, Manila.

Strom, P.O.K., J. Dale, K.M. Einarsen, J. Kasseroff, V. Zankich, P. Eilefsen, G. Gestsson, K.R. Mori and Y. Ishino. 1972. Deep sea fishing development, the Philippines fish capture. A report prepared for the Government of the Philippines by FAO/UNDP, Rome. FL:SF/Phil. 70/537 Technical Report 1. 73 p.

Sudrajat, A. and U. Beck. 1978. Report on the southern South China Sea demersal trawl survey, June and July, 1978. Mar. Fish. Res. Rep. (Spec. Rep.)/Contrib. Demersal Fish. Proj., Jakarta 4: 81-140.

Sujastani, T., P. Martosubroto and D. Pauly. 1976. A review of the demersal fishery of Indonesia in the Malacca Strait, based on recent surveys and catch and effort data, p. 23-47. *In* Report of the workshop on the fishery resources of the Malacca Strait, 29 March-2 April 1976, Jakarta. Part I. SCS/GEN/76/2. South China Sea Fisheries Development and Coordinating Programme, Manila.

Tham, A.K. 1950. The food and feeding relationships of the fishes of Singapore Straits. Colon. Off. Fish. Publ. 1(1): 35 p.

Tiews, K. 1965. Bottom fish resources investigations in the Gulf of Thailand and an outlook on further possibilities to develop the marine fisheries in Southeast Asia Arch. Fischereiwiss. 16(1): 67-108.

Tiews, K., Editor. 1969. Proceedings of an international seminar on possibilities and problems of fisheries development in Southeast Asia, West Berlin 10-30 September 1968. West Kreuz, Druckerei, Berlin.

Tiews, K., Editor. 1976. Proceedings of an international seminar on fisheries resources and their management in Southeast Asia, West Berlin, 19 Nov.-6 Dec. 1974. West Kreuz, Druckerei, Berlin.

Tiews, K. and P. Caces-Borja. 1965. On the availability of fish of the family Leiognathidae Lacepede in Manila Bay and San Miguel Bay and on their accessibility to controversial fishing gears. Philipp. J. Fish. 7(1): 59-86.

Tiews, K., P. Divino, I.A. Ronquillo and J. Marquez. 1972a. On the food and feeding habits of eight species of *Leiognathus.* Proc. Indo-Pac. Fish. Counc. 13(3): 93-99.

Tiews, K., A. Mines and I.A. Ronquillo. 1972b. On the biology of *Saurida tumbil* (Bloch, 1801) in Philippine waters. Proc. Indo-Pac. Fish. Counc. 13(3): 100-120.

Tiews, K., P. Sucondhamarn and A. Isarankura. 1967. On the changes in the abundance of demersal fish stocks in the Gulf of Thailand from 1963/1964 as a consequence of the trawl fisheries development. Contrib. Mar. Fish. Lab., Bangkok 8. 39 p. (In Thai and English)

Tseng, C.T. 1962. An investigation of the trawling grounds of the southern China Sea: report of catch by area. Rep. Taiwan Fish. Res. Inst. No. 28. 11 p. (In Chinese). *Also* IPFC Occas. Pap. 1967, 67/11, 24 p. (In English)

Ushioku, S. 1935. Report on the survey of trawling grounds by "Hijun Maru". Nos. 1-3. (In Japanese) (mimeo)

Vadhanakul, S., M. Eiamsa-ard and N. Kuantanom. 1975. Report on the survey of species composition of scrap fish in the Gulf of Thailand by M.V. Pramong II, 1974. Demersal Fish. Rep. 1. 13 p. (mimeo)

van Rosendaal, A.M. and P.N. van Kampen. 1909. Verslag van de verrichtingen van het Onder zoekings-vaartuig "Gier . . . etc. Meded. Visscherij-station Batavia (3), Buitenzard.

Vicente, D.A. 1980. Fisheries research in the Philippines: an annotated bibliography of bibliographies. J. Fish. Aquaculture (Philipp.) 1(2): 1-218.

Villoso, E.P. 1981. Catch composition, stocks and depth distribution of fishes caught by otter trawl in Samar Sea. Paper presented at the Third Fisheries Forum for 1981. Philippine Council for Agriculture and Resources Research, Los Baños, Laguna.

von Westernhagen, H. 1973. The natural food of the rabbitfish *Siganus oramin* and *S. striolata.* Mar. Biol. 22: 367-370.

von Westernhagen, H. and H. Rosenthal. 1975. Rearing and spawning siganids (Pisces: Teleostei) in a closed seawater system. Helgoländer Wiss. Meeresunters. 27: 1-18.

Warfel, H.E. and P.R. Manacop. 1950. Otter trawl explorations in Philippine waters. Fish. Wildl. Serv. (U.S.) Res. Rep. 25. 49 p.

Watanabe, Y. and S.M. Tan. 1978. Distribution of some fishes in trawling areas surveyed by R/V Changi in the Sunda Shelf, p. 107-120. *In* Report of the workshop on the demersal resources of the Sunda Shelf, 31 October-6 November, 1977. Penang, Malaysia. Part II. SCS/GEN/77/13. South China Sea Fisheries Development and Coordinating Programme, Manila.

Weber, W. 1976. The influence of hydrographical factor on the spawning time of tropical fish, p. 269-281. *In* K. Tiews (ed.) Proceedings of the international seminar on fisheries resources and their management in Southeast Asia, West Berlin, 19 Nov.-6 Dec. 1974. West Kruez, Druckerei, Berlin.

Weber, W. and A. Jothy. 1977. Observations on the fish *Nemipterus* spp. (Family: Nemipteridae) in the coastal waters of East Malaysia. Arch. Fischereiwiss. 28: 109-122.

Wetchagarun, K. et al. 1967. A report on the survey and analysis of trawl craft and gear in the Gulf of Thailand, p. 56-147. *In* Annual Report, 1967. Demersal Fish. Invest. Unit, Dept. Fish., Thailand.

Widodo, J. 1976. A check-list of fishes collected by Mutiara IV from November 1974 to November 1975. Mar. Fish. Res. Rep. (Spec. Rep.)/Contrib. Demersal Fish. Proj., Jakarta. 1: 47-77.

Wyrtki, K. 1961. Physical oceanography of the Southeast Asian waters. Naga Rep. 2: 1-195.

Yeh, S.H., H.L. Lai and H.C. Lin. 1977. Age and growth of lizard fish, *Saurida tumbil* (Bloch) in the East China Sea and the Gulf of Tonkin. Acta Oceanogr. Taiwanica 7: 134-145.

Discussion of Paper by Mr. Simpson

The discussion was concerned largely with the problems of obtaining reliable fisheries statistics, the working-up of trawl survey data and the supply of scientists.

Examples were given of the problems involved in the routine collection of data on catches and fishing effort, and the various causes of inaccurate and incomplete information or even inflated figures. It was suggested that the routine annual collection of these data should be replaced, in some cases, by intensive studies of specific fisheries at longer intervals. However, it was generally agreed that even with intensive studies, it is still necessary to have annual information on the amount of fish removed by fishing from the various fisheries. Also, in many fisheries, data on the fishing effort or the catch per unit of fishing are extremely useful. It was stressed that it is essential that data on effort are related to the catch generated by the effort in question, and that effort data for which this requirement is not met are useless for stock assessment purposes.

Concerning the use of catch and effort data, it was further pointed out that changes in the size, power and efficiency of fishing units and changes in the area of fishing and in fishing strategy are all capable of substantially changing the species composition and size of the catch. Therefore, a comprehensive knowledge of the practical aspects of the fisheries is extremely important.

It was suggested that an important reason for poor statistics is that there is often no direct user of the data collected. However, the more important aspect of this problem is that there should be ready communication between those using the data and those collecting it. It was suggested that, as the fisheries laboratories are the main users of the data, the group responsible for collecting catch and effort data and the research laboratory should be under the direction of the same or closely associated Directors, thus facilitating two-way communication. This is the case in India and in the USA.

With regard to the availability of survey data in the countries of Southeast Asia, it was reported that a start has been made in the Philippines to locate and analyze the large quantity of unused raw data obtained from past research-vessel exploratory trawling. This work is being conducted by a Philippine biologist working at ICLARM (as a part of a training program) and it was agreed that where large amounts of raw data are necessary to reach conclusions such work is much cheaper and more worthwhile than organizing new research to get more data that again might not be used. It is important to reduce these avoidable losses of data.

In Thailand and Indonesia, attempts are being made to computerize the large quantity of hand-recorded survey data now available, but this is making slow progress. While most countries want to work up their own data they welcome assistance in doing this. In fact, a lot of joint working up of data has been done through regional South China Sea Programme workshops and there also are evaluations of the resources by visiting scientists.

Turning to the supply of fisheries scientists in Southeast Asian countries the participants agreed that the problem appears to be primarily one of inadequate numbers of research leaders with adequate ability, training and experience rather than a shortage of fisheries biologists *per se*. In some countries, a major problem is the low salaries of government employees which makes it impossible for the government laboratories to recruit and retain qualified scientists, especially when competing against universities and the private sector.

Studying Single-Species Dynamics in a Tropical Multispecies Context*

DANIEL PAULY

International Center for Living Aquatic Resources Management
MCC P.O. Box 1501, Makati, Metro Manila, Philippines

PAULY, D. 1982. Studying single-species dynamics in a tropical multispecies context, p. 33-70. *In* Pauly, D. and G.I. Murphy (eds.) Theory and management of tropical fisheries. ICLARM Conference Proceedings 9, 360 p. International Center for Living Aquatic Resources Management, Manila, Philippines and Division of Fisheries Research, Commonwealth Scientific and Industrial Research Organisation, Cronulla, Australia.

Abstract

Some methods available for the study of the population dynamics of tropical fish are reviewed; emphasis is given to questions of growth, mortality and recruitment, and to relatively new approaches, such as ageing by means of daily otolith structures, detailed analysis of length-frequency data and to comparative studies.

A package of methods is discussed which is based on a set of 3 new, simple computer programs written in BASIC for the analysis of length-frequency data (ELEFAN I, II and III).

Introduction

This paper is a discussion of methods for the investigation of the major aspects of the population dynamics of single-species tropical stocks, both freshwater and marine. The reason for the presentation of this contribution at a workshop devoted to the theory and management of tropical multispecies stocks is compelling: fishery biologists modeling the dynamics of multispecies systems will often work with single-species stocks as elements of the multispecies system. They will want to know as much as possible of the autecology and biology of single-species stocks.

On the other hand, the builders of tropical models have at their disposal much less information on the various species—even on those of commercial importance—than the model builders working in the temperate areas of the world.

There are three major reasons for this:

- there are many more species—even species of commercial importance—in tropical than in temperate systems
- much less has been published, in absolute terms and on a per-species basis on tropical than on temperate fishes
- much of the work done on tropical stocks is poorly documented and often remains unpublished (see Pauly 1979a).

---

*ICLARM Contribution No. 106.

This contribution is addressed to the use of simple and rapid methods, and parameter estimates obtained from comparative methods.

Also, emphasis is given to methods which can be used to reduce and interpret previously published, but underutilized data (such as length-frequency data) and to methods in which certain biological features of tropical fishes, such as their short life cycles or their prolonged spawning seasons are turned from a liability to an asset.

Russel's axiom (Russel 1931) is still the best method of identifying, in qualitative terms, the elements that determine the yield of a fishery. This axiom states that four factors determine the dynamics of a closed population, namely:

— Recruitment of fishes into the usable stock (R)
— Growth of fishes that are part of the usable stock (G)
— Capture by fishing (C)
— Natural death (mainly by predation) (M)

Thus, we have

$$B_2 = B_1 + (R + G) - (M + C) \qquad \ldots 1)$$

where B_1 and B_2 are the stock sizes at the beginning and the end, respectively, of the time period to which the increments (R and G) and the decrements (M and C) apply. (See also Fig. 1.)

The four explicit items of Russel's axiom will be used to structure this contribution, which also covers another, implicit term of this axiom, namely the intrinsic rate of increase (r_m) of populations, corresponding to $R (G - M)/\bar{B}$.

Fish Growth

The Generalized von Bertalanffy Growth Formula

The model most commonly used in fishery biology to express the growth of fishes is the von Bertalanffy Growth Formula (VBGF) which has, for growth in length the form

$$L_t = L_\infty \left(1 - e^{-K(t - t_o)}\right) \qquad \ldots 2)$$

and

$$W_t = W_\infty \left(1 - e^{-K(t - t_o)}\right)^3 \qquad \ldots 3)$$

for growth in weight, where L_∞ and W_∞ are asymptotic sizes, or the mean sizes the fish would reach if they were to grow indefinitely, where K is a growth constant, and where t_o is the "age" the fish would have at length or weight zero if they had always grown according to the equation, while L_t and W_t are the predicted sizes at age t. Both equations can be derived from Pütter's (1920) continuity relationship

$$\frac{dw}{dt} = Hw^d - kw^m \qquad \ldots 4)$$

which states that the growth rate of fish may be conceived as the difference between two processes with opposite tendencies, one building up body substances (anabolism), the other breaking down body substances (catabolism), both processes being proportional to some power (d, m) of body weight (w). Equation (4) can be integrated in two ways:

— either by setting d = 2/3, m = 1 and by assuming that weight growth is isometric. This leads to what will be called the "special VBGF" (equations 2 and 3) called here "special" because they refer to the special case of d = 2/3 and m = 1)
— or by allowing d and m to take a wide range of values. This leads to what will be called here the "generalized VBGF".

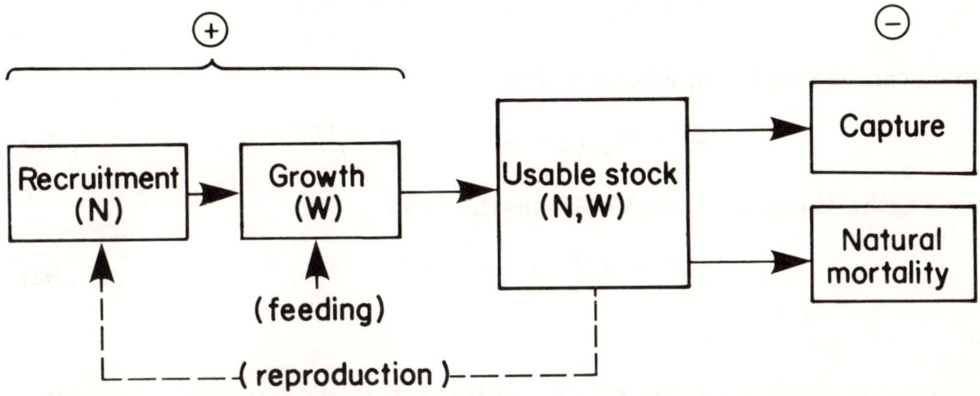

Fig. 1. Main factors investigated in stock-assessment: recruitment, growth (positive factors), capture and natural mortality (negative factors) and stock size. (N refers to numbers, W to weight). Feeding and reproduction, although also contributing to stock dynamics, are generally not considered in single-species stock assessment. Modified from Ricker (1975).

The integration of the special VBGF is well-documented (von Bertalanffy 1934, 1938, 1951; Beverton and Holt 1957; Gulland 1969), while the integration of the generalized VBGF was discussed in detail in Pauly (1979b). Here, only the definitions used in the course of that integration are presented, together with those equations pertaining to important properties of the generalized VBGF.

The substances needed for anabolism have to enter into the body across some surface (S) that increases with fish length (L) according to the relationship

$$S = p \cdot L^a \qquad \ldots.5a)$$

where a is a power that has a range of values discussed below and p is a proportionality constant.

Catabolism, on the other hand, which occurs in all cells of the fish body, may be considered directly proportional to weight (W), which is itself related to length (L) by the relationship

$$W = q \cdot L^b \qquad \ldots.5b)$$

where q is a proportionality constant and b is a power whose range of possible values lies between 2.5 and 3.5 (Carlander 1969) and b > a. Other definitions used in the course of the integration are

$$K = \frac{k}{3} \quad \ldots 6)$$

and

$$D = b - a \quad \ldots 7)$$

with the integration of expression (4) resulting for weight growth in

$$W_t^D = W_\infty^D \left(1 - e^{-KD\ 3/b\ (t - t_o)}\right)^b \quad \ldots 8a)$$

which can be simplified, when b = 3 to

$$W_t = W_\infty \left(1 - e^{-KD\ (t - t_o)}\right)^{1/D} \quad \ldots 8b)$$

For lengths, the generalized VBGF has the form

$$L_t^D = L_\infty^D \left(1 - e^{-KD\ (t - t_o)}\right) \quad \ldots 9a)$$

or

$$L_t = L_\infty \left(1 - e^{KD\ (t - t_o)}\right)^{1/D} \quad \ldots 9b)$$

It will be noted that these versions of the generalized VBGF (expressions (7) to (9) reduce to the corresponding form of the special VBGF (expression (2) and (3)) when D = 1, i.e., when a = 2 and b = 3.

A major difference between the special and generalized VBGF for length is that the latter has an inflexion point whose position on the time axis (t_i) is given by

$$t_i = t_o - \frac{\log_e D}{K \cdot D} \quad \ldots 10)$$

Thus, t_i moves toward t_o when D comes close to 1, with no inflexion point left point when D = 1.

Von Bertalanffy proposed his equations (expressions (2) and (3)) on the assumption that d = 2/3 generally holds true ("2/3 law of metabolism") and that the anabolism of fishes ought to be proportional to their metabolic rate (that is, to oxygen consumption). He also demonstrated that metabolism is indeed proportional to 2/3 of the weight of his experimental guppies, then assumed his "2/3 law of metabolism" applied to all fishes.

Actually, it can be demonstrated that the "2/3 law of metabolism" does not apply to all fishes—in fact, it seems to strictly apply only in the case of the very guppies which von Bertalanffy investigated. In other fishes (in fishes capable of reaching larger sizes), the power linking weight and metabolism (or that linking weight and gill surface area, which is equivalent) is generally close to 0.8, with values of up to 0.9 in tuna and presumably other large, active fishes (Muir 1969; Pauly 1981 and see Fig. 2).

Thus, to qualify as a "physiological" formula, the VBGF has to be used in conjunction with values of "d" in equation (4) ranging between 0.5 (as reported for small cyprinidontids by Winberg 1961) and 0.9 and higher (but always < 1).

The generalized VBGF has a number of properties which make it superior to the special VBGF. Among those, two may be listed here:

— fitting the generalized VBGF to a set of size-at-age data generates asymptotic size values that closely correspond to the largest fish reported from a given stock (Pauly 1981; Gaschütz et al. 1980), which demonstrates that values of L_{max} or W_{max} can generally be used as preliminary estimates of asymptotic size (Fig. 3).

— the values of K estimated from growth data on the basis of the generalized VBGF correspond more closely to the catabolism of fishes than the values of K generated by fitting the special VBGF (Pauly 1981).

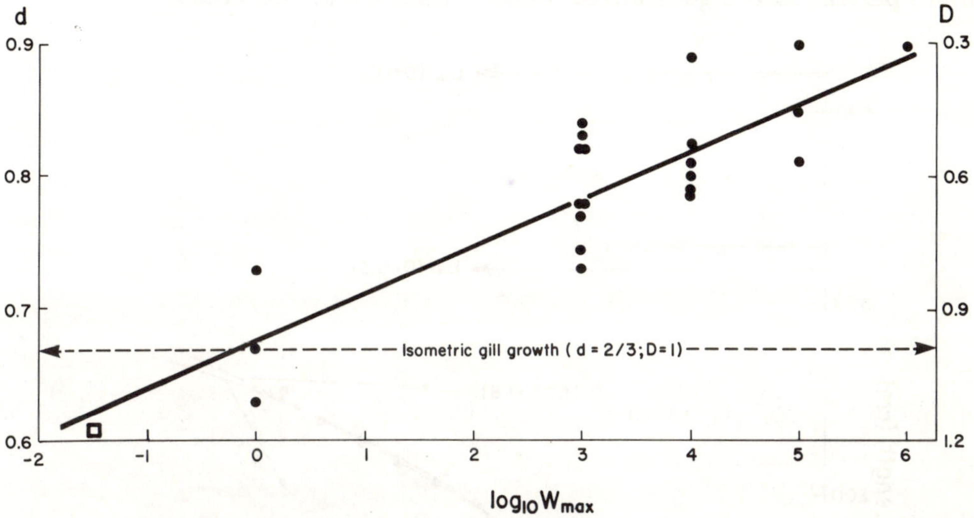

Fig. 2. Values of the power of weight (d) in proportion to which the oxygen consumption or the gill surface of fishes increase, plotted against the maximum weight commonly reached in various species of fishes (in grams). Low values of d ($\approx 2/3$) to cyprinodonts, high values (0.85 to 0.90) to tuna; intermediate values refer to various fishes such as salmonids, cyprinids, tilapias, cods and others. Based on Table V in Pauly (1981). The relationship between D and d is $D = 3(1 - d)$. [The open square in the lower left corner refers to the Philippine dwarf goby *Mistichthys luzonensis*, is based on data in Te Winkel (1935) and was added subsequently to computing the regression line, which is based on the dots only.]

Fitting the generalized VBGF to a set of growth data is as easy as fitting the special VBGF because reasonable estimates of D can be obtained independently of growth data from the relationship (in Fig. 2):

$$D \approx 3 \cdot [1 - (0.6742 + 0.03574 \log W_{max})] \qquad \ldots . 11)$$

which was established by Pauly (1981) on the basis of data on the metabolism and/or the gill surface area and the maximum weight reached by a number of fishes. Thus, once D has been estimated, L_∞ and K can be estimated by means of a Ford-Walford Plot, i.e., a plot of L_{t+1}^D on L_t^D or

$$L_{t+1}^D = \text{intercept} + \text{slope} \cdot L_t^D \qquad \ldots 12)$$

where

$$L_\infty = \left(\frac{\text{intercept}}{1 - \text{slope}}\right)^{1/D} \qquad \ldots 13)$$

and

$$K = \frac{\log_e \text{slope}}{D} \qquad \ldots 14)$$

A simple, versatile computer program written in BASIC was presented by Gaschütz et al. (1980) which can be used to fit any type of size-at-age data with the generalized VBGF, including seasonally oscillating length growth data as discussed below. In the following, references to growth curves pertain to the generalized VBGF unless otherwise noted.

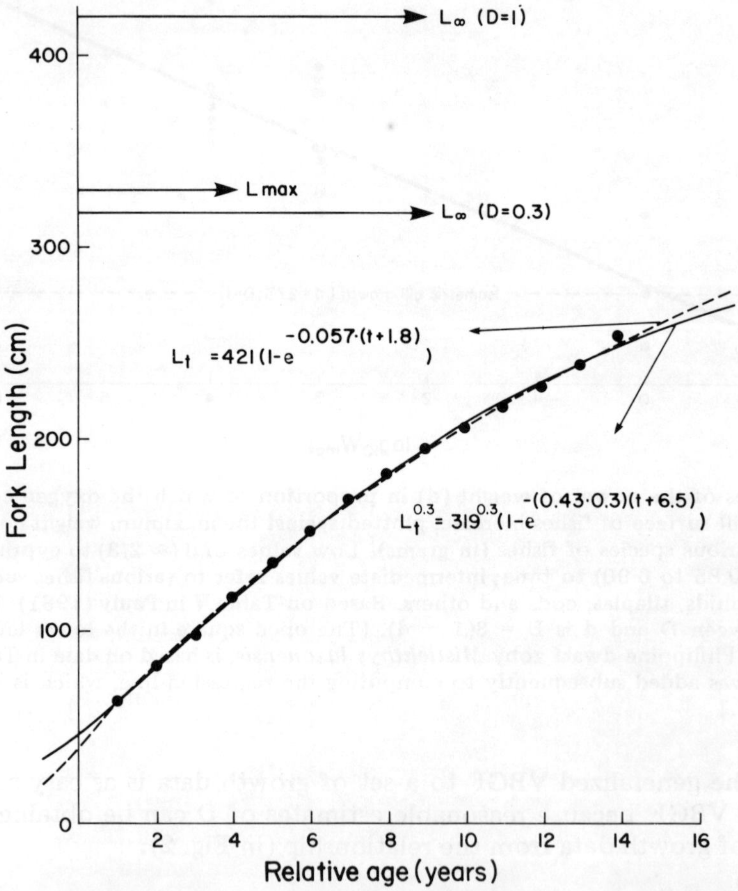

Fig. 3. Sella's (1929) length-at-age data on bluefin tuna *(Thunnus thynnus)* fitted with both the special and the generalized VBGF. Note that $L_\infty \approx L_{max}$ when using the generalized VBGF. The L_{max} value stems from Tiews (1963) and the value of d = 0.9 (hence D = 0.3) stems from Muir (1969) (see text).

Seasonally Oscillating Growth

While the various versions of the VBGF assume constant environmental conditions, it has been demonstrated by many authors that fish live in an environment with seasonally oscillating features (temperature, food availability) and that, therefore their growth also oscillates seasonally (Ursin 1963; Pitcher and MacDonald 1973; Lockwood 1974; Daget and Ecoutin 1976; Cloern and Nichols 1978). A versatile model suited to expressing the seasonally oscillating length growth of fishes was presented by Pauly and Gaschütz (1979) which is based on the VBGF and becomes, in its generalized form

$$L_t = L_\infty \left(1 - e^{-[KD(t - t_o) + C\frac{KD}{2\pi} \sin 2\pi (t - t_s)]}\right)^{1/D} \quad \ldots 15)$$

where L_∞, K, D, t_o, L_t and t are parameters as defined above, while the new parameters C and t_s refer to the intensity of the (sinusoid) growth oscillations of the model and to the onset of the first oscillation with regard to t = 0, respectively. It is a property of the parameter C that it takes a value of 1 when the growth rate (dL/dt) has exactly one zero value per year (generally in the coldest month of the year), a value of zero when the fluctuations are nil, in which case equation (15) reduces to equation (9), and intermediate values when the seasonal oscillations are sufficient to reduce, but not to halt growth in length (Fig. 4).

In Table 1 growth parameters are presented for 10 fishes the length-at-age data of which were fitted with equation (15) by means of the computer program presented by Gaschütz et al. (1980). Table 1 also contains estimates of the temperature range ΔT (= difference between the highest and lowest average mean monthly temperature to which the fish are exposed in the course of a year) for the 10 fish (stocks) in question.

As may be seen from Fig. 5, the values of C for the various fish (stocks) strongly correlate with ΔT, suggesting that

- seasonal length growth oscillations are caused by temperature fluctuations or by another environmental parameter which itself correlates with temperature (e.g., food availability, or foraging behavior, or both).
- slight seasonal fluctuations of temperature (or of correlated parameters) such as occur in the tropics are sufficient to generate seasonally oscillating growth curves, and a seasonally oscillating growth model should be used to fit growth data pertaining to intervals of less than one year (e.g., when fitting growth data obtained by reading daily otolith structures or when analyzing length-frequency data sampled at less than yearly intervals, see below).

A model similar to the one used here to simulate the seasonally oscillating length growth of fish was developed by Shul'man (1974) to depict the seasonally oscillating growth in weight of Black Sea fishes. Shul'man's approach seems particularly suitable for incorporation in seasonally oscillating metabolic models of fish growth in the tropics.

Table 1. Parameters of seasonally oscillating growth curves in 10 fish (stocks) arranged in order of increasing annual range of water temperature (ΔT).

No.	Species	Area	Growth parameters[a] L_∞	K	t_o	t_s	C	ΔT (°C)[b]	Source of growth data
1	*Pseudotolithus elongatus*	Sierra Leone	52.0	0.366	−0.098	−0.132	0.40	3*	Longhurst (1963, Fig. 10)
2	*Tilapia nigra* (untagged)	Kenyan ponds	28.2	1.95	−0.151	−0.010	0.20	4	Van Someren and Whitehead (1960, Fig. 1)
3	*Tilapia nigra* (tagged)	Kenyan ponds	27.9	1.64	−0.253	−0.005	0.45	4	Van Someren and Whitehead (1960, Fig. 1)
4	*Macrorhamphosus scolopax*	Meteor Bank	20.0	0.462	−0.507	0.482	0.90	4*	Ehrich (1976, Table 3)
5	*Hemirhamphus brasiliensis*	Florida	32.5	0.587	−1.024	0.253	0.69	6*	Berkeley and Houde (1978, Fig. 5)
6	*Trisopterus esmarki*	North Sea	17.8	1.06	−0.179	0.189	1.04	7*	Gordon (1977, Fig. 6)
7	*Salmo trutta*	Scottish river	22.2	0.411	−0.294	−0.005	1.24	10	Egglishaw (1970, Fig. 5)
8	*Salmo salar*	Baltic Sea	14	0.324	−0.101	−0.242	1.14	12*	Thurow (1973, Table 1)
9	*Tilapia nigra* ♂	open-air aquarium	16.9	1.24	−1.24	0.427	1.25	13	Cridland (1965, Fig. 1)
10	*Tilapia nigra* ♀	(Uganda)	18.5	0.956	−1.54	0.449	1.42	13	Cridland (1965, Fig. 1)

[a] All values refer to D = 1.
[b] The temperature data were obtained either from the paper that also gave the growth data, or from an oceanographic atlas (*).

Aging by means of markings on skeletal structures

The standard techniques for aging fishes by means of markings on their scales, otoliths or other hard structures are documented by vast amount of literature, much too large to be reviewed here. Mohr (1927, 1930, 1934) gave a comprehensive bibliography of the earlier literature, which was also reviewed by Graham (1928). Bibliographies of the more recent literature on fish growth are given in Menon (1950), Bagenal (1974) and Pauly (1978). Brothers (1980) reviewed and assessed methods that are particularly well-suited for use in the tropics, namely, aging by means of daily otolith structures and analysis of length-frequency data.

While the methods reviewed by Brothers (1980) for the analysis of length-frequency data are rather conventional ("Petersen method" and "modal class

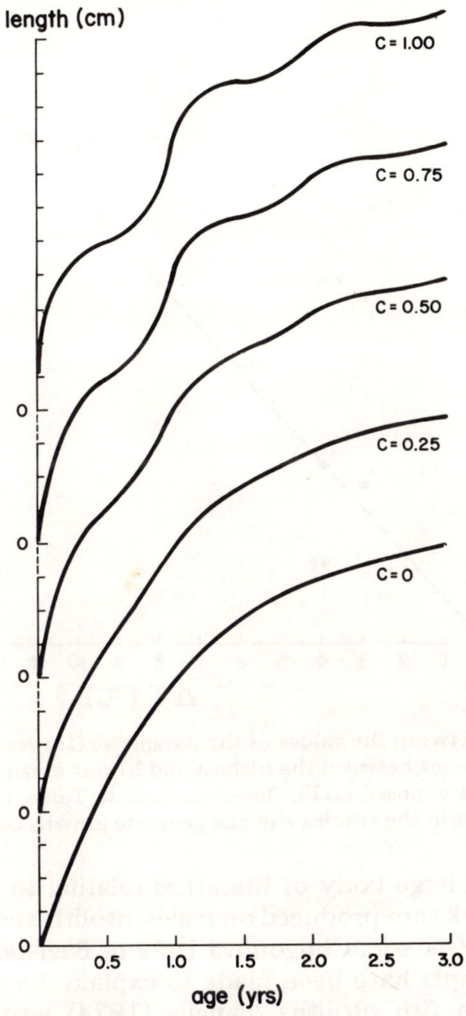

Fig. 4. Seasonally oscillating growth curves based on equation 15, with L_∞ = 25 cm, K = 1, t_o = 0, t_s = 0 and D = 1, with values of C ranging from 0 to 1.

progression analysis", see below), his review of aging by means of daily otolith structures (rings) leads to the conclusions that:

— reading daily otolith rings in tropical fishes, at least in relatively short-lived ones, is as easy as reading annual rings in temperate fishes, although it may be more time consuming,
— the overwhelming majority of commercially important tropical fish families display daily otolith rings, and
— there is no reason why aging by means of daily otolith rings should not become a standard technique for use in the tropics, especially for the validation of growth data obtained from analysis of length frequencies.

Authors reporting work conducted on daily growth rings in tropical and other fish are, among others: Brothers (1980), Brothers et al. (1976), Worthmann (1980), Gjøsaeter and Beck (1980), Methot and Kramer (1979), Le Guen (1976), Panella (1971, 1974), Steffensen (1980), Ralston (1976), Struhsaker and Uchiyama (1976) and Taubert and Cable (1977).

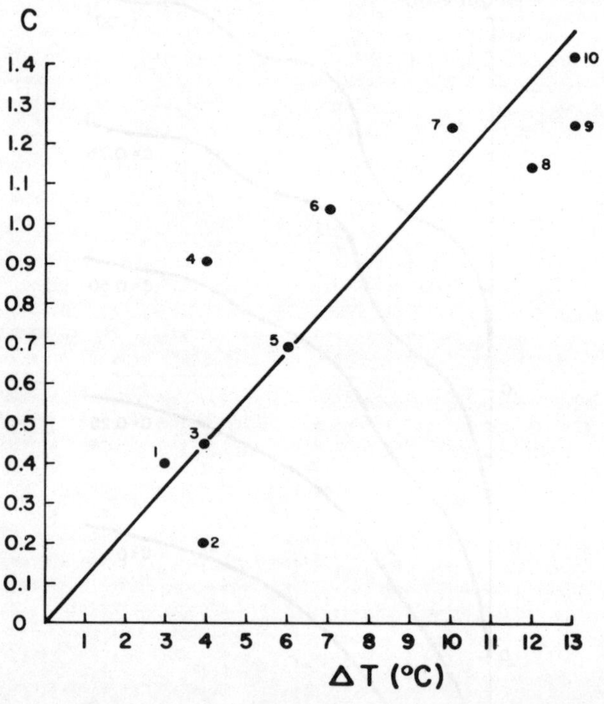

Fig. 5. Relationship between the values of the parameter C in seasonally oscillating growth curves, and the difference between the highest and lowest mean monthly temperatures to which the fishes were exposed (ΔT). Based on data in Table 1. Note that ΔT values of 2 to 3°C, such as occur in the tropics still can generate growth oscillation (see also Fig. 4).

While there is a large body of literature relating to possible reasons why annual growth checks are produced on scales, otolith and bones (and cartilage in elasmobranchs) (see e.g., Chugonova 1959 or Bagenal 1974 and references therein), few attempts have been made to explain the occurrence of marked daily structures on fish otoliths. Panella (1974) writes for example, that "whether rhythmic growth is related to diurnal migratory behavior, rhythmic feeding, or activity and rest is difficult to say."

An explanation for rhythmic growth may be provided by the hypothesis proposed earlier (Pauly 1979b, 1981) that anabolism in fishes is essentially limited by oxygen availability to the tissues (see above derivation of the generalized VBGF). This hypothesis would suggest that fish, which can store little oxygen, can be either active (e.g., feeding, escaping predators, defending territories, etc.) and diverting all available oxygen to these activities, *or* resting, in which case they can *only then* divert oxygen to anabolism. The diurnal changes in activity patterns would thus be sufficient to generate "pulses" of anabolism, the latter activity being, it must be emphasized, quite distinct from the incorporation of food into the metabolic pool (Iles 1974).

Growth as estimated from length-frequency data

The methods conventionally used for the analysis of length-frequency data have been introduced by Petersen (1892) and can be reduced to two basic techniques:

- the "Petersen Method" *(sensu stricto)*, that is the attribution of relative ages to the peaks a length-frequency sample, and
- the "Modal Class Progression Analysis" (George and Banerji 1964), that is the linking up of the peaks of length-frequency samples sequentially arranged in time by means of growth segments.

With the first method, the problem consists of identifying those peaks representing broods spawned at known or assumed time intervals. The method generally involves the separation of the length-frequency samples into normally or otherwise distributed subsets by graphical methods, such as those proposed by Harding (1949), Cassie (1954), Tanaka (1956), or by means of computer programs such as NORMSEP (Abrahamson 1971) or ENORMSEP (Yong and Skillman 1971).

The "modal class progression analysis", on the other hand, has its major problems in the identification of those peaks which should be connected (by growth lines) with each other (Pauly 1978).

Both methods are highly subjective, and an attempt was made only recently to combine them into one single, "integrated" method which to a certain extent improves the reliability of growth estimates based on the analysis of length-frequency data (Pauly 1980a).

However, the need for a rapid, yet reliable and *objective* method for the analysis of length-frequency data led to a radically new approach in the analysis of length-frequency data, and such an approach was presented in Pauly and David (1981).

Briefly, this new method (called Electronic Length-Frequency Analysis or ELEFAN I) is based on the following steps:

- objective identification (definition) of the peaks and the troughs separating peaks of a (set of) length-frequency sample(s)
- attribution to the peaks of a certain number of positive "points"
- attribution to troughs separating the peaks of a certain number of negative "points"
- iterative identification of those growth parameters generating a growth curve which, by passing through most peaks and avoiding most troughs, accumulates the highest number of "points" and thus best explains the specific structure of a (set of) length-frequency sample(s).

The procedure is illustrated in Fig. 6.

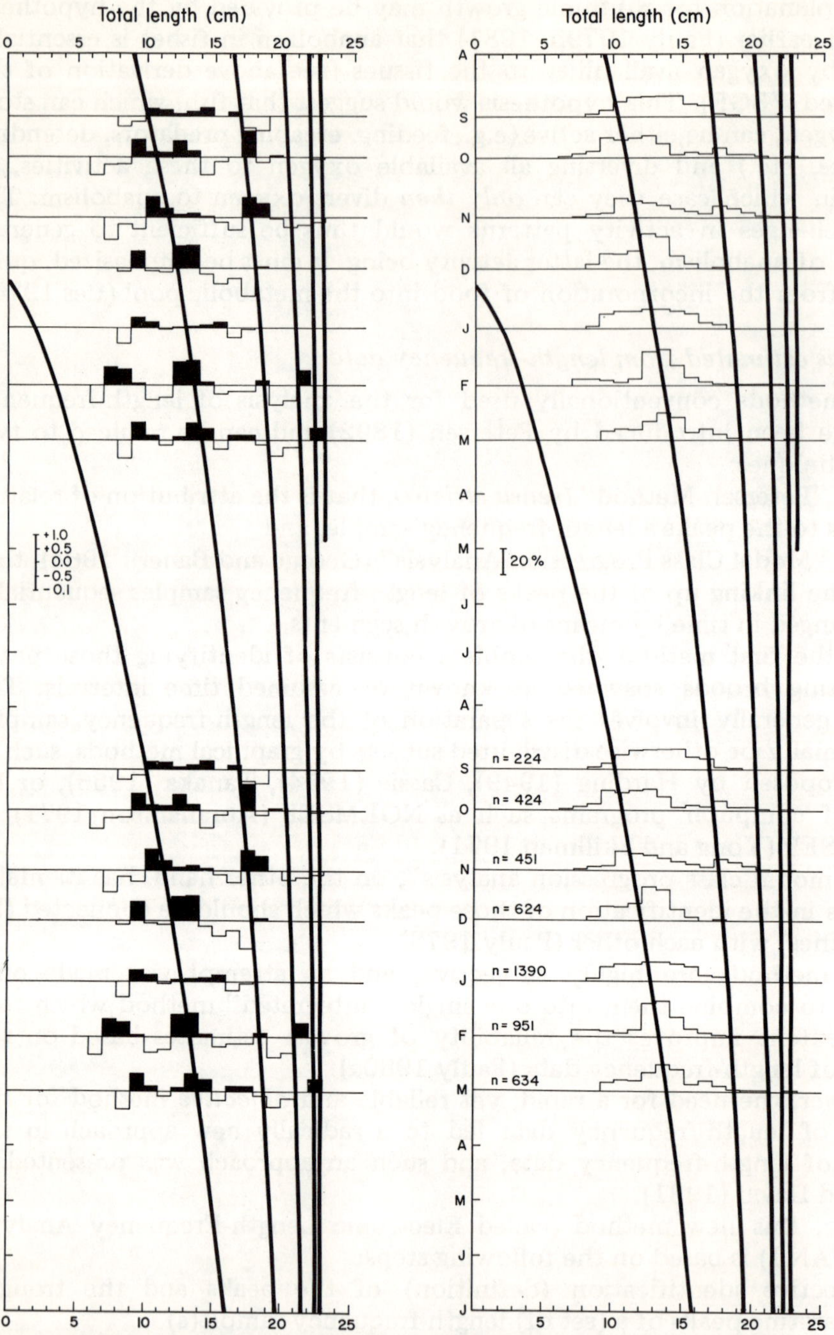

Fig. 6. Estimation of growth parameters by means of ELEFAN I. A) Showing a "restructured" length-frequency sample, where peaks have become positive points (black) and troughs have become negative points (white). These "points" are computed and used internally by ELEFAN I to identify the growth curve which passes through the largest number of positive points, and avoids negative points as much as possible. B) Showing the original length-frequency data, with superimposed growth curve. The curve, which was traced in a purely objective fashion "explains" 62% of the peaks in the length-frequency data (as defined by ELEFAN I).

Given the assumptions that the sample(s) used represent(s) the population investigated, and that the growth of the fish in question conforms to the VBGF, (seasonalized or not), the method can be used to derive growth parameters that are reproducible (i.e., any worker will get the same result from the same set of data). Moreover, an estimator is given of the proportion of the peaks in a (set of) sample(s) that are "explained" by the growth parameters obtained at the end of the iteration process. This estimator is the ratio of a sum called "Explained Sum of Peaks", or ESP, referring to the number of "points" explained by a given growth curve, divided by another sum called "Available Sum of Peaks", or ASP, which refers to the total number of points "available" in a (set of) length-frequency samples). (See Pauly et al. 1980 for details). The method which can be readily implemented on microcomputers, is fast, reliable, and objective. Ideally, the results (growth parameter estimates) should be checked against the results obtained by reading daily structures of the otoliths of a few fishes, which would also serve to obtain estimates of t_o (which is not estimated by ELEFAN I), by solving equation (9) or (15) for this parameter.

Mortality

Total Mortality

A basic equation used in fishery biology for expressing the mortality of fish is

$$N_t = N_o e^{-Zt} \qquad \ldots.16)$$

where N_o and N_t are fish numbers at time zero and t, respectively and where Z is the total mortality affecting the stock. Also, we have

$$Z = F + M \qquad \ldots.17)$$

which states that total mortality is the sum of fishing mortality (F) plus natural mortality (M). A major task of fishery biologists working on a specific stock is the splitting up of Z into its component parts, F and M; this will be discussed further below, after methods for estimating Z have been briefly reviewed.

One of the simplest methods to assess total mortality is to estimate Z from the mean size is the catch, as suggested by Beverton and Holt (1956). The method was discussed in Pauly (1980a) and Munro (this vol.).

Another method of estimating Z is to construct catch curves, i.e., plots of the natural logarithm of fish numbers against their age, where Z is the slope, with sign changed, of the "descending" part of the curve (Ricker 1975). Where fishes both are relatively long-lived and can be aged by means of annual rings on scales, otoliths or bones, catch curves can be constructed and interpreted quite straightforwardly, e.g., as described by Robson and Chapman (1961), or Ricker (1975).

Presented here are the three steps of a method which allow for the derivation of catch curves from length-frequency data, as generally collected from tropical fishes:

Table 2. Showing how to construct a catch curve from a length-frequency sample of *Sillago sihama* from Manila Bay (data from Ziegler 1979).

Class limits, in cm Lower	Upper[b]	Mid-range	N[a]	t_1[c]	t_2[c]	Δt	$N/\Delta t$	% ($N/\Delta t$)	$\log_e$ (% $N/\Delta t$)	Mean relative age (years)[d]
6	7	6.5	0.893	0.421	0.505	0.084	10.6	0.207	−1.58	0.462
7	8	7.5	7.60	0.505	0.595	0.090	84.4	1.66	0.504	0.549
8	9	8.5	21.1	0.595	0.690	0.095	222	4.31	1.46	0.641
9	10	9.5	56.5	0.690	0.792	0.102	554	10.8	2.38	0.740
10	11	10.5	75.6	0.792	0.902	0.110	687	13.4	2.59	0.846
11	12	11.5	81.6	0.902	1.021	0.119	686	13.3	2.59	0.960
12	13	12.5	122	1.021	1.151	0.130	938	18.2	2.90	1.08
13	14	13.5	121	1.151	1.294	0.143	846	16.4	2.80	1.22
14	15	14.5	79.5	1.294	1.453	0.159	500	9.73	2.28	1.37
15	16	15.5	45.9	1.453	1.632	0.179	386	5.00	1.61	1.54
16	17	16.5	32.5	1.632	1.836	0.204	159	3.09	1.13	1.73
17	18	17.5	29.2	1.836	2.075	0.239	122	2.38	0.867	1.95
18	19	18.5	14.6	2.075	2.361	0.286	51.0	0.992	−0.008	2.21
19	20	19.5	7.37	2.361	2.720	0.359	20.5	0.399	−0.918	2.53
20	21	20.5	3.49	2.720	3.201	0.481	7.26	0.141	−1.96	2.94
21	22	21.5	1.09	3.201	3.931	0.730	1.49	0.029	−3.54	3.52
22	23	22.5	1.05	3.931	5.500	1.569	0.699	0.013	−4.35	4.51

[a]The sample was obtained by adding up several samples in which frequencies were expressed in %. This procedure reduces the effects of seasonal recruitment pulses on the catch curve.

[b]The upper class limits, obviously should read 6.999, 7.999, 8.999, etc., but are rounded up for computing t_2 (see below).

[c]t_1 and t_2 are the relative ages corresponding to the lower and upper class limit, respectively, and are computed from:

$$t_{1,2} = \left\{\log_e \left[1 - (L_{t_{1,2}}/L_\infty)^{1/D}\right]\right\} / -KD$$

with $L_\infty = 23.5$, $K = 0.70$ and $D = 1$.

[d]Mean relative age is the relative age corresponding to the class-mid-range.

Pooling of length-frequency data. Since the overall age structure of a population in short-lived fishes is essentially determined by "pulses" of recruitment, a first step should be to attempt to smooth out the recruitment pulses by pooling data obtained at regular intervals during a period of one year. Also, to prevent a larger sample from unduly affecting the total sample, the various samples should be converted to percentage length-frequency samples prior to adding to obtain a single, overall sample (see also Munro, this vol.).

Construction of catch curve proper. Since the growth in length of fish is not linear, and generally slows down as length increases, there is a tendency for older size groups to contain more age groups than younger size groups. As suggested by J. Gulland (pers. comm.), this can be compensated for by dividing the number of fish in each size group (e.g., cm-class) by the time it takes the fish to grow through a size group (Δt). The catch curve equation thus becomes

$$\log_e (N/\Delta t) = a + bt \qquad \ldots . 18)$$

where t is the age corresponding to midlength of each length class, $b = -Z$, N = the number of fish in each length class and Δt = the time needed for the fish to grow through a length class. (See Table 2 and Fig. 7 for an example).

Identification of biased data points. The conversion of length-frequency samples to a catch curve by means of growth parameter values can involve fishes the individual lengths of which are very close to L_∞, in which cases unrealistically high "ages" are generated. Thus, it is imperative that a scattergram be drawn of the points to use for the computation of Z, in order to select for points really belonging to the descending part of the curve, and contained within a reasonable age-span (Fig. 7).

A versatile program, called ELEFAN II and written in BASIC is available from the author which performs the tasks outlined in 1) to 3), thus allowing for values of Z to be obtained from length-frequency data (Pauly et al. 1981). This program, which requires a (set of) representative length-frequency sample(s) and a pair of growth parameters (L_∞ and K) can be run with the length-frequency data also used in ELEFAN I for the estimation of L_∞ and K, thus allowing for growth and mortality parameters to be estimated from the same set of data. The construction of catch curves from length-frequency data can also be done with "pencil and paper" methods, as outlined by Pauly (1980a).

Another method of estimating Z from catch data akin to the use of a catch curve is the construction of a "pseudo-catch curve" (Pauly 1980b). This method is based on the interrelationships between body size and water depth on one hand, and catch rates and water depth on the other hand. The method is particularly suited to the derivation of estimates of Z from data that were gathered for purposes other than straightforward estimation of Z.

A completely different set of methods for the estimation of Z, F and M has been reviewed in Gulland (1969), Ricker (1975), Jones (1977) and elsewhere which involves the tagging and recapture of fish. Tagging programs,

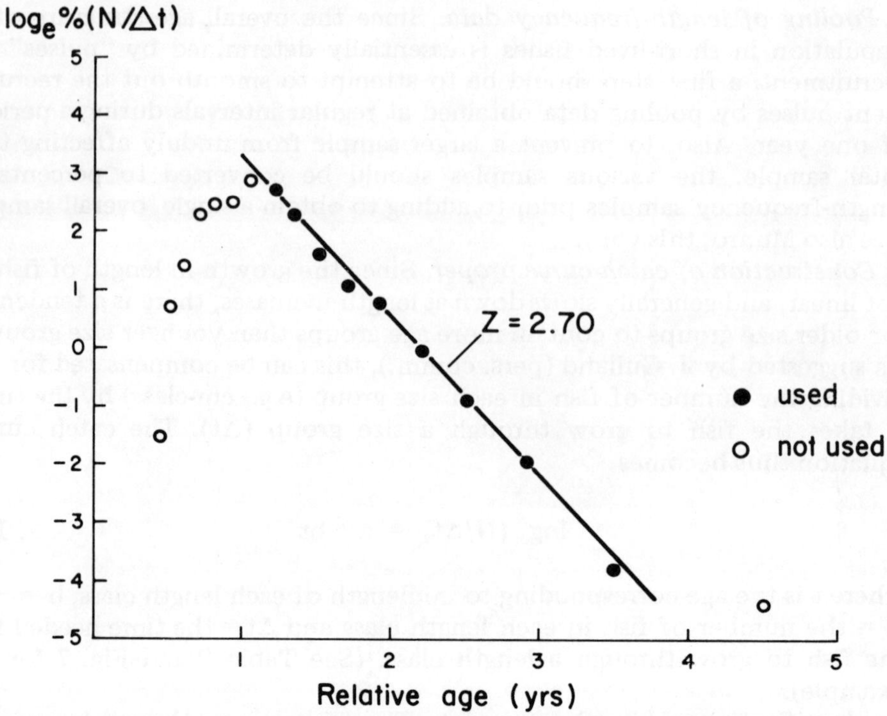

Fig. 7. A length-converted catch curve. Based on length-frequency data in Fig. 6B and on growth parameters (L_∞ = 23.5, K = 0.7 and D = 1) as estimated by means of ELEFAN I. The term "relative age" refers to the fact that t_o is set equal to zero when converting length to age. Note that for the computation of Z, one point, too far to the right was omitted, along with the ascending part of the curve. See also Table 2 and text.

which can provide mortality estimates in addition to estimates of growth, migrations, and data on stock size and identity have been conducted in conjunction with many tropical stocks. Table 3 summarizes the results of studies conducted in the tropics and subtropics. Tagging/recapture programs are generally quite costly (e.g., because of the need to advertise the program widely, in order to obtain a sufficiently large number of returns). It needs to be established whether tagging programs for more than a few key species should be conducted in multispecies stock investigations.

Natural Mortality

Natural mortality (M) is a parameter that is generally extremely difficult to estimate, and typically, natural mortality estimates of tropical fish have been obtained from estimates of total mortality in stocks known, or assumed to be unfished (e.g., Thompson and Munro 1978; Weber and Jothy 1977; Pauly and Martosubroto 1980). In a few cases, however, it has been possible to obtain time series of values of Z from the same stock, and to plot these against their corresponding values of effort, with M being obtained from the intercept of the line fitted to these data. Ricker (1975) gives the rationale of the method, which also provides an estimate of the catchability coefficient (q) of the gear in question (see Fig. 8 for an example from the tropics).

Tropical multispecies stocks are often exploited, while time series of Z and effort are generally unavailable, for which reasons M most often cannot be estimated by any of the conventional methods.

Following an earlier attempt by Beverton and Holt (1959) and others to identify predictors of M based on comparative studies, a compilation was undertaken of 175 estimates of M, ranging from freshwater to marine and from polar to tropical fishes. It was then shown (Pauly 1980c) that M can be predicted from a knowledge of the growth parameters of a given stock, and of its mean environmental temperature (e.g., as estimated from an oceanographic atlas), the empirical relationship linking these being

$$\log_{10} M = 0.0066 - 0.279 \log_{10} L_\infty + 0.6543 \log_{10} K + 0.4634 \log_{10} T \quad \ldots. 19)$$

where M is the exponential rate of natural mortality, on a yearly basis, L_∞ the asymptotic length of the fish in the stock, in cm, K their growth coefficient (on a yearly basis) and T the mean environmental temperature for the stock in question, in °C. The equation has a multiple correlation coefficient of R = 0.847, which is significant (with 171 degrees of freedom, only 0.275 is needed for significance at P = 0.01). (L_∞ and K of equation (19) refer to D = 1).

As discussed in Pauly (1980c), where the data used for the derivation of this equation are documented, the equation produces reasonable estimates of M in most stocks, the only exceptions as far as the tropics are concerned being those values of M pertaining to Clupeidae and possibly to other strongly schooling fishes, for which mortality is slightly over-estimated by the equation. Thus, in such cases, multiplying estimates of M obtained from equation (19) by a factor of 0.8 or so might be appropriate.

The values of M obtained by means of equation (19) are "reasonable" in the sense that they cannot be very different from the true values, as opposed e.g.. to estimates based on a plot of Z on effort, which at times can produce completely erroneous values of M, including negative ones (Ricker 1975 p. 174). On the other hand, it is known that the natural mortality of fishes varies with age (Boiko 1964) and most probably also with predator abundance (Pauly 1980d, 1982; Munro, this vol., Jones, this vol.). The estimates of M provided by equation (19) should thus be considered as first approximations pertaining to the exploited phase of a population, under average conditions only.

Fishing Mortality

Of the various methods used to estimate fishing mortality, four may be listed here:

— tagging/recapture studies
— subtraction of M from Z
— swept-area method in the case of trawlable demersal stocks
— Virtual Population Analysis (VPA) or Cohort Analysis

Tagging/recapture studies are not discussed further in this paper, and the reader is referred to the reviews mentioned above and to the references listed in Table 3.

Table 3. Selected tagging-recapture or marking studies conducted on tropical and subtropical fishes.[a]

Species	Family	Area	Main results	References
Trianodon obesus	*Carcharhinidae*	Johnston Island, Central Pacific	growth data, movements	Randall (1977)
Galeorhinus australis	*Carcharhinidae*	South-East Australian Waters	growth parameters, M&F	Grant et al. (1979)
Various shark spp.	—	Eastern Central Pacific	growth data, movements	Kato and Carvallo (1967)
Beryx splendens	*Bercydae*	Japan	growth parameters	Ikenouye and Masazawa (1968)
Priacanthus tayenus	*Priacanthidae*	Gulf of Thailand	growth data, movements	Chomjurai and Bunnag (1970)
Chrysophrys auratus	*Sparidae*	South-East Australian Waters	growth parameters	Sanders and Powell (1979)
Pagrus major	*Sparidae*	Japan	population size, M&F	Kato and Yamada (1975)
Tilapia esculenta	*Chichlidae*	Lake Victoria (Uganda)	growth parameters, by sex	Garrod (1963)
Amphiprion spp.	*Pomacentridae*	Eniwetok Atoll, Pacific	growth data, behavior	Allen (1975)
Acanthurus triostegus	*Acanthuridae*	Hawaii	growth data, movements	Randall (1961)
Balistes capriscus	*Balistidae*	Togo Shelf, West Africa	F, movements	Beck (1974)
Rastrelliger neglectus	*Scombridae*	Gulf of Thailand	M&F, growth parameters	Hongskul (1974)
Scomber australasicus	*Scombridae*	North of Taiwan	movements, growth data	Chang and Wu (1977)
Thunnus albacares	*Scombridae*	Eastern Central Pacific	migrations	Bayliff (1979)
Istiophorus platypterus	*Istiophoridae*	off Eastern Florida	tagging mortality	Jolley and Irby (1979)
Various teleostean spp.	—	Florida, U.S.A.	new tagging method	Thresher and Gronell (1978)
Various teleostean spp.	—	Virgin Island, Caribbean	behavior, growth data	Randall (1962, 1963)

[a]This list is not exhaustive and is presented only for illustrative purposes, and to give an entry into the literature.

The detailed analysis of length-frequency data, as outlined above, based either on the programs ELEFAN I and ELEFAN II, or on "paper and pencil methods" (Pauly 1980a) provides estimates of growth parameters and of Z, while the growth parameters, combined with an estimated value of mean environmental temperature can be used to provide an estimate of M from equation (19), which is then subtracted from Z to obtain F.

This method allows for a quick assessment of whether a stock is overfished or not, based on the assumption that the value of F which optimizes yield should be similar to M, or

$$E_{opt} \approx \frac{F}{F+M} \approx 0.5 \qquad \ldots.20)$$

where E_{opt} is the exploitation rate which optimizes the yield from a given stock (Gulland 1971).

Another method of estimating fishing mortality in trawl fisheries is the "swept area method", as treated in Gulland (1969). The method can be summarized in one equation, namely

$$F = \frac{a \cdot X_1}{A} \qquad \ldots.21)$$

where "a" is the total area "swept" by the combined effects of all gears of a fleet (of trawlers), A is the total area inhabited by the stock in question, and X_1 is the escapement factor, i.e., the fraction of the fish in the path of the gear that are actually retained by the net.

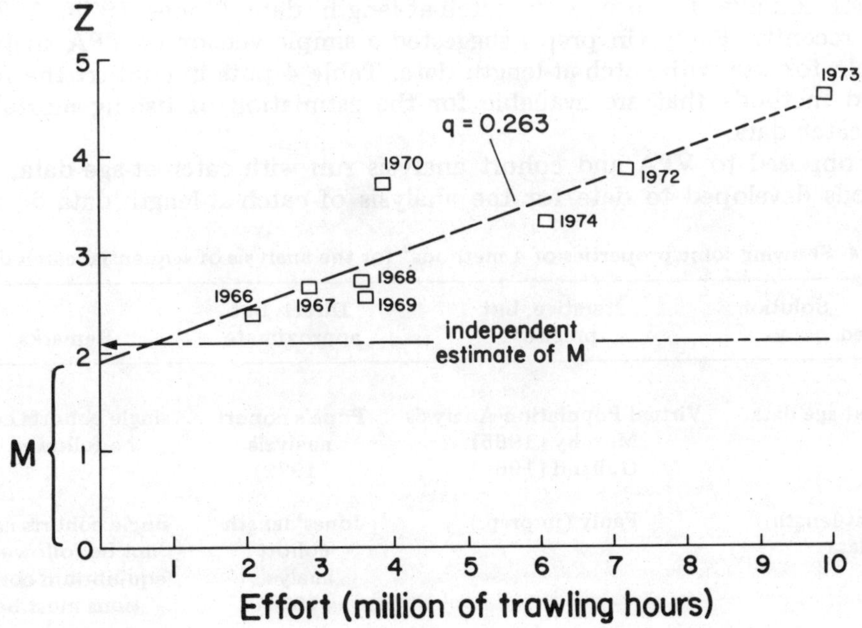

Fig. 8. Example of a plot of total mortality on effort for a tropical stock (*Selaroides leptolepis*, Gulf of Thailand). Note that estimated M (intercept) corresponds closely with value of M estimated independently, from equation (19). [The values of Z and f stem from Pauly (1980d).]

Values for the escapement factor ranging between 1 and 0.4 have been proposed and used by various authors working in tropical multispecies stocks (Isarankura 1971; Saeger et al. 1976; SCSP 1978; Gulland 1979); these estimates are commonly based on assumptions of various kinds. Pauly (1980d) demonstrated that, for the Gulf of Thailand trawl fishery, values of F obtained independently i.e., by estimating Z from mean sizes in the catch in 6 species of fish, and M by means of an empirical equation similar to equation (19) could, on the average, be made equal to values of F obtained from equation (20) by setting $X_1 = 0.5$ (and $X_2 = 0.5$, see below).

The area "swept" by the gear during one unit of effort (a′) is estimated from

$$a' = t \cdot v \cdot h \cdot X_2 \qquad \ldots 22)$$

where t is the time spent trawling, v the velocity of the trawler over ground when trawling, h is the length of the trawl's head rope and X_2 a fraction expressing the width of the area swept by the gear divided by the head rope length. Values of X_2 ranging between 0.4 and 0.66 are commonly used in Southeast Asian waters, 0.5 probably being the best guess (Pauly 1980d; see above for X_1).

Wheeler and Ommaney (1953) have attempted to adapt the swept-area method to handlining for large coral fishes. (See also Gulland 1979).

Virtual Population Analysis (VPA) and the related method of cohort analysis which can be used to estimate F from catch-at-age data have recently been reviewed by Jones (1981), who also suggested a version of cohort analysis suitable for use with catch-at-length data (Jones 1974, 1981). More recently, Pauly (in prep.) suggested a simple version of VPA analysis suitable for use with catch-at-length data. Table 4 puts in context the four related methods that are available for the estimation of fishing mortality from catch data.

As opposed to VPA and cohort analysis run with catch-at-age data, the methods developed to date for the analysis of catch-at-length data do not

Table 4. Showing some properties of 4 methods[a] for the analysis of sequential catch data.

Data required / Solution	Iterative, but precise*	Direct, but approximate	Remarks
catch-at-age data	Virtual Population Analysis Murphy (1965) Gulland (1965)	Pope's cohort analysis (1972)	single cohorts can be followed
catch-at-length data	Pauly (in prep.)	Jones' length cohort analysis (1974)	single cohorts cannot be followed; equilibrium conditions must be assumed or generated by averaging

[a]Simple programs, implemented on HP 67/97 programmable calculators are available from the author for all four of these methods.

allow for the fate of single cohorts to be investigated, i.e., length "cohort analysis provides estimates of the *average* number attaining each length during a year, as well as the *average* numbers present in a length group at any particular moment." (Jones 1981) [emphasis added, DP].

With this problem in mind, Pope et al. (in prep.) have developed a BASIC program (ELEFAN III) which allows for the use of catch-at-length data in VPA. In this program, the growth parameters of the stock in question (e.g., as obtained from ELEFAN I) are used in conjunction with the catch data arranged sequentially in time to define, in a first step, series of cohorts "sliced" from the catch-at-length data. Then VPA's are run for each of these (arbitrary) cohorts, which leads to the identification of recruitment peaks, i.e., which allow for the definition of "real" cohorts, composed of fish recruited at approximately the same time. The program then allows, in a final step, for running VPA's on whichever cohorts have been identified as "real", the crucial advantage of this approach over the two length-structured models in Table 4 being that an equilibrium situation need not occur, i.e., the decay of single cohorts can be followed over time. The program, it may be mentioned, can accommodate seasonally oscillating growth in length, and may thus be used also with temperate fishes.

Stock Size, Production and Intrinsic Rate of Increase

Estimating Stock Size

A plethora of methods are available to estimate the stock size of fish, a partial list being:

— tagging/recapture methods (see above)
— acoustic methods
— egg and larval surveys
— swept-area method
— VPA and cohort analysis
— De Lury's method
— interrelationship between catch, fishing mortality and standing stock

Acoustic methods are not reviewed here, and the reader is referred to Forbes and Nakken (1972).

Egg and larval survey, in the tropics, generally cannot be used to estimate (spawning) stock size because of the extremely short development time of most eggs (most often less than 10 hours, see Delsman 1972), which renders quasi-synoptic surveys of the spawning grounds extremely difficult to conduct.

The swept area method, which was discussed above in connection with the estimation of F can obviously also be used to estimate stock sizes (B), i.e.,

$$B = \frac{(\overline{c/f}) \cdot A}{a' \cdot X_1} \qquad \ldots 23)$$

where $\overline{c}/f$ is the mean catch-per-effort obtained during a survey (or within one stratum of a survey), and A, a′ and X_1 are defined as in equation 21 and 22).

VPA and cohort analysis (including the versions for use with catch-at-length data) can be used to estimate absolute population sizes; in fact this

Table 5. Successive sample sizes of reef eels *(Kaupichthys hyoproroides)* from an isolated Bahamian patch reef.[a]

Samples	No. of fish collected	Effort[b]
A	5	1
B	4	1
C	3	1
D + E	1	2

[a]Based on data in Smith (1973, Table 5, Station 1).
[b]The unit effort is "11 fluid ounces of emulsified rotenone applied from a plastic squeeze bottle."

is one of the reasons why the methods have become so popular. The reader is thus referred to the discussion of these methods, presented above in conjunction with the estimation of fishing mortality.

The two last methods for stock size estimation seem particularly suited for use in coral reef fish populations. The first of these is De Lury's method of plotting catch-per-effort on cumulative effort in such cases where the reduction of a population by fishing is so rapid that the effects of recruitment or natural mortality can be neglected.

In such cases, e.g., when sampling is done with repeated rotenone applications, the plot yields estimates of initial population generally higher than those obtained by simply adding up all sampled individuals (i.e., by assuming that all fish were removed). An example of this approach is given in Table 5 and Fig. 9.

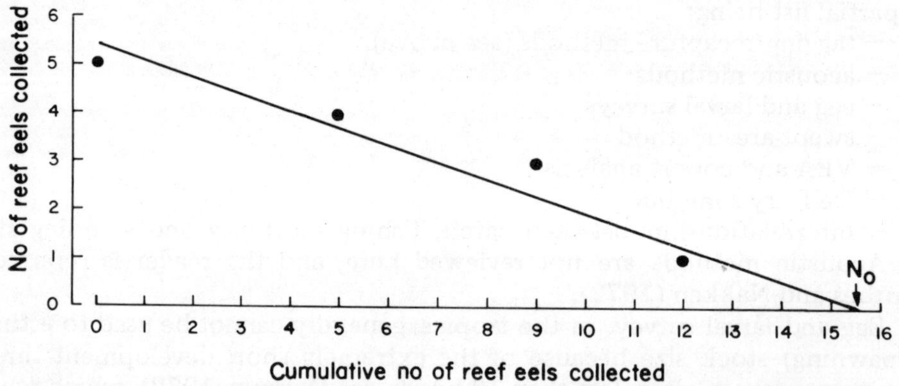

Fig. 9. Example of a DeLury plot from tropical waters: successive catches and cumulative catches of reef eels in a Bahamian reef patch. Based on data in Table 5. See Ricker (1975) for details on the method.

As documented by various authors (see reviews by Ehrlich 1975, Goldman and Talbot 1976, Russel et al. 1978) coral reef fish communities are extremely difficult to sample quantitatively. Thus, in reef stock assessment, emphasis should possibly be given to sampling catches and performing detailed analysis of these data to obtain estimates of F by means of one of the methods outlined above, and to estimate mean standing stock size ($\overline{B}$) from the relationship

$$\overline{B} = \frac{C}{F} \quad \ldots 24)$$

where C is the annual catch from the reef in question, and F the estimated fish mortality. This equation, incidentally, was used by Sekharan (1974) to estimate the standing stock of mackerels *(Rastrelliger kanagurta)* and sardines *(Sardinella longiceps)* of the West Coast of India.

Production

Production (P) is defined as "the total elaboration of fish tissue during any time interval Δt, including what is formed by individuals that do not survive to the end of Δt" (Ivlev 1966).

Allen (1971) has shown that production, in equilibrium situations, can be estimated from

$$P = Z \cdot \overline{B} \qquad \ldots .25)$$

where $\overline{B}$ is the mean annual standing stock and Z is the total mortality, as estimated by one of the methods outlined above. (Other, more sophisticated methods are available to estimate production, see Chapman 1978 a, b).

Equation (26) shows that production can be also defined as the sum of the catch plus the weight of all fish that died of natural causes or

$$P = (F \cdot \overline{B}) + (M \cdot \overline{B}) \qquad \ldots .26)$$

Regrettably, little information is available on the production of tropical marine fishes; what little data there are on tropical fish production stems from freshwater systems (Durand 1978; Welcomme 1979). Thus, it remains to demonstrate that production can be turned into a concept useful for tropical marine stock assessment, e.g., for comparative studies of various systems (see Marten and Polovina, this vol.).

Intrinsic Rate of Increase

The intrinsic rate of increase (r_m) of a population or stock is a parameter which only ecologists use explicitly (see e.g., Odum 1971, or Ricklefs 1973). However, this parameter is implicit in the parabolic surplus-yield or Schaefer model (Schaefer 1954, 1957) i.e.,

$$MSY = \frac{r_m \cdot B_\infty}{4} \qquad \ldots .27)$$

where Maximal Sustainable Yield is defined as the product of the intrinsic rate of increase of a population, times the carrying capacity of its habitat (B_∞) divided by 4 (see also Ricker 1975, p. 315).

Thus, when an estimate is available of the virgin biomass of a given population (B_v, or B_o in Gulland 1971) and when it is legitimate to set $B_\infty \approx B_v$ (it is *not* always the case, see Pauly 1979a or May et al. 1979), all that is needed to obtain a preliminary estimate of MSY is a value of the intrinsic rate of increase (r_m) of the population.

Blueweiss et al. (1978) partly based on earlier data in Fenchel (1974) and Smith (1954) have shown that r_m, in animals and various small organisms is closely related to "adult" body size, and presented a log/log plot of r_m against mean adult body weight ($\overline{w}$) spanning 22 orders of magnitude (!)

and ranging from virus (T-phage) to deers (Fig. 10). Blueweiss et al. (1978) also suggest that the product of r_m and $\overline{w}$ "is an estimate of the maximum possible rate of production" which seems to be a point worth investigating further in future production studies.

As the data used by Blueweiss et al. (1978) contained only two fish (Blueweiss pers. comm.) namely *Gadus morhua* and *Sardinops caerula* (No. 11 and 12 respectively on Fig. 10), 6 new fish stocks have been added onto their graph as well as 4 estimates of r_m and $\overline{w}$ pertaining to whales, which extends the graph to a span of 24 orders of magnitude (Fig. 10). The data used and mode of computation of the r_m and $\overline{w}$ values are presented in Table 6. Quite clearly, r_m and $\overline{w}$ values are strongly correlated; this pertains both to the data pairs of Blueweiss et al. (1978) as well as to the 10 new pairs added here. In fact, there was no need to recompute the regression equation presented by Blueweiss et al. (1978) which states, when r_m is expressed on a daily basis, that

$$r_m = 0.025 \cdot \overline{w}^{-0.26} \quad \ldots 28)$$

($r^2 = 0.90$). Combining equation (27) and (28) and converting to the year as time unit we obtain

$$MSY \approx 2.3 \cdot \overline{w}^{-0.26} \cdot B_v \quad \ldots . 29)$$

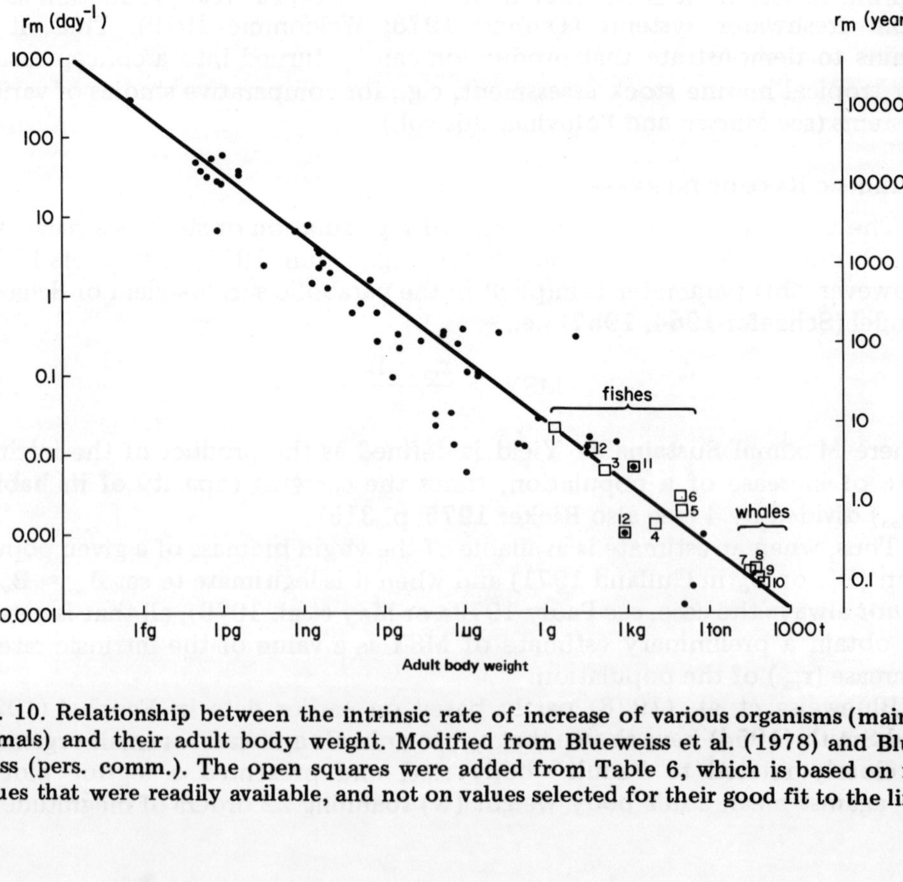

Fig. 10. Relationship between the intrinsic rate of increase of various organisms (mainly animals) and their adult body weight. Modified from Blueweiss et al. (1978) and Blueweiss (pers. comm.). The open squares were added from Table 6, which is based on r_m values that were readily available, and not on values selected for their good fit to the line.

Table 6. Estimates of the intrinsic rate of increase (r_m) obtained from fished stocks[a], as compared with those obtained from mean adult weight ($\overline{w}$)[b] and the empirical plot of r_m on $\overline{w}$ given by Blueweiss et al. (1978).

No.	Species	B_∞ or N_∞	MSY (per year)	r_m	$\overline{w}$	$\hat{r}_m$
1	*Lebistes reticulatus*	32 g	63.3	7.91	2 g	7.62
2	*Tilapia mossambica*	10 kg	11.25	4.50	50 g	3.30
3	*Saurida undosquamis*[c]	—	—	2.30	170 g	2.40
4	*Gadus morhua*	6,800 t	85.0	0.500	10 kg	0.83
5	*Hippoglossus hippoglossus*	156 million lb	28.6	0.723	90 kg	0.47
6	*Thunnus obesus*	300 thousand tonnes	82.5	1.10	90 kg	0.47
7	Humpback whale	95 thousand indiv.	3.0	0.126	35 t	0.100
8	Sei whale	150 thousand indiv.	5.0	0.133	50 t	0.091
9	Fin whale	375 thousand indiv.	10.0	0.107	70 t	0.083
10	Blue whale	150 thousand indiv.	3.5	0.093	100 t	0.076

[a]Calculations are based on the equation $r_m = (MSY \cdot 4 / B_\infty$ or N_∞), and B_∞ (or N_∞) and MSY data given in Silliman and Gutsell (1958), Silliman (1975), Garrod (1977, Fig. 39), Rothschild and Suda (1977, p. 322 and Fig. 68), Allen and Chapman (1977, Table 55), for all stocks except no. 3.

[b]"Adult body size" ($\overline{w}$) is not defined in Blueweiss et al. (1978). The definition used here is $\overline{w} = W_m + W_{max}/2$, where W_m is the mean weight at first maturity of the ♀♀ and W_{max} the maximum weight reached by the oldest ♀♀ of a stock. The W_m and W_{max} values stem from, or were computed from data in Norman and Fraser (1963), Schubert (1952), Mathiesen and Olsen (1968), Garrod (1977), Silliman and Gutsell (1958), Silliman (1975) and Gosline and Brock (1965).

[c]The value of r_m in *S. undosquamis* was estimated from a logistic population growth curve based on Fig. 5B in Ben-Yami and Glaser (1974).

which can be used to obtain first approximations of MSY when virgin stock size and mean weight ($\overline{w}$, in g) of the adult of that stock are known, $\overline{w}$ being defined as the mean of the weight at first maturity (W_m) and the maximum weight of the fish in the stock (W_{max}).

The results obtained by means of this equation generally compare well with those obtained using Gulland's (1971) widely-used relationship

$$MSY \approx \tfrac{1}{2} \cdot M \cdot B_v \qquad \ldots .30)$$

An application of equation (29) follows:

Thompson and Munro (1978) gave the following data for the Caribbean grouper *Epinephelus guttatus:* natural mortality = 0.68; TL_{max}, in cm = 53.7 (corresponding to W_{max} = 2,324 g); approximate weight at first maturity = 243 g. From these data, "adult body weight" ($\overline{w}$) as defined above was estimated to be 1,284 g, which, when inserted into equation (29) and assuming virgin stock (B_v) of one arbitrary unit, provides an MSY estimate of 0.36 per year. Equation (30), on the other hand, used in conjunction with M = 0.68 and the same virgin stock size of one arbitrary unit gives MSY = 0.34, which is very close to the first estimate.

Obviously, the advantage of equation (29) over equation (30) is that no estimate of natural mortality is required, which makes the application of equation (29) to little-investigated stocks rather easy. In practice, however, the best procedure will be to use both equations whenever possible because two independent estimates are always better than one.

Recruitment

Establishment of Stock-Recruitment Relationships

Although recruitment is an extremely important parameter, its estimation and the establishment of "stock-recruitment relationships" are extremely difficult. This is true particularly for pelagic stocks which are both more difficult to sample quantitatively and which generally display greater year-to-year variability in abundance than demersal stocks (see contributions in Parrish 1973).

Very little work has been done on the stock-recruitment relationships of tropical fishes (but see Murphy, this vol. and contributions in Saville

Table 7. Data for the establishment of a stock-recruitment relationship in the false trevally *Lactarius lactarius* from the Gulf of Thailand (adapted from Pauly 1980d).

Year	Virgin stock	1963	1966	1967	1968	1969	1970	1971	1972
Fishing mortality	0	0.22	0.64	0.87	1.09	1.12	1.18	1.92	2.23
Yield-per-recruit[a]	—	3.27	5.81	6.21	6.29	6.28	6.27	5.55	5.17
Spawning stock[b]	2,660	2,087	1,277	4.22	444	191	29.8	37.8	4.00
Recruits (millions)	—	239	292	138	202	908	15.5	55.5	8.90

[a]In grams.
[b]In thousand tonnes.

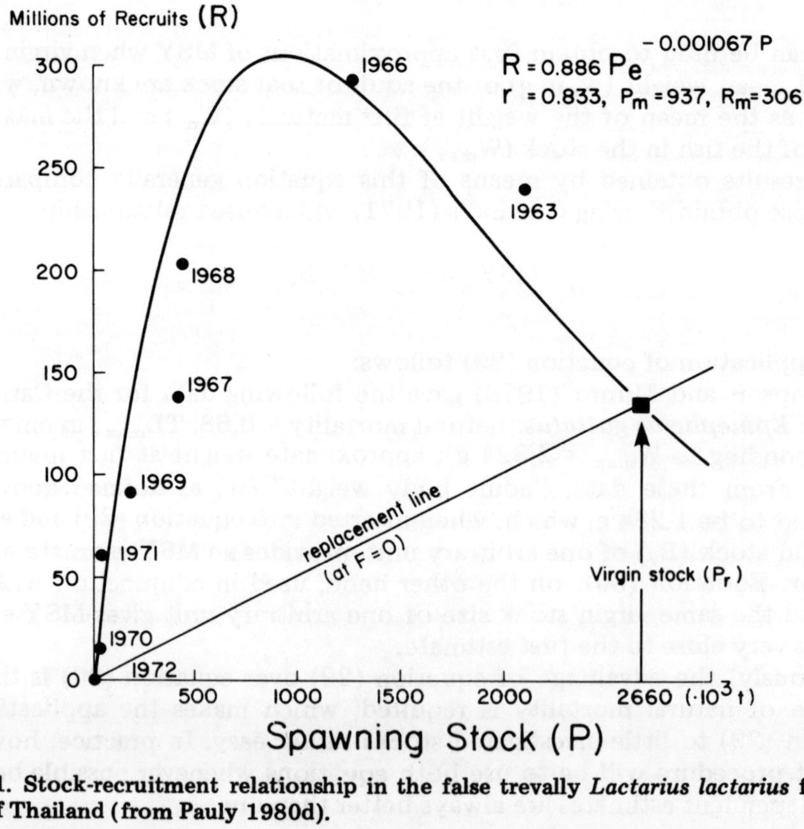

Fig. 11. Stock-recruitment relationship in the false trevally *Lactarius lactarius* from the Gulf of Thailand (from Pauly 1980d).

1980, and Sharp 1980) one of the main reasons for this being the lack of data that could be used for this purpose.

This has prompted several authors to use very indirect measures of recruitment (see e.g., Munro, this vol., or Pauly 1980d, 1982). The method used by the latter author involves the following steps:

a) estimation of growth and natural mortality parameters, using one of the methods above
b) computation of yield-per-recruit for each level of fishing mortality (i.e., for each year)
c) division of the calculated yield-per-recruit into the catch to obtain recruit numbers
d) plotting of estimated recruit numbers on parental stock or on predator stock or both.

The main drawbacks of the methods are steps (b) and (c) which violate important assumptions of the yield-per-recruit model of Beverton and Holt (1957), such as the assumption of an equilibrium situation in which a certain level of F has been operating long enough for the multi-aged population to have stabilized. Thus, when using this method to derive recruitment estimates, a bias is introduced in computing the yield-per-recruit. The magnitude of this bias increases with the inter-year changes in F, and with the life span of the fish in question.

Many tropical fishes, however, have short life spans and high rates of natural mortality (Qasim 1973b; Banerji and Krishnan 1973; Pauly 1980c). For these reasons, the biomass of a cohort generally peaks a few months after hatching, which results, when fishing is constant over the year, in a year's catch roughly corresponding to the actual yield-per-recruit for that year multiplied by the number of recruits produced in that year, to the exclusion of the remnants of older cohorts. This would then justify the operation described above.

Table 7 and Fig. 11, adapted from Pauly (1980d) shows results obtained by applying this approach to the false trevally, *Lactarius lactarius* from the Gulf of Thailand.

Murphy (this vol.) discusses some of the problems associated with this method and suggests an alternative explanation to some results obtained by applying it.

Recruitment Patterns

In addition to the numbers of recruits produced within a given time interval (e.g., a year) an important aspect of recruitment is its structure within that interval, i.e., whether recruitment is discrete or continuous. Tropical fishes are reported by many authors to have very protracted spawning seasons (e.g., Qasim 1973a; Weber 1976; Nzioka 1979; Goldman 1980), but the question has rarely been asked whether this protracted spawning season is in actuality matched by a correspondingly long "recruitment season" (see Sharp 1980 for the concept of recruitment "windows", open during only brief periods of the year).

Clearly, recruitment even in tropical fishes must oscillate seasonally, since year-round recruitment would generate size frequency distributions lacking peaks and troughs. Conversely, by projecting a set of length-frequency

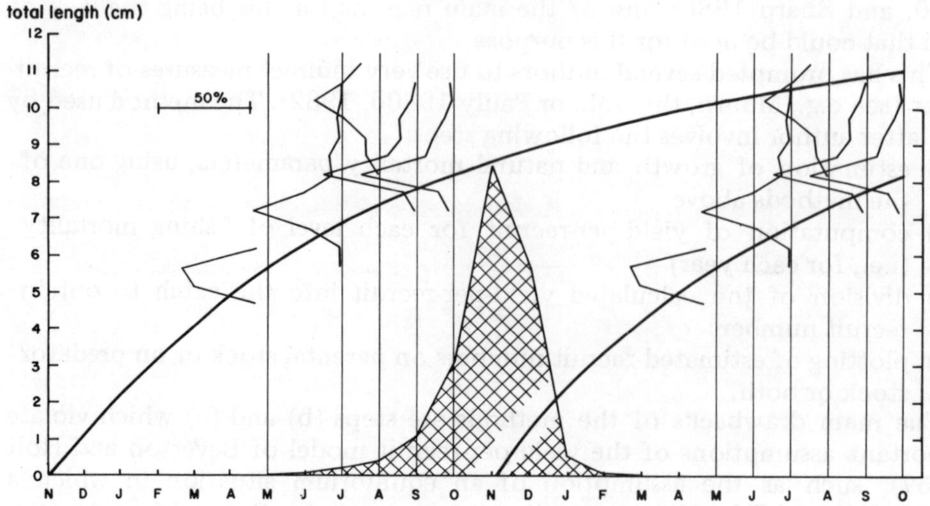

Fig. 12. Recruitment pattern in *Leiognathus bindus*. Note that the time scale implies that t_o = 0, which most certainly doesn't apply. Thus, only the *shape* of the recruitment pattern is considered, not its position on the time axis. The growth parameters used for projecting the length-frequency data are L_∞ = 12.2 cm, K = 1.3 and D = 1 (ESP/ASP = 0.804) as estimated by means of ELEFAN I from data in Balan (1967).

data backward onto the time axis (by means of growth parameters) down to zero length, one can obtain a frequency distribution reflecting the "pattern of recruitment" of a stock (see Fig. 12). A procedure generating such "patterns of recruitment" has been incorporated into the ELEFAN II program. One first result obtained with this new approach is that, although it seems continuous (i.e., occurring over the whole year), recruitment to the stocks hitherto investigated is often distributed normally, with one or two peaks per year. The position of the peak(s) when adjusted to real time by means of a value of t_o thus indicates the time(s) of the year when recruitment is most intense, and this should generally correspond to the "peak" spawning season(s).

A standard deviation can be calculated for recruitment patterns, which is related to the width of the annual recruitment "window". Therefore, the standard deviation of recruitment patterns can be used to classify these according to the recruitment strategy they express (Fig. 13). However, the standard deviations of recruitment patterns are also related to the longevity of the investigated fishes, i.e., to the degree of blurring that occurs because the individual fishes, as they grow older, increasingly deviate from their predicted length for age. This effect can be easily countered by plotting the standard deviations of fishes with one single, very distinct spawning season (open dots in Fig. 14) against their estimated longevity. The resulting regression line (dotted in Fig. 14) is then shifted upward and used to separate fishes with one recruitment season (fish nos. 1-6, 11 in Figs. 13 and 14) from those with two recruitment seasons (fish nos. 8, 10, 12-14 in Figs. 13 and 14), leaving one fish (no. 9) with an undefined status. This analysis is very preliminary and will be expanded and validated when more data on recruitment patterns become available.

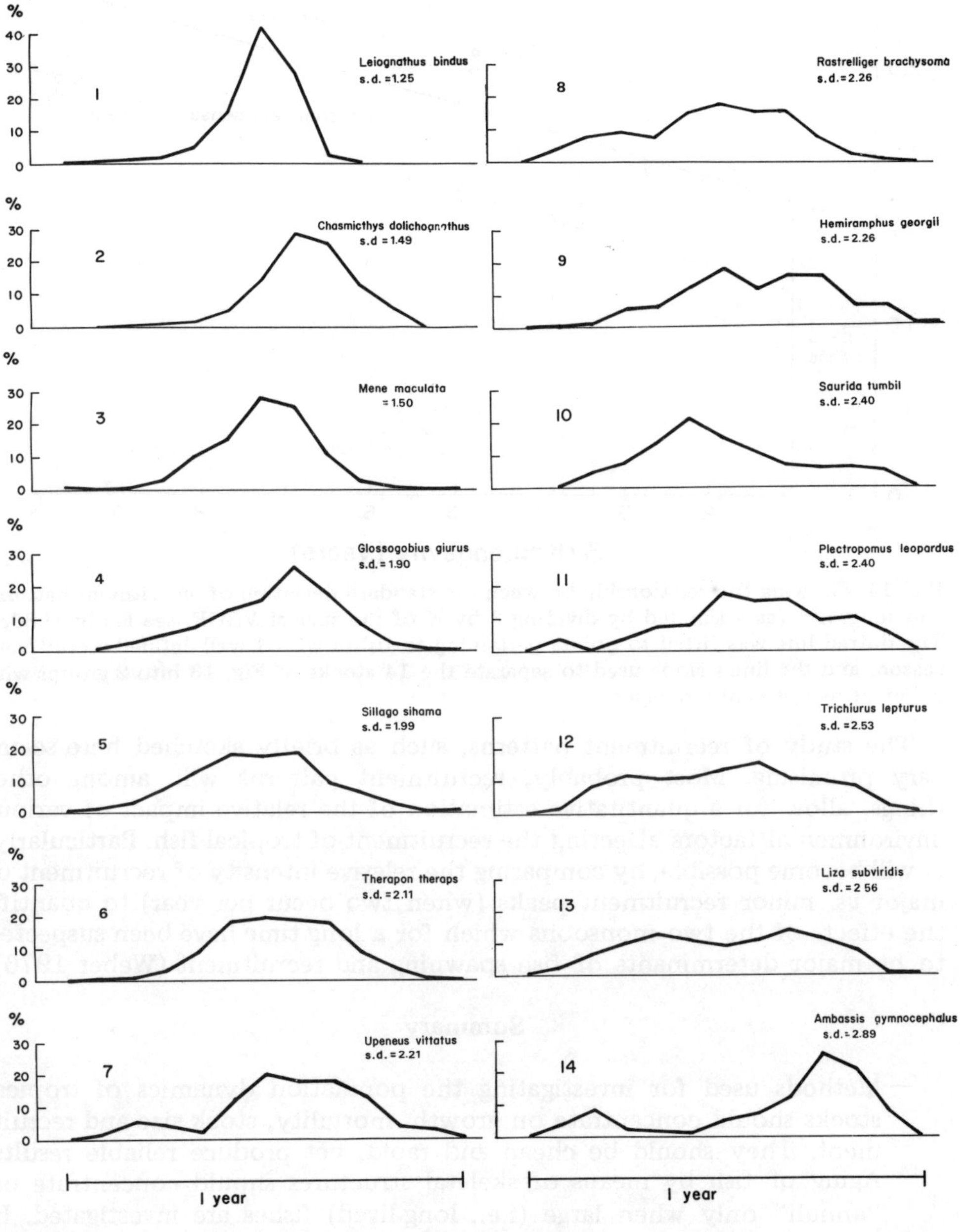

Fig. 13. Recruitment patterns in 14 stocks of fishes whose growth parameters were estimated by means of ELEFAN I. Note increase in standard deviation from *Leiognathus bindus* (s.d. = 1.25) to *Ambassis gymnocephalus* (s.d. = 2.89), with corresponding shift from 1 to 2 recruitment seasons. Source of length-frequency data: Balan 1967 (1), Goeden 1978 (11), Tamura and Honma 1977 (2), Marquez 1960 (4) and Ziegler 1979 (3, 5, 6, 7, 8, 9, 10, 12, 13, 14).

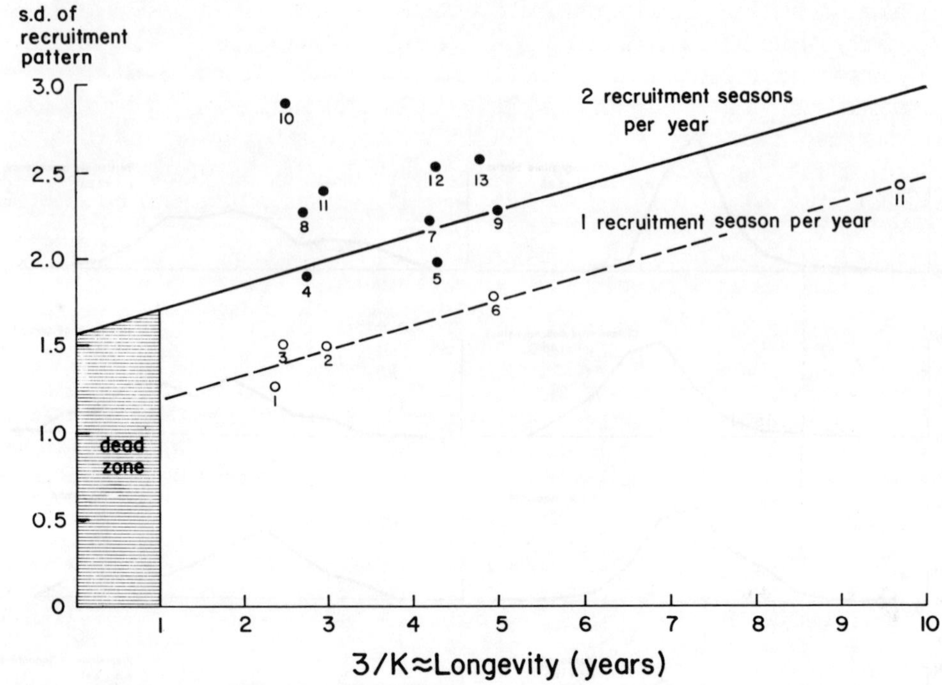

Fig. 14. Showing the relationship between the standard deviation of recruitment patterns and longevity (as estimated by dividing 3 by K of the special VBGF, see Pauly 1980c). The dotted line was fitted to points pertaining to fishes with 1 well-defined recruitment season, and the line's slope used to separate the 14 stocks of Fig. 13 into 2 groups with different recruitment strategies.

The study of recruitment patterns, such as briefly sketched here seems very promising. Most probably, recruitment patterns will, among other things, allow for a quantitative estimation of the relative impact of various environmental factors affecting the recruitment of tropical fish. Particularly, it will become possible, by comparing the relative intensity of recruitment of major *vs.* minor recruitment peaks (when two occur per year) to quantify the effects of the two monsoons which for a long time have been suspected to be major determinants of fish spawning and recruitment (Weber 1976).

Summary

- Methods used for investigating the population dynamics of tropical stocks should concentrate on growth, mortality, stock size and recruitment. They should be cheap and rapid, yet produce reliable results.
- Aging of fish by means of skeletal structures should concentrate on "annuli" only when large (i.e., long-lived) fishes are investigated. In fishes with short life span (i.e., in fishes that remain small) aging should be done by using daily otolith structures. As this method is very tedious, it should be used mainly to validate and complement growth estimates based on the analysis of length-frequencies.
- The von Bertalanffy Growth Formula may be used for modelling the growth of tropical fish, but workers should be aware of the assumptions involved in the derivation of the formula's two versions, and of season-

ally oscillating growth patterns, which may occur even in the tropics.

— Length-frequency data can be used to reliably estimate the values of the growth and mortality parameters of well-sampled populations. The work can be considerably facilitated by using new computer programs designed for these purposes (ELEFAN I and II). The second of these programs can also be used to generate "patterns of recruitment".
— When combined with catch data, length-frequency data can be used to generate catch-at-length data which can be used for Cohort or Virtual Population Analysis, to which several new approaches are available.

Acknowledgments

It is my pleasure to acknowledge the assistance of ICLARM Intern José Ingles, University of the Philippines, College of Fisheries, who calculated all examples based on ELEFAN I and II. Thanks are also due to Roberto Bugay for drafting the figures, and to Drs. Richard Neal, ICLARM and John Gulland, FAO for reviewing preliminary versions of this paper.

References

Abrahamson, N.J. 1971. Computer programs for fish stock assessment. FAO. Fish. Tech. Pap. 101. pag. var.

Allen, G.R. 1975. Anemonefishes. T.F.H. Publ. 2nd ed., Jersey City, U.S.A., 351 p.

Allen, K.R. 1971. Relation between production and biomass. J. Fish. Res. Board Can. 28: 1573-1581.

Allen, K.R. and D.G. Chapman. 1977. Whales, p. 335-358. *In* J.A. Gulland (ed.) Fish population dynamics. Wiley Interscience, New York.

Bagenal, T.. Editor. 1974. Ageing of fish. Unwin Bros. Ltd., Old Woking, Surrey, England.

Balan, V. 1967. Biology of the silver-belly *Leiognathus bindus* (Val.) of the Calicut Coast. Ind. J. Fish. 10(1): 118-134.

Banerji, S.K. and T.S. Krishnan. 1973. Acceleration of assessment of fish populations and comparative studies of similar taxonomic groups, p. 158-175. *In* Proc. Symp. Living Res. Seas Around India, Spec. Publ., Cen. Mar. Fish. Res. Inst., Cochin. 748 p.

Bayliff, W.H. 1979. Migrations of yellowfin tuna in the Eastern Pacific Ocean as determined from tagging experiments initiated during 1968-1974. Inter-Amer. Trop. Tuna Comm. Bull. 17: 447-506.

Beck, U. 1974. Bestandskundliche Untersuchungen an einigen Fischarten der Grundschleppnetzfischerei auf dem Schelf von Togo (Westafrika) M.S. Thesis, Hamburg University, 126 p.

Ben-Yami, M. and T. Glaser. 1974. The invasion of *Saurida undosquamis* (Richardson) into the Levant Basin—an example of biological effects of interoceanic canals. Fish. Bull. 72(2): 359-373.

Berkeley, S.A. and E.D. Houde. 1978. Biology of two exploited species of halfbeaks, *Hemirhamphus brasiliensis* and *H. balao* from Southeast Florida. Bull. Mar. Sci. 28: 624-644.

Bertalanffy, L. von. 1934. Untersuchungen über die Gesetzlichkeiten des Wachstums I. Roux' Archiv Entwicklungsmech. 131: 613-652.

Bertalanffy, L. von. 1938. A quantitative theory of organic growth (Inquiries in growth laws II) Hum. Biol. 10: 181-213.

Bertalanffy, L. von. 1951. Theoretische Biologie. Zweiter Band: Stoffwechsel, Wachstum. A. Franke A.G. Verlag, Bern

Beverton, R.J.H. and S.J. Holt. 1956. A review of methods for estimating mortality rates in fish populations, with special reference to sources of bias in catch sampling. Rapp. P.-v. Reún. Cons. Int. Explor. Mer 140: 67-83.

Beverton, R.J.H. and S.J. Holt. 1957. On the dynamics of exploited fish populations. U.K. Min. Agric. Fish., Fish. Invest. (Ser. 2) 19: 533 p.

Beverton, R.J.H. and S.J. Holt. 1959. A regiew of the lifespans and mortality rates of fish in nature and their relation to growth and other physiological characteristics. Ciba Found. Colloq. on Ageing 5: 142-180.

Blueweiss, L., H. Fox, V. Kudzma, D. Nakashima, R. Peters and S. Sams. 1978. Relationships between body size and some life-history parameters. Oecologia (Berl.) 37: 257-272.

Boiko, E.G. 1964. Evaluating natural mortality in the Azov zander [*Stizostedion lucioperca*]. Fish. Res. Board Can. Transl. Ser. (541) 15 pp. + Appendices.

Brothers, E.B., 1980. Age and growth studies on tropical fishes, p. 119-136. *In* S. Saila and P. Roedel (eds.) Proceedings of the International Workshop on Tropical Small Scale Fishery Stock Assessment, University of Rhode Island, September, 1979, Kingston, 198 p.

Brothers, E.B., C.P. Mathew and R. Lasker. 1976. Daily growth increments in otoliths from larval and adult fishes. Fish. Bull. (U.S.) 74(1): 1-8.

Carlander, K. 1969. Handbook of freshwater fishery biology. Vol. I. Iowa State University Press, Ames.

Cassie, R.M. 1954. Some use of probability paper in the analysis of frequency distributions. Aust. J. Mar. Freshwater Res. 5: 513-522.

Chang, K-H. and W-L. Wu. 1977. Tagging experiments on the spotted mackerel *(Scomber australasicus)* in Taiwan. Bull. Inst. Zool. Academic Sinica 16(2): 137-139.

Chapman, D.W. 1978a. Production in fish populations, p. 5-25. *In* S.D. Gerking (ed.) Ecology of freshwater fish production, John Wiley, New York.

Chapman, D.W. 1978b. Production, p. 202-217. *In* T. Bagenal (ed.) Methods for assessment of fish production in fresh waters. IBP Handbook No. 3. Blackwell Sci. Publ., Oxford.

Chomjurai, W. and R. Bunnag. 1970. Preliminary tagging studies of demersal fish in the Gulf of Thailand. *In* J.D. Marr (ed.) The Kuroshio, a symposium on the Japan current. East-West Center Press, Honolulu.

Chugunova, N.I. 1959. Handbook for the study of age and growth of fishes. Akademiya Nauk Press, Moscow. 164 p. (English version, 1963. Age and growth studies in fish. Office of Technical Services, Washington, D.C. 132 p.)

Cloern, J.E. and F.H. Nichols. 1978. A von Bertalanffy growth model with a seasonally varying coefficient. J. Fish. Res. Board Can. 35: 1479-1482.

Cridland, C.C. 1965. Laboratory experiments on the growth of *Tilapia* spp. The reproduction of *Tilapia esculenta* under artificial conditions. Hydrobiol. 18(3): 177-184.

Csirke, J. 1980. Recruitment in the Peruvian anchovy and its dependence on the adult population, p. 307-313. *In* A. Saville (ed.) The assessment and management of pelagic fish stocks, I.C.E.S., Rapp. P.-V. Réun. 177. 517 p.

Daget, J. and J.M. Ecoutin. 1976. Modèles mathématiques de production applicables aux poisson tropicaux subissant un arrêt prolongé de croissance. Cah. O.R.S.T.O.M. ser. Hydrobiol. 10: 59-69.

Delsman, J.C. 1972. Fish eggs and larvae from the Java Sea. Linnaeus Press, Amsterdam.

Durand, J.R. 1978. Biologie et dynamique des populations d'*Alestes baremoze* (Pisces, Characidae) du bassin tchadien. Trav. Doc. O.R.S.T.O.M. (98) 332 p.

Egglishaw, J. 1970. Production of salmon and trout in a stream in Scotland. J. Fish. Biol. 2: 117-136.

Ehrich, S. 1976. Zur Taxonomie, Ökologie und Wachstum von *Macrorhamphosus scolopax* (Linnaeus, 1758) (Pisces, Syngnathiformes) aus den subtropischen Nordostatlantik. Ber. Dtsch. Wiss. Komm. Meeresforsch. (n'F.) 24(4): 251-266.

Ehrlich, P.R. 1975. The population biology of coral reef fishes. Ann. Rev. Ecol. System. 6: 221-247.

Fenchel, T. 1974. Intrinsic rate of natural increase: the relationship with body size. Oecologia 14: 317-326.

Forbes, S.T. and O. Nakken. 1972. Manual of methods for fisheries resource survey and appraisal. Part 2. The use of acoustic instruments for fish detection and abundance estimation. FAO Man. Fish. Sci. 5: 138 p.

Garrod, D.J. 1963. The application of a method for the estimation of growth parameters from tagging data at unequal time intervals. North Atlantic Marking Symposium, Woods Hole (1961) ICNAF Special Publication (4): 258-261.

Garrod, D.J. 1977. The North Atlantic Cod, p. 216-242. *In* J.A. Gulland (ed.) Fish population dynamics. Wiley Interscience, New York.

Gaschütz, G., D. Pauly and N. David. 1980. A versatile BASIC program for fitting weight and seasonally oscillating length growth data. I.C.E.S. CM 1980/D:6 Statistics Cttee., 23 p.

George K. and S.K. Banerji. 1964. Age and growth studies on the Indian mackerel Rastrelliger kanagurta (Cuvier) with special reference to length frequency data collected at Cochin. Ind. J. Fish. 11(2): 621-638.

Gjøsaeter, J. and I-M. Beck. 1980. Age determination of fish species from Mozambique using daily rings in otoliths FI: Noz./77/001, Mimeo, FAO, Rome. 10 p.

Goeden, G.B. 1978. A monograph of the coral trout, *Plectropomus leopardus* (Lacépède). Queensland Fisheries Serv. Res. Bull. (1) 42 p.

Goldman, B. 1980. "Letter to the editor". ICLARM Newsl. 3(2): 22.

Goldman, B. and F.H. Talbot. 1976. Aspects of the ecology of coral reef fishes. Chapter 4. *In* O.A. Jones and R. Endean (eds.) Biology and Geology of Coral Reefs. Vol. III, Biology 2: 125-154. Academic Press, New York.

Gordon, I.D.H. 1977. The fish populations of the inshore waters of the west of Scotland. The biology of the Norway pout *(Trisopterus esmarkii)*. J. Fish. Biol. 10: 417-430.

Gosline, W.A. and V.E. Brock. 1965. Handbook of Hawaiian fishes. Univ. of Hawaii Press, Honolulu.

Graham, M. 1928. Studies of age determination in fish. Part II. A survey of the literature. U.K. Min. Agric. Fish., Fish. Invest. (Ser. 2) 11(3): 50 p.

Grant, C.J., R.L. Sandland and A.M. Olsen. 1979. Estimation of growth, mortality and yield per recruit of the Australian school shark *Galeorhinus australia* (Macleay), from tag recoveries. Aust. J. Mar. Freshw. Res. 30: 625-637.

Gulland, J.A. 1965. Estimation of mortality rates. Annex to Rep. Arctic Fish. Working Group, I.C.E.S. C.M. 1965(3): 9 p.

Gulland, J.A. 1969. Manual of methods for fish stock assessment. Part I. Fish Population analysis. FAO Man. Fish. Sci. No. 4. 154 p.

Gulland, J.A. 1971. The fish resources of the oceans. FAO/Fishing News (Books), Ltd., Surrey.

Gulland, J.A. 1979. Summary of workshop, p. 1-37. *In* Report of the FAO/IOP Workshop on the fishery resources of the western Indian Ocean south of the equator. IOFC/DEV/79/45.

Harding, J.P. 1949. The use of probability paper for the graphical analysis of the polymodal frequency distributions. J. Mar. Biol. Ass. U.K. 28: 141-153.

Hongskul, V. 1974. Population dynamics of Platu *Rastrelliger neglectus* (van Kampen) in the Gulf of Thailand. Proc. Indo-Pac. Fish. Counc. 15(3): 297-350.

Iles, T.D. 1974. The tactics and strategy of growth in fishes, p. 331-345. *In* F.R. Harden Jones (ed.) Sea fisheries research. Elek Scientific Books. London. 510 p.

Ikenouye, H. and H. Masazawa. 1968. An estimation on parameters of growth equation basing on the results of tagging experiments of the Japanese Alfonsin fish. Bull. Jap. Soc. Sci. Fish. 34(2): 97-102.

Isarankura, A. 1971. Assessment of stocks of demersal fish off the West Coasts of Thailand and Malaysia. IOFC/DEV/71/20.

Ivlev, V.S. 1966. The biological productivity of waters. J. Fish. Res. Board Can. 23: 1727-1759.

Jolley, J.W. and E.W. Irby. 1979. Survival of tagged and released Atlantic sailfish (*Istiophorus platypterus:* Istiophoridae) determined with acoustical telemetry. Bull. Mar. Sci. 29(2): 155-169.

Jones, R. 1974. Assessing the long term effects of changes in fishing effort and mesh sizes from length composition data, I.C.E.S. CM 1974/F:33. Demersal Fish (Northern) Committee. 13 p.

Jones, R. 1977. Tagging: theoretical methods and practical difficulties, p. 46-66. *In* J.A.

Gulland (ed.) Fish population dynamics. Wiley Interscience.
Jones, R. 1981. The use of length composition data in fish stock assessment (with notes on VPA and cohort analysis) FAO Fish. Circ. 734, FAO, Rome. 55 p.
Kato, S. and A.H. Carvallo. 1967. Shark tagging in the Eastern Pacific Ocean, 1962-65, p. 93-109. *In* P.W. Gilbert, R.F. Matthewson and D.P. Hall (eds.) Sharks, skates and rays. Johns Hopkins Press, Baltimore.
Kato, F. and Y. Yamada. 1975. Estimation of abundance and related parameters of age I red sea bream *Pagrus major*, in Iida Bay, by the multiple tag-recapture method. Bull. Jap. Sea. Reg. Fish. Res. Lab. (26): 1-16.
Le Guen, J.C. 1976. Utilisation des otolithes pour lecture de l'âge des Scianides intertropicaux, marques saisonnaire et journalière. Cah. O.R.S.T.O.M. Ser. Oceanogr. 14(4): 331-338.
Lockwood, S.J. 1974. The use of the von Bertalanffy equation to describe the seasonal growth of fish. J. Cons., Cons. Int. Explor. Mer 35: 175-179.
Longhurst, A.R. 1963. The bionomics of the fisheries resources of the Eastern Tropical Atlantic. Col. Off. Fish. Publ. London. (20) 65 p.
Marquez, J.R.S. 1960. Age and size at sexual maturity of white goby *(Glossogobius giurus)*, a common species of fish in Laguna de Bay, with notes on its food habits. Philipp. J. Fish. 8(1): 71-99.
Mathiesen, O.A. and S. Olsen. 1968. Yield isopleths of the halibut, *Hippolossus hippoglossus*, in Northern Norway. Fisk. Dir. Skr. Ser. Havunders. 14: 129-159.
May, R.M., J.R. Beddington, C.W. Clark, S.J. Holt and R.M. Laws. 1979. Management of multispecies fisheries. Science 205: 267-277.
Menon, M.D. 1950. The use of bones, other than otoliths, in determining the age and growth-rate of fishes. J. Cons., Cons. Int. Explor. Mer 16: 311-335.
Methot, R.D. Jr. and D. Kramer. 1979. Growth of northern anchovy *Engraulis mordax* larvae in the sea. Fish. Bull. (U.S.) 77(2): 413-423.
Mohr, E. 1927. Bibliographie der Alters- und Wachstums-Bestimmung bei Fischen. J. Cons., Cons. Int. Explor. Mer 2(2): 236-258.
Mohr, E. 1930. Bibliographie der Alters- und Wachstums-Bestimmung bei Fischen. III. Nachträge und Fortsetzung. J. Cons., Cons. Int. Explor. Mer 9(2): 377-391.
Mohr, E. 1934. Bibliographie der Alters- und Wachstums-Bestimmung bei Fischen. III. Nachträge und Fortsetzung. J. Cons., Cons. Int. Explor. Mer 9(2): 377-391.
Muir, B.S. 1969. Gill size as a function of fish size. J. Fish. Res. Board Can. 26: 165-170.
Murphy, G.I. 1965. A solution of the catch equation. J. Fish. Res. Board Can. 22(1): 191-202.
Norman, J.R. and F.C. Fraser. 1963. Riesenfische, Wale und Delphine. Parey, Hamburg and Berlin.
Nzioka, R.M. 1979. Observations on the spawning seasons of East African reef fishes. J. Fish Biol. 14: 329-342.
Odum, E. 1971. Fundamentals of ecology. 3rd ed. Saunder's International Student Edition, Philadelphia, Pa.
Panella, G. 1971. Fish otoliths: daily growth layers and periodical patterns. Science (Wash.) 137(4002): 1124-1127.
Panella, G. 1974. Otolith growth patterns: an aid in age determination in temperate and tropical fishes, p. 28-39. *In* T.B. Bagenal (ed.) The ageing of fish. Unwin Brothers Ltd., Old Woking, Surrey, England.
Parrish, B.B., Editor. 1973. Fish stocks and recruitment. I.C.E.S. Rapp. P.-v. Réun. 164, 372 p.
Pauly, D. 1978. A preliminary compilation of fish length growth parameters. Ber. Inst. f. Meereskunde (Kiel University) No. 55. 200 p.
Pauly D. 1979a. Theory and management of tropical multispecies stocks: a review, with emphasis on the Southeast Asian demersal fisheries. ICLARM Studies and Reviews 1, 35 p. International Center for Living Aquatic Resources Management, Manila.
Pauly, D. 1979b. Gill size and temperature as governing factors in fish growth: a generalization of von Bertalanffy's Growth Formula. Ber. Inst. f. Meereskunde (Kiel University) No. 63. 153 p.
Pauly, D. 1980a. A selection of simple methods for the assessment of tropical fish stocks. FAO Fish. Circ. No. 729. FIRM/C701, 54 p.

Pauly, D. 1980b. The use of a pseudo-catch curve for the estimation of mortality rates in *Leiognathus splendens* (Pisces: Leiognathidae) in Western Indonesian waters. Meeresforsch. 28(1): 56-60.

Pauly, D. 1980c. On the interrelationships between natural mortality, growth parameters and mean environmental temperature in 175 fish stocks. J. Cons., Cons. Int. Explor. Mer 39(3): 175-192.

Pauly, D. 1980d. A new methodology for rapidly acquiring basic information on tropical fish stocks: growth, mortality and stock-recruitment relationships, p. 154-172. *In* S. Saila and P. Roedel (eds.) Proceedings of the international workshop on tropical small-scale fishery stock assessment, University of Rhode Island, Sept. 1979. Kingston.

Pauly, D. 1981. The relationships between gill surface area and growth performance in fish: a generalization of von Bertalanffy's theory of growth. Meeresforsch. 28: 251-282.

Pauly, D. 1982. A method to estimate the stock-recruitment relationships of shrimps. Trans. Amer. Fish. Soc. 111: 13-20.

Pauly, D. and N. David. 1981. A BASIC program for the objective extraction of growth parameters from length-frequency data. Meeresforsch. 28: 205-211.

Pauly, D. and G. Gaschütz. 1979. A simple method for fitting oscillating length growth data, with a program for pocket calculators. I.C.E.S. CM 1979/G:24. Demersal Fish Cttee., 26 p.

Pauly, D. and P. Martosubroto. 1980. The population dynamics of *Nemipterus marginatus* off Western Kalimantan, South China Sea. J. Fish Biol. 17: 263-273.

Pauly, D., N. David and J. Ingles. 1980. ELEFAN I: Users' instructions and program listing. Mimeo. pag. var.

Pauly, D., N. David and J. Ingles. 1981. ELEFAN II: Users' instructions and program listing. Mimeo. pag. var.

Petersen, J. 1892. Fiskenbiologiske forhold i Holboek Fjord, 1890-91. Beretnig fra de Danske Biologiske Station for 1890(91), 1: 121-183.

Pitcher, T.J. and P.D.M. MacDonald. 1973. Two models for seasonal growth in fishes. J. Appl. Ecol. 10: 599-606.

Pope, J.G. 1972. An investigation of the accuracy of virtual population analysis using cohort analysis. Int. Comm. Northwest Atl. Fish. Res. Bull. 9: 65-74.

Pope, J.G., N. David and D. Pauly. 1982. ELEFAN III: Users' instructions and program listing. Mimeo, pag. var.

Pütter, A. 1920. Studien über physiologische Ähnlichkeit. VI. Wachstumsähnlichkeit. Pflugers Arch. Gesamte Physiol. 180- 298-340.

Qasim, S.Z. 1973a. An appraisal of the studies on maturation spawning in marine teleosts from the Indian waters. Indian J. Fish. 20(1): 166-181.

Qasim, S.Z. 1973b. Some implications of the problems of age and growth in marine fishes from the Indian waters. Indian J. Fish. 20(2): 351-371.

Ralston, S. 1976. Age determination of a tropical reef butterflyfish utilizing daily growth rings in otoliths. Fish. Bull. (U.S.) 74(4): 990-994.

Randall, J.E. 1961. A contribution to the biology of the convict surgeonfish of the Hawaiian Islands: *Acanthurus triostegus sandvicensis* Pac. Sci. 153(2): 215-272.

Randall, J. 1962. Tagging reef fishes in the Virgin Islands. Proc. Gulf Caribb. Fish. Inst. 14: 201-241.

Randall, J.E. 1963. Additional recoveries of tagged reef fishes from the Virgin Islands. Proc. Gulf & Caribb. Fish. Inst. 15: 155-157.

Randall, J.E. 1977. Contribution to the biology of the whitetip reef shark *(Trianodon obesus)* Pac. Sci. 31(2): 143-164.

Ricker, W.E. 1975. Computation and interpretation of biological statistics of fish populations. Bull. Fish. Res. Board Can. 191. 382 p.

Ricklefs, R.E. 1973. Ecology. Chiron Press, Portland, Oregon.

Robson, D.S. and D.G. Chapman. 1961. Catch curves and mortality rates. Trans. Amer. Fish. Soc. 90(2): 181-189.

Rothschild, B.J. and A. Suda. 1977. Population dynamics of tuna, p. 309-334. *In* J.A. Gulland (ed.) Fish population dynamics. Wiley Interscience, New York.

Russel, F.S. 1931. Some theoretical considerations on the "overfishing" problem. J. Cons., Cons. Int. Explor. Mer 6(1): 3-27.

Russel, B.C., F.H. Talbot, G.R.V. Anderson and B. Goldman. 1978. Collection and sampling of reef fishes, p. 329-345. *In* D.R. Stoddart and R.E. Johannes (eds.) Coral reefs: research methods. UNESCO, Paris.

Saeger, J., P. Martosubroto and D. Pauly. 1976. First report of Indonesian German demersal fisheries project (Results of a trawl survey in the Sunda Shelf Area). Marine Fisheries Research Reports/Contributions of the Demersal Fisheries Project No. 1, Jakarta. 46 p.

Sanders, M.J. and D.G.M. Powell. 1979. Comparison of the growth rates of two stocks of snappers *(Chrysophrys auratus)* in South-east Australian waters using capture-recapture data. N.Z. J. Mar. Freshw. Res. 13(2): 279-284.

Saville, A., Editor. 1980. The assessment and management of pelagic fish stocks. I.C.E.S. Rapp. P.-v. Réun. 177. 517 p.

Schaefer, M. 1954. Some aspects of the dynamics of populations important to the management of the commercial marine fisheries. Inter. Amer. Trop. Tuna Comm. Bull. 1(2): 27-56.

Schaefer, M.B. 1957. A study of the dynamics of the fishery for yellowfin tuna in the eastern tropical Pacific Ocean. Inter-Amer. Trop. Tuna Comm. Bull. 2: 247-268.

Schubert, K. 1952. Das Gewicht der Wale. Die Fischwirtschaft (6-9) 5 p.

SCSP. 1978. Report on the workshop on the demersal resources of the Sunda Shelf, Part I. SCS/GEN/77/12. 44 p.

Sekharan, K.V. 1974. Estimates of the stocks of oil sardine and mackerel in the present fishing grounds off the west coast of India. Indian J. Fish. 21(1): 177-182.

Sella, M. 1929. Migrazioni e habitat del tonno *(Thunnus thynnus)* studiati col metodo degli ami, con osservazioni su l'accrescimento, etc. Memoria R. Comit. Talassog. Ital. 156: 1-24.

Sharp, G., Editor. 1980. Workshop on the effects of environmental variation on the survival of larval pelagic fishes. Unesco, Intergov. Oceanogr. Comm. Workshop. Report No. 28, 323 p.

Shul'man, G.E. 1974. Life cycles of fish; physiology and biochemistry. Isr. Progr. Sci. Transl. Wiley & Sons, New York.

Silliman, R.P. 1975. Selective and unselective exploitation of experimental populations of *Tilapia mossambica.* Fish Bull. (U.S.) 73(3): 495-507.

Silliman, R.P. and J.S. Gutsell. 1958. Experimental exploitation of fish population. Fish. Bull. (U.S.) 58(133): 215-252.

Smith, C.L. 1973. Small rotenone stations: a tool for studying coral reef communities. Am. Mus. Novit. (2512) 21 p.

Smith, F.E. 1954. Quantitative aspects of population growth, p. 274-294. *In* E. Boell (ed.) Dynamics of growth processes. Princeton Univ. Press, Princeton.

Steffensen, E. 1980. Daily growth increments observed in otoliths from juveniles East Baltic cod. Dana 1: 29-37.

Struhsaker, P. and J.H. Uchiyama. 1976. Age and growth of the nehu, *Stolephorus purpureus* (Pisces, Engraulidae), from the Hawaiian Islands as indicated by daily growth increment of saggittae. Fish. Bull. (U.S.) 74(1): 9-17.

Tamura, E. and Y. Honma. 1977. Histological changes in the organs and tissues of the gobiid fishes throughout their life-span—VII. Seasonal changes in four species of gobies. Bull. Jap. Soc. Sci. Fish. 43: 893-974.

Tanaka, S. 1956. A method for analyzing the polymodal frequency distribution and its application to the length distribution of porgy, *Taius tumifrons* (T. & S.) Bull. Tokai Reg. Fish. Res. Lab. 14: 1-12.

Taubert, B.Ø. and D.W. Cable. 1977. Daily rings in otoliths of three species of *Lepomis* and *Tilapia mossambica.* J. Fish. Res. Board Can. 34: 332-340.

Te Winkel, L.E. 1935. A study of *Mistichthys luzonensis* with special reference to conditions correlated with reduced size. J. Morph. 58: 463-535.

Tiews, K. 1963. Synopsis of biological data on bluefin tuna *Thunnus thynnus* (Linnaeus) 1978 (Atlantic and Mediterranean), p. 422-482. *In* Proceedings of the world scientific meeting on the biology of tuna and related species, La Jolla, California, U.S.A., 2-4 July 1962. FAO Fish Rep. (6), Vol. 2. Rome.

Thompson, R. and J.L. Munro. 1978. Aspects of the biology and ecology of Caribbean reef fishes: Serranidae (hinds and groupers). J. Fish. Biol. 12: 115-146.

Thresher, R.E. and A.M. Gronell. 1978. Subcutaneous tagging of small reef fishes. Copeia 1978(2): 352-353.

Thurow, F. 1973. Growth parameters of Baltic salmon (*Salmo salar* L.) Ber. dt. wiss. Komm. Meeresforsch. 22(4): 445-451.

Ursin, E. 1963. On the incorporation of temperature in the von Bertalanffy growth equation. Medd. fra Danmarks Fiskeri-og Havundersøgelser 4: 1-16.

Van Someren, V.D. and P.J. Whitehead. 1960. The culture of *Tilapia nigra* (Günther) in ponds. IV. The seasonal growth of male *T. nigra.* East Afr. Agric. For. J. 26: 79-86.

Weber, W. 1976. The influence of hydrographical factors on the spawning time of tropical fish, p. 269-281. *In* K. Tiews (ed.) Proceedings of an international seminar on possibilities and problems of fisheries development in Southeast Asia, West Berlin, 19 Nov.-6 Dec. 1974. West Kreuz, Druckerei, Berlin.

Weber, W. and A. Jothy. 1977. Observations on the fish *Nemipterus* spp. (Family: Nemipteridae) in the coastal waters of East Malaysia. Arch. Fischereiwiss. 28(2/3): 109-122.

Welcomme, R.L. 1979. Fisheries ecology of floodplain rivers. London, Longman.

Wheeler, J. and F.D. Ommanney. 1953. Report on the Mauritius-Seychelles Fisheries Survey, 1948-1949. Colon. Off. Fish. Publ. (London) 1(3) 148 p.

Worthmann, H. 1980. The early growth stages of the pescada *(Plagioscion monti* Soares). Arch. Fischereiwiss. 30(2/3): 145-156.

Winberg, G.G. 1961. [New information on metabolic rate in fishes] Vopr. Ikthiol. Vol. 1: 156-165. Fish. Res. Board Can. Transl. Ser. No. 362. 11 p.

Yong, M.Y. and R.A. Skillman. 1971. A computer program for analysis of polymodal frequency distribution (ENORMSEP), FORTRAN IV. U.S. Fish. Bull. 73(3): 681.

Ziegler, B. 1979. Growth and mortality rates of some fishes of Manila Bay, Philippines, as estimated from the analysis of length frequencies. M.S. thesis, Kiel University (FRG) 115 p.

Discussion of Dr. Pauly's Paper

The discussion concentrated around five points: clarification concerning the generalized VBGF, the reliability of growth parameter estimates based on ELEFAN I, the relationships relating r_m and body weight, recruitment patterns and the relationships between single-species and community dynamics.

Concerning ELEFAN I, Mr. Jones pointed out that in fish which can grow to an old age, there is a real danger of peaks overlapping each other to the point of being indistinguishable. Dr. Pauly fully agreed with this and pointed out that the method in general should not be used in conjunction with fishes known to reach ages beyond 4 to 5 years, unless the available length frequency samples are extraordinarily well-structured. Several participants suggested statistical and mathematical techniques suitable for improving ELEFAN I.

Concerning the relationship between r_m and body weight (Table 5 and Fig. 11 in Pauly's paper), Dr. Sainsbury pointed out that the identity of estimates of r_m based on numbers and with that of estimates based on biomass has never been demonstrated, and suggested that considering these estimates to be equal is "at best sloppy and at worst completely wrong." Concerning this same relationship Dr. Gulland and Dr. Marten pointed out that the scatter of the points is, for any range of body weights too large for the relationship to be of much help. Pauly held to the usefulness of the relationship, despite its shortcomings.

The relationship between single-species and community dynamics was discussed, especially by Dr. Marten. He suggested that there are systems (e.g.,

Lake Victoria) in which single-species investigations lead to results that are not confirmed when the dynamics of the overall stock is investigated. A case in point is that small fishes in Lake Victoria were assumed to have higher turnover rates than larger fishes (and hence could be expected to withstand heavy fishing) yet collapsed rather rapidly when exploitation increased. It was suggested, however, that this might have been due to the stock having been maintained "fully exploited" prior to the onset of any fishery.

Estimation of Biological and Fishery Parameters in Coral Reef Fisheries

J.L. MUNRO[a]

Fisheries Research and Survey Branch
Department of Primary Industry
Konedobu, Papua New Guinea

MUNRO, J.L. 1982. Estimation of biological and fishery parameters in coral reef fisheries, p. 71-82. *In* Pauly, D. and G.I. Murphy (eds.) Theory and management of tropical fisheries. ICLARM Conference Proceedings 9, 360 p. International Center for Living Aquatic Resources Management, Manila, Philippines and Division of Fisheries Research, Commonwealth Scientific and Industrial Research Organisation, Cronulla, Australia.

Abstract

The problems of parameter estimation in coral reef fisheries are discussed and the methods used by the author and co-workers to investigate the fisheries biology of coral reef fishes in Jamaican waters are evaluated. It is concluded that the systematic analysis of length-frequency data offers the best opportunities for producing usable estimates of fishery parameters. Daily growth rings in otoliths appear to offer excellent opportunities for determining ages of coral reef fishes and should be more extensively used.

Attention is drawn to the possibilities of attempting to derive fishery parameters from easily obtainable statistics such as mean weights of individuals in catches, catch compositions in multispecies fisheries, catch-per-effort indices and to the use of gross measures such as coastal population densities or numbers of boats per unit area of shelf in order to make progress in the assessment of tropical multispecies stocks.

Introduction

Our knowledge of the fishery biology of coral reef fishes stems primarily from three major studies: those of Bardach and co-workers in Bermuda (1957, 1958), Randall in the U.S. Virgin Islands (1962 and 1963) and Munro and co-workers (1973-1980) in Jamaica. Additionally, Nagelkerken (1979) has undertaken a major study of the small grouper, *Epinephelus (= Petrometopon) cruentatum* around Curaçao. For the Indo-Pacific region I am aware of no large-scale studies which have yet been completed. The contributions by Goeden (1978) and Lebeau and Cueff (1975) appear to be among the few studies of any of the larger species of Indo-Pacific reef fishes.

The reasons for the paucity of data are extremely simple. Coral reefs are inhabited by an enormous diversity of harvestable animals, principally

[a]Present address: ICLARM, MCC P.O. Box 1501, Makati, Metro Manila, Philippines.

fishes, which range in size from large carcharhinid sharks to small demoiselles and anchovies. In most areas where reef fisheries exist, several hundred species might normally appear in catches taken by a variety of methods to which each species will be differentially vulnerable. Of all the species encountered, no one species is overwhelmingly dominant in terms of numbers or weights. Munro (1974b, 1977) listed 35 species each of which exceeded 1.5% of the total weight of trap catches in at least one of the areas studied. Of these species, 11 had to be omitted from a multispecies assessment because the necessary parameter values could not be estimated.

In the study undertaken by Munro and co-workers (1973-80), biological data were gathered on about 70 species of reef fishes. An examination of scales and otoliths of many of the common species showed that if marks were present, they were irregularly distributed and uninterpretable on the basis of small samples. However, there were good indications that, for some species at least, interpretations could be made if sufficient time and sufficient specimens were available (Munro and Thompson 1973). Nagelkerken (1979) made age determinations for *Epinephelus cruentatum* based upon marks observed in otoliths. He concluded that about seven rings were laid down in the first year and three per year thereafter. The reasons why such a pattern should exist in nature were not clearly established. The work is an excellent example of the difficulties which face those who work on tropical reef fishes.

The Jamaican Investigations

Because of the problems of age determination, Munro and co-workers concentrated on accumulating length-frequency distributions of male and female fishes, mean weights of individuals in the catches and basic data on reproduction for as many specimens of as many species as could be obtained, within the constraints of time and of other activities. Investigations were concentrated in four particular areas; the heavily exploited Port Royal reefs near to Kingston, Jamaica; California Bank, an isolated pinnacle quite close to the South Jamaica shelf, which is quite heavily exploited; the lightly exploited area surrounding the Pedro Cays (islands) on oceanic Pedro Bank and those parts of Pedro Bank which are remote from any land and which were unexploited at the time of the investigations.

Analytical Format

The data obtained at the Port Royal reefs were the result of 28 lunar months of continuous sampling. As such, information on the relative abundance of each species was available for the full period, subject to normal chance variability or seasonal factors affecting the success of the sampling. In the case of the offshore oceanic banks, our objective was to sample each of the main areas of interest once in every ten weeks. Owing to variable weather conditions and mechanical problems this objective was not entirely fulfilled and a complete time series of data was not available for every species in every depth zone or area. An analytical format was established with the objective of fully utilizing the available data and arriving at the best possible estimates of biological parameters (growth rates, population structure and natural or fishing mortality rates applicable to each species). The basic plan is outlined below:

— Length distributions of samples obtained from different areas or depth zones or using different fishing gear (traps or hooks) were first compared in terms of modal length, mean length and range in order to establish by visual inspection whether any differences existed between such samples. If no differences existed in size distributions, it then became possible, if necessary or desirable, to combine sample frequencies and treat, for example, a particular depth zone as a unit irrespective of the actual location of the sampling stations.

— Having sorted the samples into categories according to depth, area or gear or a combination thereof, the length distributions were then arranged in a time series to determine whether evidence of growth, in the form of modal progressions, was apparent. If samples of the species in question were small, frequencies were combined as necessary on a monthly, two-monthly or three-monthly basis. Where modal progression was evident, growth curves (in terms of length) were traced and values of the parameters of the von Bertalanffy growth equation were estimated. Generally, the regression equation used was in the form

$$\log_e (L_\infty - L_t) = \log_e L_\infty + Kt_o - Kt \qquad (1)$$

in which K is the coefficient of growth and L_∞ is the asymptotic length towards which the average fish is growing and is determined by trial regressions (Ricker 1958). L_t is the modal size, and t is the time interval between estimates of modal size (usually 1, 2 or 3 months). The theoretical origin of the growth curve at time t_o is estimated from the y-axis intercept, $\log_e L_\infty + Kt_o$. Where aging was possible on the basis of scale or otolith readings, the parameters of the von Bertalanffy equation were similarly estimated.

- The annual, mean length composition of the catch of each species was estimated on the following basis:

(i) The unexploited portions of Pedro Bank, the exploited Pedro Cays area, California Bank and the Port Royal reefs were treated as separate entities, and the length-frequency distributions of the catches were summed on a monthly, bi-monthly, quarterly, four-monthly, half-yearly or annual basis in such a way that *more than* 100 fishes constituted the *smallest* summation.

(ii) The percentage frequency of each length group was calculated for each time period and the mean of the frequencies in each period then calculated for each length group to obtain the best estimate of the mean annual length-frequency distribution of the catch.

(iii) Estimates obtained in this manner clearly vary quite widely in the degree of confidence which can be attached to the resulting estimates of mean annual population length structure, according to the size of the samples, and the degree to which samples have had to be combined, e.g., an estimate of population length structure based on a sample of at least 200 fishes in every month of the year will have a very much greater validity than an estimate based on samples so small that length-frequency distributions obtained at irregularly spaced intervals during the year have merely been summed to obtain an estimate of the population length structure.

(iv) For descriptive purposes only, usable estimates of population length structure were graded into five groups between "poor" and "excellent" according to the format given in Table 1. For many of the less common species dealt with in the study the sample sizes were so small that no usable estimates could be obtained and for many others, combinations of small sample size or irregular distribution through the year placed them in the lower categories. Wherever this could be done, confidence limits were calculated, even where the limits were extraordinarily wide.

Table 1. Criteria used for assessing the representativeness of length-frequency samples.

Total sample size	Time period (months) 12 Grade	6 Grade	4 Grade	3 Grade	2 Grade	1 Grade
1 - 100	NU	NU	NU	NU	NU	NU
101 - 500	0	0	0	1	1	1
501 - 1,000	0	0	1	1	2	2
1,001 - 1,500	0	1	1	2	3	3
>1,500	0	1	2	3	4	4

NU = not usable, 0 = poor, 1 = fair, 2 = good, 3 = very good, 4 = excellent.

Total Mortality Rates

Beverton and Holt (1956) showed that if fishes were growing in accordance with the von Bertalanffy growth equation, and were subjected after recruitment to a constant mortality rate, then

$$Z = K\,(L_\infty - \overline{L})\,/\,(\overline{L} - L') \qquad (2)$$

in which Z is the coefficient of the total mortality rate, K and L_∞ are constants in the von Bertalanffy equation, L' is not smaller than the smallest length group fully represented in the length-frequency distribution of the catch and $\overline{L}$ is the mean of all lengths equal to or greater than L'.

Given values of K and L_∞ it is possible to calculate the value of Z from any reasonable representation of the annual, average, population length-frequency distribution and to compare values of Z obtained in different areas. Furthermore, if samples of a species are also available from an unexploited population, then Z = M (the coefficient of the natural mortality rate) and the fishing mortality rate, F, in exploited areas, can then be derived as F = Z — M.

In addition, in cases where the value of K is unknown, the equation in the form $Z/K = (L_\infty - L)\,/\,(L - L')$ can be used to obtain estimates of the ratios Z/K or M/K (in the case of unexploited populations) provided that the asymptotic length L_∞ can be reasonably estimated or bracketed. This is important in so far as the values of M and K appear in Beverton and Holt's (1966) yield equation only in the form of the ratio of M/K, and the actual magnitudes of M and K need not be known in order to make a preliminary assessment of the fishery potential of a species.

Also, if the values of Z/K for an exploited population and M/K for an unexploited population are known, the value of F/K and the rate of exploitation (E = F/M + F) can be estimated.

Assumptions implicit in the above equations are that throughout the exploited size range the natural and fishing mortality rates remain constant and the patterns of growth conform with the von Bertalanffy growth equation. For the species considered for which "good" estimates of growth and mortality were available, there was no evidence of any radical departures from the patterns of growth or mortality which have generally become accepted as the basis of fish population dynamics.

This formulation has not been widely used. Most investigators have preferred the more laborious method of deriving estimates of mortality rates from the age compositions of samples of catches, where data have been available. However, Le Guen (1971) has shown that estimates of Z based upon age compositions and upon length-frequency distributions of samples of *Pseudotolithus elongatus* from the Gulf of Guinea are nearly identical and, provided that growth over the exploited range is reasonably well described by the Bertalanffy growth equation, the above equation will yield accurate estimates of Z. As with any other such estimate, however, it is likely that very large samples will enhance the accuracy of the results.

Additionally, estimates of Z can be derived from the mean weights of individuals in the catches. The equation for deriving the mean weight of individuals in the catch is

$$\overline{W} = W_\infty \sum_{0}^{3} \frac{U_n Z_e^{-nK(t_c - t_o)}}{Z + nK} \tag{3}$$

in which the mean weight ($\overline{W}$) is a function of the asymptotic weight (W_∞), the growth coefficient K, the total mortality rate (Z) and the age at entry to the exploited phase (t_c). U_n is the summation variable in which $U_0 = 1$, $U_1 = -3$, $U_2 = 3$ and $U_3 = -1$ (Gulland 1969). For those species for which the growth parameters are known, estimates of Z can, therefore, be derived from the observed mean weight of individuals in catches from various areas. The accuracy of such estimates of Z is much dependent upon the estimates of $\overline{W}$ having been based upon catches which were evenly distributed throughout the year or which have been appropriately weighted to allow for seasonal changes in condition and the growth of year-classes. As in the case of equation (2), it is assumed that growth patterns conform with the von Bertalanffy growth function, that mortality rates are constant and that the stock is in equilibrium.

Natural Mortality Rates

A coral reef fish community normally contains a wide spectrum of fish species, ranging from major predators, such as sharks, mainly piscivorous species, such as groupers, snappers and jacks, omnivores, such as grunts and purely herbivorous or at least non-piscivorous groups, such as parrot fishes and surgeon fishes. Most species are resident upon the reefs from a very small size and are liable to predation by larger fishes throughout most of their lives, but the likelihood of death by predation will decrease with increasing size. Predatory species will themselves be liable to predation throughout all of their life cycle, the only likely exceptions being the largest adult sharks. Most species will, therefore, suffer great reductions in their numbers as a result of predation.

The conventional concept of trophic structure in which species A feeds on species B is therefore largely inapplicable and, with the exception of the herbivorous species, the position of a fish in the trophic pyramid is largely determined by its size. Most species contribute to the biomass of predators in accordance with their numbers, average size and their proclivity to the piscivorous habit. All species form part of the pool of prey in accordance with their numbers and average size.

Our estimates of natural mortality rates were based upon the length-frequency compositions of catches taken at unexploited parts of Pedro Bank. The mortality rates are relatively high and probably reflect abundance of predatory species in such areas. As a general case it can be suggested that the following equation describes the situation:

$$M = M_x + gP \tag{4}$$

in which M is the natural mortality from all causes, M_x is natural mortality caused by factors other than predation, P is the biomass of predators and g is the amount of mortality generated in the prey population which is caused by one unit of biomass of predators. With the exception of communities from which predators are virtually absent, the value of M_x will probably be small, as any fish which sickens as a result of disease, injury or lack of food becomes highly vulnerable to predation (see Jones, this vol.). Thus, for most natural communities, $M \cong gP$. The above equation is directly analogous to the expression, $F = qf$, used for describing fishing mortality, F, in which q is the catchability, or amount of mortality generated by one unit of fishing effort, f. Thus, the mortality rates in a population of prey can be viewed as directly proportional to the abundance or biomass of the predators. When predator and prey are concurrently exploited, the biomasses and abundance of both groups should decline exponentially in response to increases in fishing effort. The natural mortality rate of the prey species should, therefore, also decline exponentially. However, certain compensatory effects could occur:

— The numbers of the prey species might not decline as rapidly as those of the predators, because the natural mortality of prey may have decreased.
— The growth rates of the prey species might be enhanced as a result of decreased competition.
— The ratio of predators to prey might decrease, thus reducing competition between predators for the available prey and possibly enhancing growth rates of predators.
— The predators will be of smaller average size and survival of the largest prey species might be differentially improved, thus producing relative increases in the biomass of prey species.

As a result of the complexity of these interactions, the exact magnitude of the changes in natural mortality rates which might result from exploitation are not calculable at present. However, it is possible to make generalized estimates of the magnitude of the decreases in natural mortality rates in exploited areas by comparing the relative abundances of predatory species in various areas. The trap-catch rates of predatory species were, therefore, calculated and expressed as catch in kg per 1,000 Hexagonal traps (Munro 1973) set for 24 hours. Groups classed as major predators for this purpose include all species of sharks, lizard fishes *(Synodontidae)*, eels (*Gymnothorax* spp.), the trumpet fish *(Aulostomus maculatus)*, *Sphyraena barracuda*, hinds and groupers *(Epinephelus* spp., *Cephalopholis fulva, E. cruentatum* and *Mycteroperca* spp.) the tile fish *(Malacanthus plumieri)* and all species of jacks (*Caranx* and *Seriola* spp.) and snappers (*Lutjanus* spp. and *Ocyurus chrysurus*). The relative total biomasses of these species are given in Table 2. In order to estimate the relative amounts of fishing and natural mortalities in exploited areas, the natural mortality rates estimated for unexploited parts of Pedro Bank were multiplied, for each area, by the relative biomass of predators, as given in Table 2 on the assumption that mortality rates are a function of predator biomass, as described in Equation 4.

Table 2. Relative biomass of predators in various areas proximal to Jamaica.

Area	kg/1,000 H.t-n	Relative biomass
Unexploited Pedro Bank	718.3	1.00
Pedro Cays area	355.8	0.50
California Bank	188.9	0.26
S. Jamaica Shelf	161.0	0.22
Port Royal reefs (unbaited traps)	53.7	0.07 x 2 = 0.14

The adjustment of the Port Royal reefs biomass allows for the fact that unbaited traps were used (Munro 1974c).

Fishing Mortality Rates and Catchability

Fig. 1 illustrates the possible relationships between natural mortality rates (M), fishing mortality rates (F), and total mortality rate (Z), at different levels of fishing effort.

It is clear from this figure that fishing mortality rates and catchability cannot be derived from conventional techniques of regression of estimates of Z against fishing effort. Instead, estimates of F must be derived as F = Z — M for each level of fishing effort.

Estimates of catchability, q, were derived for the exploited Pedro Cays area, California Bank and the Port Royal reefs from the formula:

$$q = (Z - P_w M_o) / f \qquad (5)$$

in which Z is the total mortality rate in a given area
P_w is the relative biomass of predators in that area
M_o is the natural mortality rate at unexploited parts of Pedro Bank, and
f is the estimated fishing intensity, expressed as canoes/km² in the area in question.

The approximate fishing intensities at the sampling stations were as follows:

Pedro Cays area	0.15 canoes/km^2
California Bank	0.38 canoes/km^2
South Jamaica Shelf	0.38 canoes/km^2
Port Royal reefs	1.52 canoes/km^2

Recruitment Rates

Estimates of annual recruitment to the stocks were not available for any species, there being no statistics of landings available. Recruitment has, therefore, been expressed on a relative basis, and a recruitment index (R′) calculated which is equal to the theoretical number of recruits required to produce the observed catch rate (numbers per 1,000 Hexagonal traps set for 24 hours). Such estimates are derived simply as $R' = (C/f)Z$ for various areas.

Recruitment rates tended to be greatest at Pedro Bank, but some of the smaller species such as grunts and surgeon fishes had relatively more recruits in the exploited areas. This suggests that predation upon juveniles might be an important factor determining recruitment rates and that if the Pedro Bank stocks are exploited, the patterns of recruitment will shift in favor of the prey species.

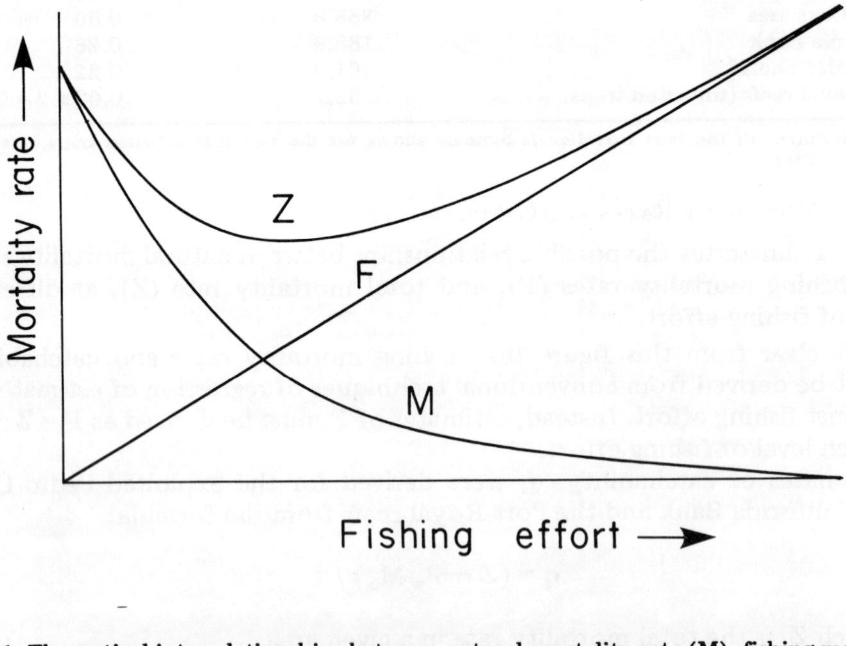

Fig. 1. Theoretical interrelationships between natural mortality rate (M), fishing mortality rate (F) and total mortality rate (Z = F + M) which will exist if natural mortality rates in an exploited community decline as a result of concurrent exploitation of predatory species and given the assumption that natural mortality rates are proportional to predator biomass (from Munro 1974a).

Discussion

The question which must now be raised is whether or not there are better ways of analyzing data in those circumstances which are common to tropical multispecies fisheries: namely, a great diversity of species, relatively small samples of each species obtained at relatively great cost and a pressing need to extract the maximum amount of information from the least amount of data. Undoubtedly, the answer is in the affirmative.

For example, Pauly and David (1981) have developed a new method for highlighting or identifying prominent modes in modal progressions and using a computer to identify objectively the growth curve most likely to fit the data (see Pauly, this vol.). This represents a very useful advance and could usefully be employed to reassess existing data.

However, recent advances in the interpretation of daily rings in otoliths (Panella 1971, 1974) might make analyses of modal progressions redundant for many species or groups, although, at the present time it might still be difficult to interpret and distinguish daily and sub-daily rings in at least some of the important groups (Ralston 1980; Moffitt 1980).

If, as seems a strong possibility, it appears that age determinations will become feasible for many tropical species from examination of daily rings in otoliths, a major unknown will be removed from much assessment work. In particular, information on changes in growth rates (if any) in response to increased rates of exploitation and decreased stock size might become quantifiable. Additionally, if the parameters of the von Bertalanffy growth equation (Equation 1) became more accurately determinable, the utility of Equations 2 and 3 might be enhanced. It is, therefore, suggested that age determinations based on daily rings in otoliths should receive a very high priority in tropical areas.

A separate question which needs to be answered concerns the degree of accuracy which should be aimed at in parameter estimation. Also, it should be assessed whether or not it is worthwhile estimating coefficients of natural mortality, catchability and growth parameters with great accuracy and precision, if, in the final analysis, it is necessary to either assume that they remain constant or, as in the case of the Jamaican studies, blandly assume that mortalities will change as suggested in Equations 4 and 5 and illustrated in Fig. 1.

A different approach to this question would be to ask how it might be possible to evaluate changes in growth and mortality coefficients in response to exploitation of a fish community. Clearly, the Jamaican studies could have produced definitive answers for some species (instead of mere suggestions for a handful of species) if the program had been of longer duration and had possessed a capacity for more intensive sampling efforts. A group of Pacific islands which are environmentally uniform but which are subjected to different levels of exploitation might be highly suited to an investigation of the changes in parameters which are generated by fishing (Munro 1980).

Finally, it is clear from the Jamaican studies that, in common with other temperate and tropical fisheries, the community composition will change in response to exploitation. I believe that it might be possible to derive estimates of fishery parameters from data on catch composition (by numbers and by weight), catch per unit of effort and fishing effort. This is because the catch composition is a function of the differing catchabilities of the species, their growth and natural mortality rates and their recruitment rates. Given information on changes in relative abundance, percentage composition of the catch and say, accurate estimates of growth parameters, it may then be possible to calculate mortality and relative recruitment rates. Alternatively, if information is available on the mean weight of individuals in the catch *and* accurate growth estimates are available then total mortality rates might be derived (Equation 2). If mortality and growth estimates are thus available, it might be possible to combine these with catch composition, effort and catch per unit of effort data to derive information on changes in rates of recruitment.

I believe that it is essential to look to the easily derived statistics, such as mean weights, catch compositions and catch per unit of effort (perhaps using standardized units of effort generated by a research vessel) and gross measures of effort, such as coastal population densities or number of boats per unit area of shelf, if we are to make progress in the assessment of tropical multispecies stocks.

References

Bardach, J.E. and D.W. Menzel. 1957. Field and laboratory observations on the growth of some Bermuda reef fisheries. Proc. Gulf Caribb. Fish. Inst. 9: 106-112.

Bardach, J.E., C.L. Smith and D.W. Menzel. 1958. Bermuda fisheries research program, final report. Bermuda Trade Development Board, Hamilton, Bermuda. 59 p.

Beverton, R.J.H. and S.J. Holt. 1956. A review of methods for estimating rates in exploited populations, with special reference to sources of bias in catch sampling. Rapp. P.-v. Réun. Cons. Int. Explor. Mer. 140: 67-83.

Beverton, R.J.H. and S.J. Holt. 1957. On the dynamics of exploited fish populations. Fish. Invest. Min. Agric. Fish. Food (G.B.) Ser. II; Sea Fish. 533 p.

Beverton, R.J.H. and S.J. Holt. 1966. Manual of methods for fish stock assessment. Part II. Tables of yield functions. FAO Fish. Rech. Pap. 38 (Rev. 1). 67 p.

Goeden, G.B. 1978. A monograph of the coral trout, *Plectromus leopardus* (Lacépède). Queensl. Fish. Serv. Res. Bull. 1. 42 p.

Gulland, J.A. 1969. Manual of methods for fish stock assessment. Part I. Fish population analysis. FAO Man. Fish. Sci. 4. 154 p.

Lebeau, A. and J.C. Cueff. 1975. Biologie et pêche du Capitaine *Lethrinus enigmaticus* (Smith) 1959 du banc de Saya de Malha (Océan Indien). Rev. Trav. Inst. Pêches Marit. Nantes 39: 415-442.

Le Guen, J.C. 1971. Dynamique des populations de *Pseudotolithus (Fonticulus) elongatus* (Bowd. 1825). Poissons–Sciaenidae. Cah. O.R.S.T.O.M. Ser. Océanogr. 9: 3-84.

Moffitt, R. 1980. A preliminary report on bottom-fishing in the Northwestern Hawaiian Islands, p. 216-225. *In* R.W. Grigg and R.T. Pfund (eds.) Proceedings of the symposium on status of resource investigations in the norwestern Hawaiian Island, April 24-25, Honolulu. University of Hawaii, Honolulu.

Munro, J.L., editor. 1973-1980. The biology, ecology, exploitation and management of Caribbean reef fishes. Scientific report of the ODA/UWI Fisheries Ecology Research Project: 1969-1973. Parts I-VI. Res. Rep. Zool. Dep., Univ. West Indies. 3.

Munro, J.L. 1973. Large volume stackable traps for offshore fishing. Proc. Gulf Caribb. Fish. Inst. 25: 121-128.

Munro, J.L. 1974a. The biology, ecology, exploitation and management of Caribbean reef fishes. Part V. Summary of biological and ecological data pertaining to Caribbean reef fishes. Res. Rep. Zool. Dep., Univ. West Indies. 3. (V) 24 p.

Munro, J.L. 1974b. The biology, ecology, exploitation and management of Caribbean reef fishes. Part VI. Assessment of the potential productivity of Jamaican fisheries. Res. Rep. Zool. Dep., Univ. West Indies 3. (VI) 56 p.

Munro, J.L. 1974c. The mode of operation of Antillean fish traps and the relationships between ingress, escapement, catch and soak. J. Cons., Cons. Int. Explor. Mer 35: 337-350.

Munro, J.L. 1977. Actual and potential fish production from the coralline shelves of the Caribbean Sea. FAO Fish. Rep. 200: 301-321.

Munro, 1980. Southwest of Hawaii, p. 325-328. *In* R.W. Grigg and R.T. Pfund (eds.) Proceedings of the symposium on status of resource investigations in the northwestern Hawaiian Island, April 24-25, Honolulu. University of Hawaii, Honolulu.

Munro, J.L. and R. Thompson. 1973. The biology, ecology, exploitation and management of Caribbean reef fishes. Part II. The Jamaican fishing industry, the area investigated and the objectives and methodology of the ODA/UWI Fisheries Ecology Research Project. Res. Rep. Zool. Dep., Univ. West Indies 3. (II) 44 p.

Nagelkerken, W.P. 1979. Biology of the graysby, *Epinephelus cruentatus*, of the coral reefs of Curaçao. Stud. Fauna Curaçao 60: 1-118.

Panella, G. 1971. Fish otoliths: daily growth layers and periodical patterns. Science 173: 1124-1127.

Panella, G. 1974. Otolith growth patterns: an aid in age determination in temperate and tropical fishes, p. 28-39. *In* T.B. Bagenal (ed.) Ageing of fish. Unwin Bros. Ltd., Old Woking, Surrey, England.

Pauly, D. and N. David. 1981. ELEFAN I, a BASIC program for the objective extraction of growth parameters from length-frequency data. Meeresforsch. 28: 205-211.

Ralston, S. 1980. An analysis of the Hawaiian offshore handline fishery: a progress report, p. 204-215. *In* R.W. Grigg and R.T. Pfund (eds.) Proceedings of the symposium on status of resource investigations in the northwestern Hawaiian Island, April 24-25, Honolulu, University of Hawaii, Honolulu.

Randall. J.E. 1962. Tagging reef fishes in the Virgin Islands. Proc. Gulf Caribb. Fish. Inst. 14: 201-241.

Randall, J.E. 1963. Additional recoveries of tagged reef fishes from the Virgin Islands. Proc. Gulf Caribb. Fish. Inst. 15: 155-157.

Discussion of Dr. Munro's Paper

The discussion of Dr. Munro's paper, which was presented by Dr. Pauly in the absence of its author, concentrated on the relative numbers of fish in the various species and the problem these numbers pose, natural mortality and its variability and on problems of stock definitions and identity.

Concerning relative numbers of fish in various species, it was regretted that Dr. Munro had not attempted to provide more than subjective criteria for the quality assessment of available length-frequency data (Table 1 in Munro's paper).

Dr. Marten mentioned that, in Hawaii, 100 species are deemed important enough to be recorded in catch statistics. Dr. Sale pointed out that in all reef assemblages he investigated the numbers of fishes in the various species were lognormally distributed; this implies that there is no inherent difference from temperate communities, and that sampling schemes developed there can be applied on reefs.

It was noted that Dr. Munro had not demonstrated changes in natural mortality, but rather assumed these changes. It was agreed that data are not available which could help test the validity of this assumption (but see Jones, this vol.).

Concerning the identity of the populations sampled, it was mentioned by several participants that studies have been conducted which suggest population homogeneity over a wide scale, while other studies suggested populations were highly localized. The differences between such populations will obviously depend on the mode of recruitment, i.e., on the mechanism for transport of larval stages.

Simple Models for Multispecies Fisheries

G. P. KIRKWOOD

Division of Fisheries Research
CSIRO Marine Laboratories
P.O. Box 21, Cronulla
N.S.W. 2230, Australia

KIRKWOOD, G.P. 1982. Simple models for multispecies fisheries, p. 83-98. *In* Pauly, D. and G.I. Murphy (eds.) Theory and management of tropical fisheries. ICLARM Conference Proceedings 9, 360 p. International Center for Living Aquatic Resources Management, Manila, Philippines and Division of Fisheries Research, Commonwealth Scientific and Industrial Research Organisation, Cronulla, Australia.

Abstract

Possibly the simplest and most studied class of mathematical models for the dynamics of multispecies fish communities comprises differential equation models relating the exploitable biomasses N_i of each species of the form

$$\frac{1}{N_i}\frac{dN_i}{dt} = f_i(N_1, N_2, \ldots N_m).$$

The general properties of models in this class are reviewed, with emphasis on their practical implications for the management of tropical multispecies fisheries.

A particularly simple model in this class has each function $f_i(.)$ linear in its arguments. This model has been termed the multispecies Schaefer model. The properties of this model and its application in tropical regions have been discussed by J.G. Pope, and his findings are reviewed.

The extent to which the multispecies Schaefer model is an adequate approximation to more general models in the class is examined and the stability properties of the models and timescales of response to perturbations discussed. It is shown that there is a large diversity in the behavior of general models in the class that defies synthesis into generally applicable properties in the absence of specific knowledge of parameter values.

Available data suggest that the prediction by the multispecies Schaefer model of a linear equilibrium catch-per-effort versus effort relationship may not hold. Some aspects of the behavior of multispecies models, formulated as an extension of generalized stock production models to a multispecies situation, analogous to Pope's extension of the single-species Schaefer model, are examined. While these models have properties similar to those of the multispecies Schaefer models, the generality of results for the latter models is lost.

In view of the lack of generally applicable model properties, the poor understanding of the underlying biological processes and the unavailability of estimates of possible model parameters, models even in this simple class appear to have relatively little to offer at present in the way of management paradigms for tropical multispecies fisheries. It is recommended that a more empirical approach to these problems be adopted.

Introduction

Consider an exploited marine ecosystem comprising m species in which the biomasses of the exploitable segments of the populations are denoted $N_1, N_2, \ldots, N_m$. In this review, we shall consider "simple" models for the dynamics of this system of the form

$$\frac{1}{N_i}\frac{dN_i}{dt} = f_i(N_1, N_2, \ldots N_m) \qquad (1)$$

In principle, this class of models includes members ranging from those where $f_i(N_1, N_2, \ldots N_m) = \alpha_i + \beta_i N_i$, i.e., where there are no interspecific interactions and very simple intraspecific interactions, to those in which the functions f_i (.) are complicated non-linear functions of each of the N_i. For large m, these models really are simple neither in terms of number of parameters, behavior, or even form. Despite this, we shall class them as simple, in order to distinguish them from the complex ecosystem models of the Andersen and Ursin (1978) variety, whose possible applications are reviewed elsewhere (Larkin and Gazey, this vol.).

There is a vast literature on the mathematical theory of ecological processes, including the equilibrium and dynamic properties of models such as (1) (see Pielou 1969; May 1975, 1976 for summaries). Even a brief trip through the foothills of this mountain of literature convinces one that plausible models with form (1) are as numerous and varied as is their behavior. If this review is to be of value, and not merely a loosely connected compendium of differing models and their properties, a firm focus needs to be set. Thus, we shall here concentrate on the *practical* implications of these models to tropical multispecies fisheries and their management.

Expanding on this theme, the first observation that may be made is that our understanding of the ways tropical multispecies marine ecosystems "work" is generally very poor. These systems contain large numbers of species; thus, even if one particular member of the class of simple models was considered to accurately portray the dynamics of the system, (and this is by no means clear, see Sainsbury, this vol.) the sheer difficulty of parameter estimation alone precludes practical application of the model. Also tropical multispecies fisheries are often undirected, with the overall biomass yield being of most importance, rather than the yield of a few commercially valuable species (except in the case when penaeid shrimps occur in the catch).

The available models are in the main empirical, and we must perforce concentrate on those properties of the models that are largely independent of specific parameter values. Thus, we shall concentrate here particularly on relations between total yield and fishing effort, and on the relative biomass levels at which MSY occurs.

Although it was not so stated explicitly, this reasoning appeared to motivate the approach taken by Pope (1976, 1979), who in the latter publication directed attention specifically to tropical multispecies fisheries. This approach may also be seen in the Report of the Expert Consultation on Management of Multispecies Fisheries (FAO 1978).

An important subclass of (1) comprises those models in which the functions $f_i(.)$ are linear in the N_i. These are the Lotka-Volterra type models, and in the single-species case correspond to the familiar Schaefer model. Pope (1976, 1979) has given an excellent review of the behavior of such models. Below, the behavior of these multispecies Schaefer models is briefly outlined; then we will revert to the full class of models in order to examine to what extent the multispecies Schaefer models are an adequate approximation to more general models.

One property of the multispecies Schaefer models is that the total catch per unit effort is a linear function of fishing effort. However, a number of data sets for tropical fisheries suggest that this relationship may be curvilinear. For single species, such a relationship is demonstrated by generalized production models described by Pella and Tomlinson (1969). These models can be generalized to the multispecies case in a manner exactly analagous to that in which the single-species Schaefer model is extended to the multispecies Schaefer model (see below). The behavior of these models is compared with that of the multispecies Schaefer models.

In the final section, the results are synthesized and discussed. In addition, the effects of treating the differential equations (1) as difference equations, and of inclusion of complicating (but realistic) variables such as time lags and age structure are briefly touched upon.

Multispecies Schaefer Models

Under the Schaefer model for a single species, the dynamics of an exploited population in terms of its biomass N and fishing mortality rate F are described by

$$\frac{1}{N}\frac{dN}{dt} = b - aN - F.$$

When F = 0, there is an unexploited stable equilibrium at

$$N_o = \frac{b}{a}$$

and if F = qf, where f denotes fishing effort, the sustainable yield at given f is

$$Y(f) = qf\,(b - qf)/a.$$

Thus the sustainable yield curve as a function of f is a parabola, with a maximum at $\frac{1}{2}N_o$. If the fishing mortality rate resulting in the taking of the maximum sustainable yield (MSY) is approximately the same as M, the natural mortality rate, we then have the oft quoted formula (Gulland 1970)

$$Y_{max} \cong \tfrac{1}{2} M N_o \qquad (2)$$

Even if the Schaefer model is assumed to be correct, formula (2) depends critically on whether $F \cong M$ at MSY level. Francis (1974) has examined this formula in the context of an underlying Schaefer model, and has shown that at MSY

$$F \geqslant M \quad \text{if} \quad R_{opt} \leqslant R_{max},$$

where R_{opt} and R_{max} are the recruitments at MSY and unexploited levels respectively.

A natural extension of the single species Schaefer model to m species is the set of equations

$$\frac{1}{N_i}\frac{dN_i}{dt} = b_i - \sum_{j=1}^{m} a_{ij} N_j - q_i f \tag{3}$$

where $i = 1, 2, \ldots, m$. If $a_{ij} = 0$ for $i \neq j$, we have simply a set of m independent single-species Schaefer models. More generally, (3) is of the familiar Lotka-Volterra form, modified to include the effects of fishing. Equation (3) forms the basis of the studies by Pope (1976, 1979) and the remainder of this section outlines results he obtained.

If there is an equilibrium at $N_1^*, N_2^*, \ldots, N_m^*$, where $N_i^* \geqslant 0$ for each i, then from (3)

$$b_i - \sum_{j=1}^{m} a_{ij} N_j^* - q_i f = 0 \qquad \text{for } i = 1, 2, \ldots, m$$

This set of equations may be written conveniently in matrix notation

$$B - A\,\underset{\sim}{N}^* - Qf = 0, \tag{4}$$

where B, $\underset{\sim}{N}^*$ and Q are (l x m) matrices and A is an (m x m) matrix. It follows that

$$\underset{\sim}{N}^* = A^{-1}\,(B - Qf) \tag{5}$$

and the total yield from the system at $\underset{\sim}{N}^*$ is

$$Y = f\,Q^T \underset{\sim}{N}^*$$

$$= f\,Q^T A^{-1} B - f^2\,Q^T A^{-1} Q \tag{6}$$

Thus for any matrix A, the total yield is a parabolic function of fishing effort f, given of course a constant Q vector and provided none of the N_i^* become zero.

This encouraging similarity to the single-species Schaefer model behavior leads one to examine under what circumstances the maximum yield occurs at half the virgin biomass. From (4), the total yield as a function of N* can be written

$$Y = \underset{\sim}{N}^{*T} B - \underset{\sim}{N}^{*T} A \underset{\sim}{N}^*$$

so that performing vector differentiation with respect to $\underset{\sim}{N}^*$ and equating to zero, we find that the maximum yield occurs at $\underset{\sim}{N}_{MSY}$, where

$$B - (A + A^T)\,\underset{\sim}{N}_{MSY} = 0$$

i.e.

$$\underset{\sim}{N}_{MSY} = (A + A^T)^{-1}\,B \tag{7}$$

Thus if A is a symmetric matrix ($a_{ij} = a_{ji}$ for all $i \neq j$)

$$\underset{\sim}{N}_{MSY} = \tfrac{1}{2}A^{-1}\,B.$$

However, from (5) the virgin unexploited biomass occurs at

$$\underset{\sim}{N}^o = A^{-1}\,B,$$

so that if A is symmetric, MSY occurs at

$$\underset{\sim}{N}_{MSY} = \tfrac{1}{2}\underset{\sim}{N}^o \tag{8}$$

In general, the ratio of the yield when each species is at half its virgin biomass to the overall maximum yield is

$$\tfrac{1}{4}\,B^TA^{-1}B \;:\; B^T(A + A^T)^{-1}B \;-\; B^T(A + A^T)^{-1}\,A(A + A^T)^{-1}B$$

This differs slightly from the expression derived by Pope (1979). It is clear that if A is nearly symmetric, or if the off diagonal elements of A are small, this ratio will be close to 1:1. The ratio will be furthest from 1:1 when A is strongly asymmetric, which will occur in particular if predator-prey interactions predominate.

Equation (7) above defines the equilibrium species biomass levels at which overall MSY occurs. On the other hand, for a fixed catchability vector Q, equation (5) describes a trajectory of equilibrium biomass levels as a function of fishing effort f, which will not in general pass through $\underset{\sim}{N}_{MSY}$. Thus, while (6) ensures that for a fixed Q the total sustainable yield is a parabolic function of f, in general the maximum sustainable yield for fixed Q will be less than the overall MSY. Further, both the maximum sustainable yield and the level of fishing effort at which it is taken will vary as the vector of catchabilities varies.

Pope (1979) also demonstrated that the overall MSY is generally less than the sum of the individual single species MSYs, this diminution being greatest when the interspecific interactions are strongest. This conclusion was based on calculating the individual species MSYs when all other species in the system are at zero biomass. However, Pope (1979) noted that in practice these MSYs will normally be assessed in the presence of other species. The other important effect of increasing strength of interspecific interactions is that the regions of equilibria in which all species biomasses are positive decrease in size.

General Models

The discussion in the previous section suggests that an approach to multispecies fisheries assessments similar to the single-species Schaefer approach may be quite fruitful. However these discussions have been based solely on a multispecies Schaefer model, and as no one would seriously suggest that this model is exactly correct, it is important to determine to what extent the model (3) is an adequate approximation to the more general models (1). This, of course, begs the question of whether models (1) are adequate, but that lies outside the scope of this review.

A frequent, but rather loose comment is that the multispecies Schaefer model is a first order Taylor series approximation to the more general models. Let us examine this, and suppose that $N_1{*}, N_2{*}, \ldots N_m{*}$ is an equilibrium point for the set of equations

$$\frac{1}{N_i}\frac{dN_i}{dt} = f_i(N_1, N_2 \ldots N_m) \qquad \text{for } i = 1, 2, \ldots m$$

Then $f_i(N_1, N_2, \ldots N_m)$ may be expanded in a Taylor series about this equilibrium, yielding to first order

$$f_i(N_1, N_2, \ldots N_m) = \sum_{j=1}^{m} (N_j - N_j{*}) \frac{\partial f_i}{\partial N_j}(N_1{*}, N_2{*} \ldots N_m{*})$$

where the superscript asterisks in the partial derivatives indicate that the derivatives are to be evaluated at the equilibrium point. Combining these two equations, we indeed find

$$\frac{1}{N_i}\frac{dN_i}{dt} = b_i - \sum_{j=1}^{m} a_{ij} N_j$$

for some constants b_i, a_{ij}. However, it is quite clear that in the first order expansion about a different equilibrium, while the linear form will be retained, the constants will be different. Thus any particular model (3) will be an adequate approximation to (1) only in a neighborhood of an appropriate point of equilibrium of (1). That these trivial comments are worth making at all is due to the fact that models such as (3) have been applied, not in small regions of the space of the N_i, but over the whole space, often with the justification that they are first order Taylor approximations to a more general model. In fact this approach is valid only if the true form of (1) is very similar to that of (3).

A second area of approximation arises if not all of the interacting species in a system are included in the model. This may arise, for example, if the model deals only with a fish community, and a number of the fish are benthic or plankton feeders. Suppose there are m fish species of biomasses N_i, $i = 1, 2, \ldots m$, and r resource species R_j, $j = 1, 2, \ldots r$, and that the true full model for the system is

$$\frac{1}{N_i}\frac{dN_i}{dt} = b_i - \sum_{j=1}^{m} a_{ij} N_j - \sum_{k=1}^{r} c_{ik} R_k - q_i f \qquad \text{for } i = 1, 2, \ldots m,$$

$$\frac{1}{R_i}\frac{dR_1}{dt} = b_i^1 - \sum_{j=1}^{m} a_{ij}^1 N_j - \sum_{k=1}^{r} c_{ik}^1 R_k \qquad \text{for } i = 1, 2, \ldots r.$$

Then at equilibrium for a particular value of f,

$$b_i - \sum_{j=1}^{m} a_{ij}N_j - \sum_{k=1}^{r} c_{ik}R_k - q_i f = 0 \qquad i = 1, 2, \ldots m$$

$$b_i^1 - \sum_{j=1}^{m} a_{ij}^1 N_j - \sum_{k=1}^{r} c_{ik}^1 R_k = 0 \qquad i = 1, 2, \ldots r.$$

Then the second set of equations may be solved for R_i in terms of the N_i, and substituted into the first set of equations, leading to

$$b_i'' - \sum_{j=1}^{m} a_{ij}'' N_j - q_i f = 0$$

for some b_i'', a_{ij}''. Thus, models of the form (3) may still be appropriate even if some species have been omitted, but again this holds only in the neighborhood of the original equilibrium. This topic is dealt with in more detail by Schoener (1974), although he derives the approximation slightly differently.

An important aspect of the behavior of models (1) is their stability, particularly in the neighborhood of an equilibrium point. An equilibrium point is said to have neighborhood stability if the system returns over time to the equilibrium point after an initial small perturbation from that equilibrium. This obviously is an essential property if an equilibrium yield for a given fishing effort is to be in any real sense a "sustainable" yield. Neighborhood stability is governed by the behavior of equations (8), where if we put $x_i = N_i - N_i^*$ as an initial perturbation from the equilibrium at N_i^* for each i, we have in matrix notation the linearized approximation

$$\frac{d\underset{\sim}{x}(t)}{dt} = A\,\underset{\sim}{x}(t)$$

Here $\underset{\sim}{x}(t)$ is the (m x 1) column vector of the initial perturbations $x_i(t)$, and A is the (m x m) "community matrix" (Levins 1968) with elements

$$a_{ij} = \frac{\partial\, N_i f_i\,(N_1, N_2, \ldots N_m)}{\partial N_j}$$

evaluated at the equilibrium point $N_1^*, N_2^*, \ldots N_m^*$. May (1975) shows that the equilibrium point will have neighborhood stability if and only if all eigen values of the community matrix A have negative real parts.

Unfortunately, while this criterion may be tested for any given model, it depends on the precise form of the matrix A, and little can be said in general of the neighborhood stability of equilibria of models (1) without a knowledge of their exact form. However, we clearly wish the system (as opposed to the models) to have this property, if we are to be able to rely on yield-effort curves. It is worth noting at this point that two-species models tend to behave rather better and have better stability properties than models with more than two species. It is unfortunate that the two-dimensional book page acts as a strong incentive to use two-species models as examples (as indeed will be done later in this review). Suffice it to say that generalizations from two-multispecies models should be made with caution.

Another important topic is the speed, or time scale of transitions to the equilibrium points. In the main, the transient behavior of the models like (1) is much more complex than their static and equilibrium behavior; however, two observations may be made. The first concerns the speed of return to an equilibrium after an initial perturbation, assuming neighborhood stability of the equilibrium.

A relative measure of the speed of return of the system to an equilibrium following a small perturbation is given by the "characteristic return time" (May et al. 1974; Beddington et al. 1976). This measure, which is the reciprocal of the magnitude of the real part of the dominant eigenvalue (May et al. 1978), typically increases as the rate of exploitation increases (Beddington and May 1977). However, Shepherd and Horwood (1979) have shown that to some extent this depends on the manner in which the perturbations are introduced. Despite this, it is generally agreed that the resilience of a system decreases with increasing harvesting (May et al. 1978; May et al. 1979). More recently, Horwood and Shepherd (1981) have shown that if the system and the perturbing noise are modelled more realistically, the sensitivity of the system to noise increases monotonically as exploitation increases, and that the return time corresponds poorly with this sensitivity. Thus the conclusion holds, even if the basis for the original arguments is somewhat shaky. Similar conclusions have been reached by Sissenwine (1977) and Doubleday (1976).

Of possibly more practical importance in the present context are the time scales of responses to larger changes. This is particularly important when attempting to interpret yield-effort curves. All the results discussed here are couched in terms of equilibria and sustainability. However, if the fishing effort changes from f_1 in one year to f_2 in the next year, the yield in the second year will normally not reflect the sustainable yield corresponding to that level of effort, due to time lags in response of the system. This phenomenon has been recognized, of course, for a long time, and means of allowing for this are well known (Gulland 1961, 1969; Fox 1975). Essentially these techniques involve averaging of the catches per unit effort over the duration of a year class in the fishery, which is related in turn to the life span of the fish, and its "intrinsic growth rate" (the parameter r in the standard single-species Schaefer model). Similar considerations apply in the multispecies case, but the situation is complicated by the presence of more than one species, and by their interactions. It can be stated, however, that the response of each species to changes occurs on a time scale set roughly by the reciprocal of the intrinsic growth rate ($1/b_i$ in equations (3), for example) of the single species in the system. Most importantly, the system as a whole responds, again roughly, on the longest time scale of each of the component species. This is seen clearly in the figures in May et al. (1979).

Lastly, we must advert at least briefly to several types of behavior of models (1) and to some extent of the real systems they attempt to mimic, that cast some doubt on the simple picture that has been drawn so far. To this stage, we have blithely been speaking of equilibria as if they were equilibrium points. That even the simplest of models can have not an equilibrium point, but rather a stable limit cycle, is well known, and seemingly similar behavior is exhibited in real systems (May 1975). Another finding is that both models and real systems may also have more than one stable state. The review of Sutherland (1974) demonstrates that "multiple stable points are an undeniable reality." The existence of multiple equilibria for models (1) also has been amply demonstrated (e.g., Clark 1976; Gilpin and Case 1976). The evidence for existence of multiple stable states has been reviewed by May (1977).

Finally, we should also note that models such as (1), especially discrete time and stochastic versions of them, can exhibit a bewildering multiplicity of odd behaviors. It is known that even relatively simple models, with at least *prima facie* plausibility, can exhibit quite chaotic behavior (e.g., Guckenheimer et al. 1977), which has even led some authors to invoke the currently fashionable catastrophe theory (Thom 1972; Zeeman 1977). Population levels in real systems can also give the appearance of fluctuating chaotically. However it is extremely difficult to establish whether this genuine behavior is generated by the mechanisms in the models (Poole 1977). Clark (1976) gives a fine review of some of these aspects of model properties.

Whether particular models (1) actually exhibit behavior similar to that described in the preceding two paragraphs depends entirely on their precise form and parameter values, and few general statements can be made. On the other hand, there is ample evidence that real biological systems do exhibit behavior similar to that described above. In practical terms, we are not in a position to know the true form of the appropriate model. Thus these complicating properties should be taken as warning signals and in the presence of uncertainty we can only counsel caution. Especially in a developing fishery, a prudent approach would have management goals being set initially very conservatively, as has been suggested in a slightly different context by Kirkwood (1981) and others.

Multispecies Extension of Generalized Production Models

Pope's (1976, 1979) analysis of multispecies Schaefer models suggests that the sustainable yield-effort curve should be parabolic, provided the individual species catchabilities remain constant. In turn, this implies that the equilibrium catch-per-effort should be linearly related to the effort. However, the typical observed plot of total catch-per-effort against effort is concave rather than linear; see, for example, Gulland (1972, Fig. 13c) and Liu et al. (1978) for Gulf of Thailand data.

There is a major problem in interpreting these data, as they represent plots of an annual time series of catches-per-effort against effort, and the observed relationship will almost certainly differ from the "true" equilibrium relationship. This difference occurs because of the delayed response of the system to changes in effort, and as discussed above, these responses should occur on a time scale similar to the longest time spent in the fishery by any species. Nevertheless, the available evidence suggests, if anything, a concave relationship between equilibrium catch-per-effort and effort rather than a linear one.

In the single species case, such a relationship is predicted by generalized production models of the form

$$\frac{1}{N}\frac{dN}{dt} = b - aN^{m-1} \quad (10)$$

introduced by Pella and Tomlinson (1969). When m = 2, this model (10) reduces to the Schaefer model. If m > 2, the sustainable yield versus effort curve has a peak to the left of half the unexploited equilibrium level, and the sustainable catch-per-effort against effort curve is concave.

Thus, models exhibiting this property, and possibly those of this form, may be more appropriate as an initial starting point than the Schaefer model. When m approaches 1 in (10), the Pella and Tomlinson model becomes equivalent to the model proposed by Fox (1970), and in fact this model was applied by Liu et al. (1978).

In this section, we briefly examine the general behavior of models which extend (10) to the multispecies case in an analogous manner to the extension of the Schaefer model to models (3). As in (3), suppose N_i denotes the biomass of species i. We assume that

$$\frac{1}{N_i}\frac{dN_i}{dt} = b_i - \sum_{j=1}^{n} a_{ij}N_j^{m-1} - q_i f \quad (11)$$

for i = 1, 2, . . . n. Then if we define, at the equilibrium of this set of equations $N_1^*, N_2^*, \ldots N_n^*$ for given f, $\underset{\sim}{N}^{*m-1}$ to be the (l x n) matrix with elements $N_1^{*m-1}, N_2^{*m-1}, \ldots N_n^{*m-1}$, and define matrices B, A and Q as in equation (4), we have at equilibrium

$$\underset{\sim}{N}^{*m-1} = A^{-1}(B - Qf) \quad (12)$$

The total yield from the system at $\underset{\sim}{N}^*$ is

$$Y = fQ^T \underset{\sim}{N}^* \quad (13)$$

In view of (12), equation (13) unfortunately does not simplify to a convenient function of f as in equation (6) obtained for the Schaefer model. However, it is clear that (13) does not in general define a parabola in f, and the MSY does not occur at $\frac{1}{2}\underset{\sim}{N}_o$, even if A is a diagonal matrix, as it would be if there were no interspecies interactions. Naturally, this is to be expected. To gain an idea of the types of behavior of models such as (11), we need to resort to specific examples.

We examine a two-species competitive community with "weak" and "strong" interactions identical (when m = 2) to those used as examples by Pope (1976). For the weakly competitive community, the dynamics are described by

$$\frac{1}{N}\frac{dN_1}{dt} = 0.43 - 0.000143\,N_1{}^{m-1} - 0.0000143\,N_2{}^{m-1} - F_1$$

$$\frac{1}{N}\frac{dN_2}{dt} = 1.10 - 0.001\,N_2{}^{m-1} - 0.00005\,N_1{}^{m-1} - F_2 \qquad (14)$$

while for the strongly competitive community,

$$\frac{1}{N_1}\frac{dN_1}{dt} = 0.43 - 0.000143\,N_1{}^{m-1} - 0.000118\,N_2{}^{m-1} - F_1$$

$$\frac{1}{N_2}\frac{dN_2}{dt} = 1.10 - 0.001\,N_2{}^{m-1} - 0.000266\,N_1{}^{m-1} - F_2 \qquad (15)$$

Contours of the equilibrium yield as a function of F_1 and F_2 are shown, for $m = 3$, 2 and 1.5, in Figs. 1, 3, 5 for the weak competition case, and in Figs. 2, 4, 6 in the strong competition example. Figs. 3 and 4 are identical to Figs. 2 and 3 in Pope (1976).

Turning first to Figs. 3 and 4, we see as suggested by Pope (1976), that the contours of equal equilibrium yield are ellipses within the region in which both equilibrium species biomasses are positive. This region is smaller when the interspecies interactions are stronger, and the overall MSY is also lower with stronger interactions. For the weak interaction example, the sustainable yield when both species are at half their equilibrium biomasses is close to the overall MSY.

From the earlier discussion, the examples when $m = 3$ (Figs. 1 and 2) may be more appropriate for tropical fisheries data than those when $m = 2$. Whenever $m \neq 2$, the contours of equal total yield are no longer ellipses, as is seen in Figs. 1 and 2. Within the region where both species coexist (identical to those in Figs. 3 and 4 respectively), the contours are more widely separated near the origin than they are for high values of fishing mortality F_1 and F_2. This implies that when $F_1 = q_1 f$ and $F_2 = q_2 f$, the yield curve as a function of f is no longer parabolic, as it was when $m = 2$. In this case the slope of the yield curve is smaller in the left hand limb than in the right hand limb, which implies a concave catch-per-effort against effort curve. Also for different values of q_1, q_2, the shape of this curve differs, as well as its maximum. Again the overall MSY is lower when interactions are higher.

In the single-species generalized production model, MSY occurs at a fraction $(1/m)^{1/m-1}$ of the unexploited population biomass. In Fig. 1 we see that the yield when each species is at this fraction of its virgin biomass is relatively close to the overall MSY when interactions are weak. Also shown on Fig. 1 is the yield when each species is at ½ its virgin biomass. Again this is not far from the overall MSY, and it is also more conservative, in the sense that the overall fishing effort is lower than at MSY, while the biomass levels are higher.

The third set of figures (Figs. 5 and 6) correspond to $m = 1.5$. For single species, this may be appropriate for top predators (May et al. 1979), and by analogy, such a multispecies set of equations may be appropriate for a com-

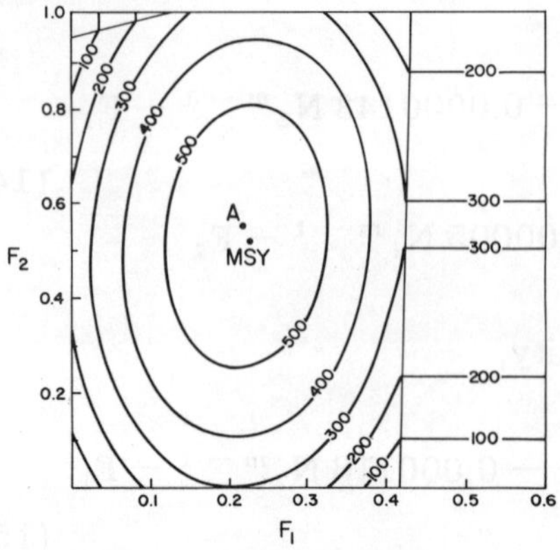

Fig. 1. Weak competition, m = 3.0.

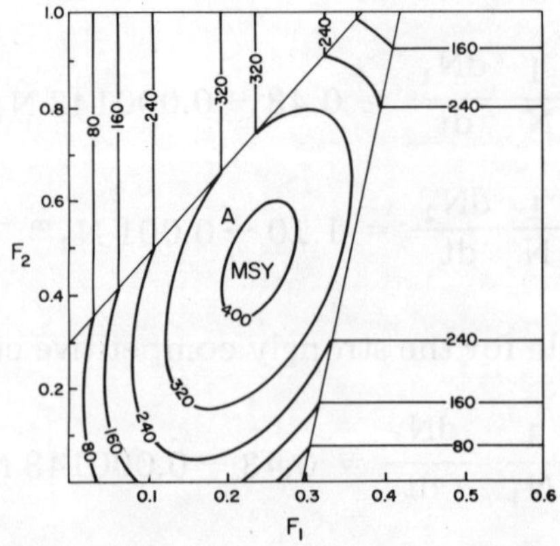

Fig. 2. Strong competition, m = 3.0.

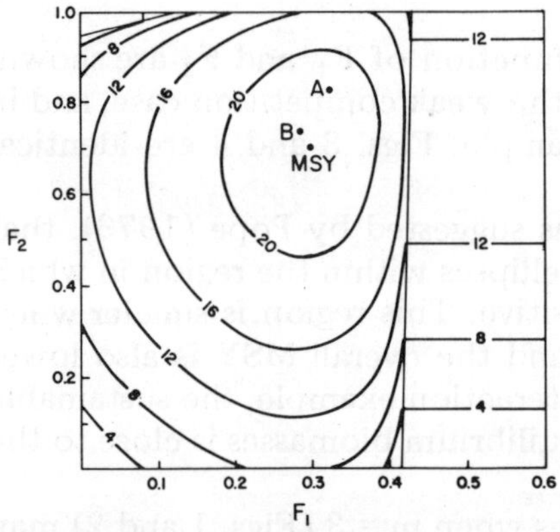

Fig. 3. Weak competition, m = 2.0.

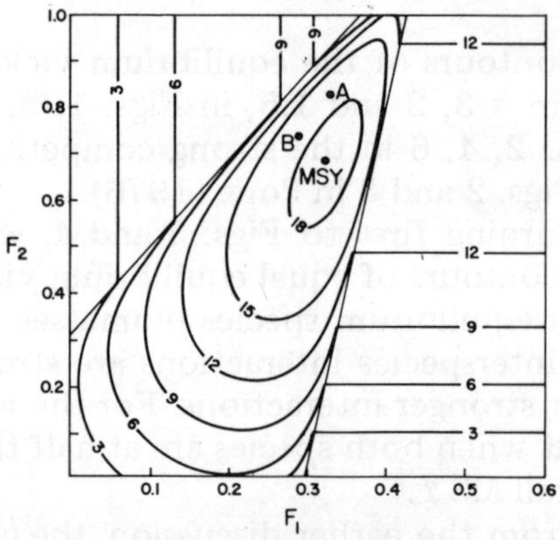

Fig. 4. Strong competition, m = 2.0.

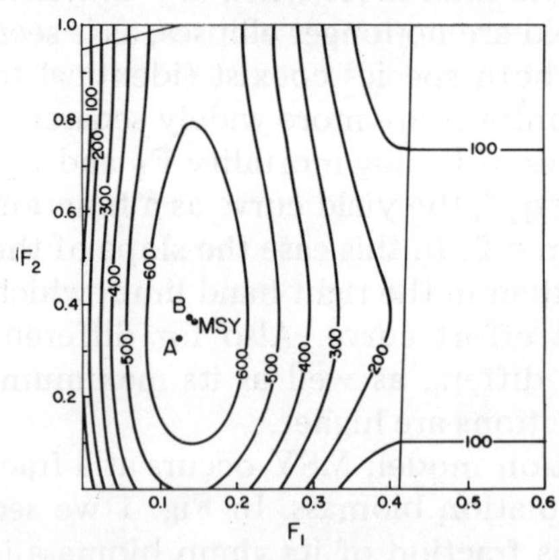

Fig. 5. Weak competition, m = 1.5.

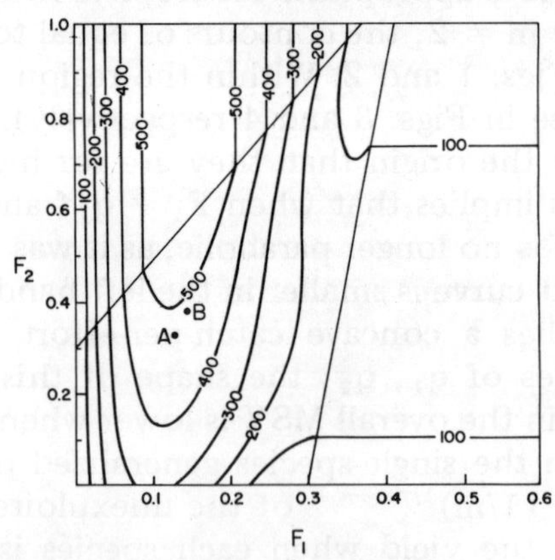

Fig. 6. Strong competition, m = 1.5.

Contours of total sustainable biomass yield as a function of fishing mortality rates F_1 and F_2 for a two-species competitive community, whose dynamics are described by equation (14) for weak competition and (15) for strong competition, and for different values of m. The point of maximum sustained yield (MSY) as indicated as are the sustainable yields at ½ virgin biomass (A) and at $(1/m)^{1/m-1}$ virgin biomass (B).

munity of top predators. In this case the general features are a mirror image of the general features of Figs. 1 and 2. However. Fig. 6 has a special feature of interest in that no MSY is indicated. In fact, the maximum yield within the region of coexistence occurs right on the boundary, where the biomass of species 2 becomes zero, and the overall MSY is a yield of 576 taken solely from species 1 with $F_1 = 0.143$. This provides an excellent illustration of the fact that for particular combinations of catchabilities, the total-yield-effort curve may still be rising at a stage where one of the species becomes extinct.

Discussion

The major aim of this review of models (1) and their implications for management of tropical multispecies fisheries was to determine whether an approach based on examination of the relationship between total biomass yield and fishing effort or relative species biomass levels was feasible. The work of Pope (1976, 1979) outlined above certainly suggests that such an approach merits close attention. Using a multispecies Schaefer model, Pope showed that for any constant vector of catchabilities, the relation between total yield occurred at species biomasses of close to one half their unexploited levels if the interactions were symmetric or weak.

Examination of the extent to which the multispecies Schaefer model is an adequate approximation to the more general models leads to the conclusion that the global results obtained by Pope only hold when the multispecies Schaefer model is the correct model. This occurs because, while the form of the model is appropriate in the neighborhood of points of equilibrium of management models, the parameter values change as the point of equilibrium changes. Unfortunately, most approximate models have a relatively restricted region of validity; it being unfortunate in the sense that we really wish to apply these models over a large region.

In general, the behavior of models in the class (1) ranges from the regular behavior described by Pope to seemingly chaotic behavior. Given the above expressed caveats on use of approximate models, simple examinations of the models (1) *per se* will not lead to widely applicable management paradigms, unless some of the models can be excluded from consideration. This is unfortunately difficult even in principle, as some of the models with irregular behavior are specifically based on what was felt to be biologically more reasonable assumptions than those inherent in the well-behaved models. In practice, our knowledge of tropical multispecies fish communities is insufficient to exclude more than a very few models in class (1).

There are, however, a few general conclusions that can be drawn from our knowledge of these models. These include:

- the overall MSY for the system is generally less than the sum of the individual single-species MSYs, and the amount by which it is less increases as the strength of interspecies interactions increases;
- the sensitivity of the system to environmental fluctuations increases as the level of exploitation increases;
- the models that do exhibit 'unexpected' and irregular behavior normally are those that include strong and nonlinear interspecies interactions. If these interactions are weak, the behavior of the models is more regular.

Since the discussion of the models did not generate widely applicable management guidelines, it appears that we must resort to an even more empirical approach, and here the prospects may be somewhat brighter. At least in the medium term, and for moderate levels of exploitation, exploited fish communities do seem to have fairly regular behavior. Thus if, for example, for each different level of fishing effort there appears to be an equilibrium yield that is stable in the sense of the neighborhood stability of some of the models, then in principle a sustainable yield-effort curve could be drawn empirically. In such a case, it is important that this curve is not immediately forced to fit a convenient functional form. It may be tempting to fit a parabola in view of the results obtained by Pope. However, the discussions on observed data and the multispecies extension of the generalized production models of Pella and Tomlinson (1969) suggest not only that this may not be the most appropriate curve to fit, but more importantly that the form of the curve will differ for differing sets of catchabilities. If such an empirical relationship can be found, then rational conclusions may be drawn from it, and in time it may even be possible to adjust the individual catchabilities and head towards the overall MSY using methods similar to those suggested by Pope (1979).

An empirical approach to the provision of management advice may not only provide answers to the urgent management problems of the present, but in conjunction with ongoing research, may also lead to a better understanding of the underlying mechanisms of the ecosystem. With a better knowledge of the true mechanisms, the range of appropriate models may be sufficiently reduced to allow them to be of great practical value.

References

Andersen, K.P. and E. Ursin. 1978. A multispecies analysis of the effects of variations of effort upon stock composition of eleven North Sea fish species. ICES Symposium (1975) on the changes in the North Sea fish stocks and their causes. Rapp. P.-V. Réun. Cons. Int. Explor. Mer. 112: 286-291.

Beddington, J.R., C.A. Free and J.H. Lawton. 1976. Concepts of stability and resilience in predator-prey models. J. Anim. Ecol. 45: 791-816.

Beddington, J.R. and R.M. May. 1977. Harvesting natural populations in a randomly fluctuating environment. Science 197: 463-465.

Clark, C.W. 1976. Mathematical bioeconomics. John Wiley and Sons, New York.

Doubleday, W.G. 1976. Environmental fluctuations and fisheries management. Int. Comm. Northw. Atlant. Fish. Sel. Pap. 1: 141-150.

FAO. 1978. Some scientific problems of multispecies fisheries. Report of the Expert Consultation on Management of Multispecies Fisheries (1977). FAO Fish. Tech. Pap. No. 181.

Francis, R.C. 1974. Relationship of fishing mortality to natural mortality at the level of maximum sustainable yield under the logistic stock production model. J. Fish. Res. Board Can. 31: 1539-1542.

Fox, W.W. Jr. 1970. An exponential surplus-yield model for optimising exploited fish populations. Trans. Amer. Fish. Soc. 99: 80-88.

Fox, W.W. Jr. 1975. Fitting the generalised stock production model by least-squares and equilibrium approximation. Fish. Bull. 73(1): 23-37.

Gilpin, M.E. and T.J. Case. 1976. Multiple domains of attraction for competition communities. Nature 261: 40-42.

Guckenheimer, J., G. Oster and A. Ipaktchi. 1977. The dynamics of density dependent population models. J. Math. Biol. 4: 101-147.

Gulland, J.A. 1961. Fishing and the stocks of fish at Iceland. Fish. Invest. Minist. Agric. Fish. Food (G.B.) Ser. 2. 23(4). 52 p.

Gulland, J.A. 1969. Manual of methods for fish stock assessment. Part 1. Fish population analysis. FAO Man. Fish. Sci. 4. 154 p.

Gulland, J.A. 1970. The fish resources of the oceans. FAO Fish. Tech. Pap. No. 97. 425 p.

Gulland, J.A. 1972. Population dynamics of world fisheries. Washington Sea Grant Publications, University of Washington, Seattle.

Horwood, J.W. and J.G.Shepherd. 1981. The sensitivity of age-structured populations to environmental variability. Math. Biosci. 57: 59-82.

Kirkwood, G.P. 1981. Allowing for risks in setting catch limits based on MSY. Math. Biosci. 53: 119-129.

Levins, R. 1968. Evolution in changing environments. Princeton University Press, Princeton, N.J.

Liu, H.C., H.L. Lai and S.Y. Yeh. 1978. General review of demersal fish resources in the Sunda shelf and the Australian waters. Acta Oceanographica Taiwanica 8: 109-140.

May, R.M. 1975. Stability and complexity in model ecosystems. (2nd ed.) Princeton University Press, Princeton, N.J.

May, R.M., Editor. 1976. Theoretical ecology: principles and applications. Blackwell, Oxford.

May, R.M. 1977. Thresholds and breakpoints in ecosystems with a multiplicity of stable states. Nature 269: 471-477.

May, R.M., G.R. Conway, M.P. Hassell and T.R.E. Southwood. 1974. Time delays, density dependence, and single species oscillations. J. Anim. Ecol. 43: 747-770.

May, R.M., J.R. Beddington, J.W. Horwood and J.G. Shepherd. 1978. Exploiting natural populations in an uncertain world. J. Math. Biosci. 42: 219-252.

May, R.M., J.R. Beddington, C.W. Clark, S.J. Holt and R.M. Laws. 1979. Management of multispecies fisheries. Science 205: 267-277.

Pella, J.J. and P.K. Tomlinson. 1969. A generalized stock production model. Inter. Amer. Trop. Tuna Comm. Bull. 13: 421-496.

Pielou, E.C. 1969. An introduction to mathematical ecology. Wiley-Interscience, New York.

Poole, R.W. 1977. Periodic, pseudoperiodic, and chaotic population fluctuations. Ecology 58: 210-213.

Pope, J.G. 1976. The effect of biological interaction on the theory of mixed fisheries. Int. Comm. Northw. Atlant. Fish. Sel. Pap. 1: 157-162.

Pope, J.G. 1979. Stock assessment in multispecies fisheries. South China Sea Fisheries Development and Coordinating Programme SCS/DEV/79/19. FAO, Manila.

Schoener, T.W. 1974. Resource partitioning in ecological communities. Science 185: 27-39.

Shepherd, J.G. and J.W. Horwood. 1979. The sensitivity of exploited populations to environmental "noise", and the implications for management. J. Cons. Int. Explor. Mer. 38(3): 318-323.

Sissenwine, M.P. 1977. The effect of random fluctuations on a hypothetical fishery. Int. Comm. Northw. Atlant. Fish. Sel. Pap. 2: 137-144.

Sutherland, J.P. 1974. Multiple stable points in natural communities. Amer. Nat. 108: 859-873.

Thom, R. 1972. Stabilité structurelle et morphogénèse. W.A. Benjamin, New York.

Zeeman, E.C. 1977. Catastrophe theory, 1972-1977. Addison-Wesley Publ. Corp., Reading, Mass.

Discussion of Paper by Dr. G. Kirkwood

Discussion centered on the following points:

— Although the possibility of multiple equilibria and limit cycles was noted, much of the analysis related to static equilibria. The point was raised that the inclusion of explicit time dependence in the parameters, which surely is present, would suggest that limit-cycle behavior may be the norm, rather than the exception. While agreeing that this was indeed likely to be the case, and that strictly constant equilibria would not be expected, Kirkwood noted that the importance of these fluctuations depended on the magnitude of the amplitude of the oscillations relative to the mean value.

— In view of the large numbers of species caught, it was recognized that in the models at least some grouping of species was necessary. In answer to a query as to whether this grouping could extend to dealing with only two groups, predators and prey, Kirkwood noted that as a general rule, two-species models tended to be somewhat better behaved than models with more than two species. However, he felt that in our current state of ignorance such an approach could well be very useful.

— The multispecies generalizations of the Pella and Tomlinson general production model allowed the possibility of a yield effort curve that was flatter topped and declined much less steeply than was the case in the multispecies Schaefer model. A number of examples were cited in which this behavior was observed, and it was felt that these models could be useful. In particular, a number of discussants felt that the Schaefer model predictions of a relatively rapid collapse as effort increased past that providing MSY have not been borne out in practice.

Another means of producing such behavior was suggested to be the inclusion of a refuge for some species. It was agreed that this was a useful suggestion, particularly if such a model was analytically tractable.

— It was felt important that models should not be set up merely to estimate, for example, the appropriate level of next year's quota. Rather models of multispecies fisheries should be set up to allow provision of advice on the much wider range of management problems and options that currently exist.

— It was noted that the production models as described deal with the single dimension of fishing effort, and that an additional dimension that may have a substantial effect is that of size at first capture. In particular, the levels of effort at which species successions take place could change considerably according to the size at first capture.

— Agreeing with the views expressed by Kirkwood that a more empirical approach was warranted involving close examination of the observed catch-effort data for tropical multispecies fisheries, Simpson reiterated his concern on the need to improve existing catch and effort information. There was considerable discussion on the interpretation of observed catch-effort relationships, particularly on the likely reversibility or otherwise or trends in catch-per-effort as effort first increased and then decreased. It was felt that these trends may be reversible only over a restricted region or range of parameters. It was noted that at least on a geological time scale, it was the nature of biological systems not to return to their previous state. The possibility of alteration of the gene pool and its effects were raised.

An often held view, at least of fishermen, that historically high abundances of a species were the norm, and that low current abundances represented severe crashes or species switching was also cited. It was felt that at least in some cases the lower levels were the norm, with only occasional large catches, usually during the development or expanding phase of a fishery on that species.

Application of the Three Bays Ecosystem Model to Fisheries Management

M.K. JAMES
K.P. STARK
Department of Civil and Systems Engineering
James Cook University of North Queensland
Townsville, Queensland
4811, Australia

JAMES, M.K. AND K.P. STARK. 1982. Application of the three bays ecosystem model to fisheries management, p. 99-121. *In* Pauly, D. and G.I. Murphy (eds.) Theory and management of tropical fisheries. ICLARM Conference Proceedings 9, 360 p. International Center for Living Aquatic Resources Management, Manila, Philippines and Division of Fisheries Research, Commonwealth Scientific and Industrial Research Organisation, Cronulla, Australia.

Abstract

The theory and management of a multispecies tropical fishery is considered in the context of a systems framework developed under the Three Bays Project. The ecosystem model developed as part of that Project is extended and applied to the problem of determining the response of the fishery to different types and levels of fishing effort. Extension of the model involved the disaggregation of compartments representing fish biomass variables, the consideration of recruitment and age structure, and the modelling of selective fishing mortality.

The potential value of the model as an aid in determining sustainable yields and changes in species composition of the catch is demonstrated by computer simulations of a hypothetical multispecies fishery.

Introduction

The coastal waters near Townsville comprise three roughly comparable bays which are increasingly used for shipping, recreation, fishing and disposal of urban and industrial waste. Questions naturally arise concerning the extent of unacceptable environmental effects resulting from these activities, and the need to develop rational management policies based on an understanding of the entire system and how it reacts to the pressures imposed upon it.

The extreme complexity of environmental systems and the need for the development and application of systems methodologies for the understanding and resolution of environmental conflicts have been discussed by Stark

(1980). A systems framework is required, which provides a multidisciplinary approach, and finds expression in the development of a set of appropriate models.

In recognition of the environment as a dynamic system involving the complex interaction of physical, chemical and biological processes, the "Three Bays Project" was initiated as a pilot study to determine the resources, time, and scale of operations necessary to implement within an academic setting a fully integrated multidisciplinary investigation of the Three Bays (James 1980). In this pilot study, it was decided to explore the development of several mathematical models, in an attempt to assess more clearly the difficulties involved, and to define the information requirements.

Included among these was a biological, or ecosystem model which attempts to simulate the important processes determining the transfers of biomass and nutrients among the biological and chemical components of the system. The system *processes* are modelled by mathematical equations, and the system *behavior* (in terms of observable properties) is modelled by the time-dependent solutions of those equations.

The scope of the model is represented in Fig. 1. The model is spatially integrated in the sense that the values assumed by variables are taken to represent averages over the entire domains for which the variables are relevant

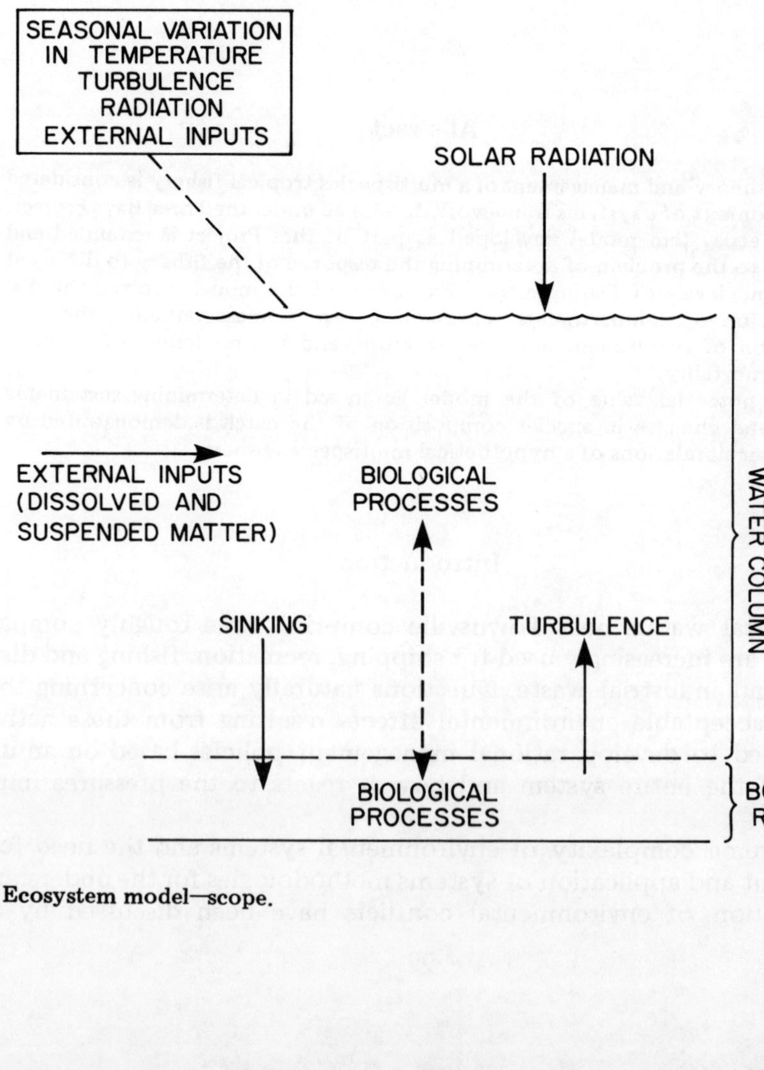

Fig. 1. Ecosystem model—scope.

(i.e., either the water column, or the bottom region). The units employed are mg of organic carbon per liter, for biomass concentration, and mg of nitrogen per liter, for inorganic nutrient.

Model Structure

The system is characterized by a very large number of species. For example, approximately 180 fish species are present. Therefore, species have not been individually modelled, but have been grouped according to biological character and dominant behavior. The model then simulates group behavior. This has resulted in a total of twenty system state variables. Of these, fourteen represent living components and the remainder non-living chemical components, as shown in Table 1.

Structure of the model is most easily represented in terms of the interaction matrix (Fig. 2). An entry in a cell indicates a flow of matter from the compartment associated with the row, to the compartment associated with the column. The actual entry indicates the type of process involved in that flow (e.g., feeding, mortality, etc.).

Table 1. Ecosystem model—species aggregations and state variables.

Species Aggregated into Compartments	
1	Phytoplankton in water column
2	Phytoplankton in bottom region
3	Zooplankton (mainly copepods)
4	Crab larvae
5	Mollusc larvae
6	Polychaete larvae
7	Crabs
8	Molluscs
9	Polychaetes
10	Benthic-feeding fish
11	Plankton-feeding fish
12	Nekton-feeding fish
13	Pelagic bacteria
14	Benthic bacteria
15	Pelagic P.O.M.*
16	Benthic P.O.M.*
17	Pelagic D.O.M..*
18	Benthic D.O.M.*
19	Pelagic dissolved nutrients
20	Benthic dissolved nutrients

*P.O.M.: Particular Organic Matter
D.O.M.: Dissolved Organic Matter

			1	2	3	4	5	6	7	8	9	10	11	12	13	14	15	16	17	18	19	20
*	1	Phytos (1)		s	f	f	f	f					f				m		e,m			
*	2	Phytos (2)	t						f	f	f	f						m		e,m		
	3	Zooplankton			f	f		f					f				d,m		e,m		e	
	4	Crab Larvae			f	f	f	f	mat	f	f		f				d,m		e,m		e	
	5	Mollusc Larvae			f	f	f	f		mat f	f		f				d,m		e,m		e	
	6	Polychaete Larvae			f	f	f	f		f	mat f		f				d,m		e,m		e	
	7	Crabs				b			f			f						d,m		e,m		e
	8	Molluscs					b		f	f		f						d,m		e,m		e
	9	Polychaetes						b	f		f	f						d,m		e,m		e
*	10	Fish (a)										f		f				d,m		e,m		e
*	11	Fish (b)										f		f			d	m	e	m	e	
*	12	Fish (c)										f		f			d	m	e	m	e	
	13	Pelagic Bacteria			f	f	f	f					f			s	m		m		e	
	14	Benthic Bacteria							f	f	f	f			t			m		m		e
	15	Pelagic P.O.M.			f	f	f	f					f		f			s				
	16	Benthic P.O.M.							f	f	f	f				f	t					
	17	Pelagic D.O.M.	f		f	f	f	f					f	f	f							
	18	Benthic D.O.M.		f					f	f	f	f				f			t			
	19	Pelagic Nutr.	f																			
	20	Benthic Nutr.		f																	t	

*Phytos (1) : Phytoplankton in water column
Phytos (2) : Phytoplankton in bottom region
Fish (a) : Benthic feeders
Fish (b) : Planktonic feeders
Fish (c) : Nektonic feeders

b : birth
d : defecation
e : excretion
f : feeding
m : mortality
mat : maturation
s : sinking
t : turbulent mixing

Fig. 2. Ecosystem model—interaction matrix.

Note that nutrient cycling is included, as inanimate matter enters the food chain via uptake by certain organisms, and is released through excretion, etc., by other organisms. The overall food web is represented schematically in Fig. 3.

Mathematically, the model comprises twenty non-linear differential equations for the variables $X_1, X_2, \ldots, X_{20}$:

$$X_i = F_i(X_1, X_2, \ldots, X_{20}, t),$$

where the functions F_i involve mathematical formulations representing the following processes:

- photosynthesis
- feeding
- defecation
- metabolic loss (respiration)
- natural mortality
- excretion
- reproduction and maturation
- sinking
- turbulent resuspension
- external inputs

The formulations employed are those for which some basis has been found either in discussion with marine biologists, or in the literature.

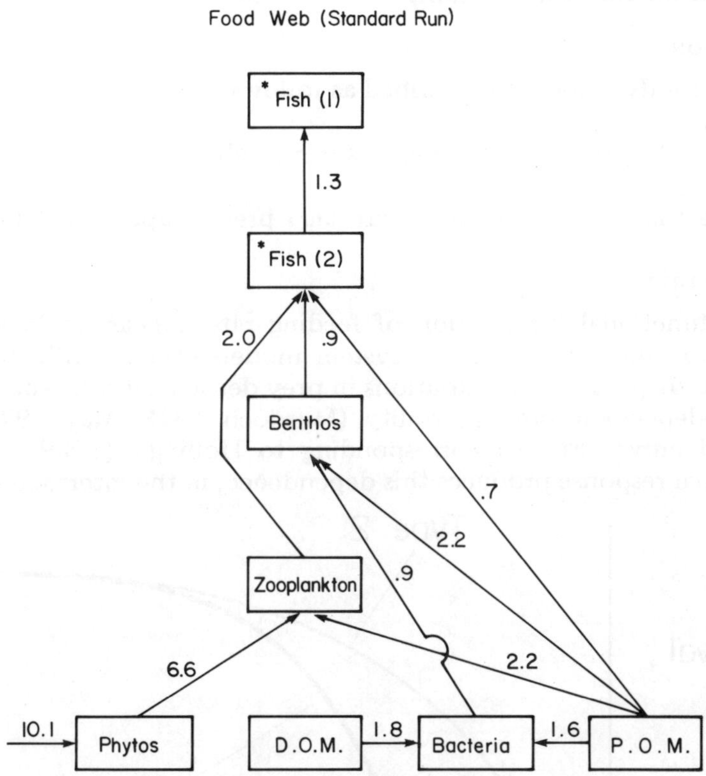

Fig. 3. Ecosystem model—food web transfers under standard conditions.

Model Processes

Phytoplankton Dynamics

Phytoplankton dynamics are modelled by the following equation:

$$\dot{X}_1 = \text{photosynthesis} - \text{respiration} - \text{organic excretion} - \text{mortality} - \text{predation} - \text{sinking} + \text{turbulent mixing.}$$

Photosynthesis is assumed to be both light-limited and nutrient (nitrogen)-limited, and governed by Michaelis-Menten saturation relationships (Parsons and Takahashi 1973, chapter 3). Insolation is a time dependent input to this process, and attenuation with depth is assumed to be exponential (e.g., Parsons and Takahashi 1973, p. 63), with an extinction coefficient which is linearly dependent on the concentration of suspended matter in the water column (Scott 1978).

Respiration loss and organic excretion are proportional to gross photosynthesis, while natural mortality and sinking loss are proportional to phytoplankton concentration. The effect of wind-generated turbulence in resuspending algae is exponentially dependent on wind-strength (Belperio 1978).

The processes of photosynthesis, respiration and mortality depend exponentially on the temperature (e.g., Kelly 1976). This dependence is also incorporated in the processes of predation, respiration and mortality discussed in the following sections.

PREDATION

Predator dynamics are described as follows:

$$\dot{X}_i = \text{feeding rate} - \text{metabolic loss} - \text{mortality} - \text{predation} - \text{defecation}$$

(Note that certain predators are also preyed upon by other predators.)

Feeding rate

The functional formulation of feeding rate appears to have vital significance for the stability of ecosystem models (May 1973). The functional response of predators to variations in prey density must be such as to lead to density-dependent prey mortality (Murdoch 1973; May 1973). Only the s-shaped curve (Fig. 4) corresponding to Holling's (1959, 1966) type 3 functional response produces this dependence, in the interval a-b.

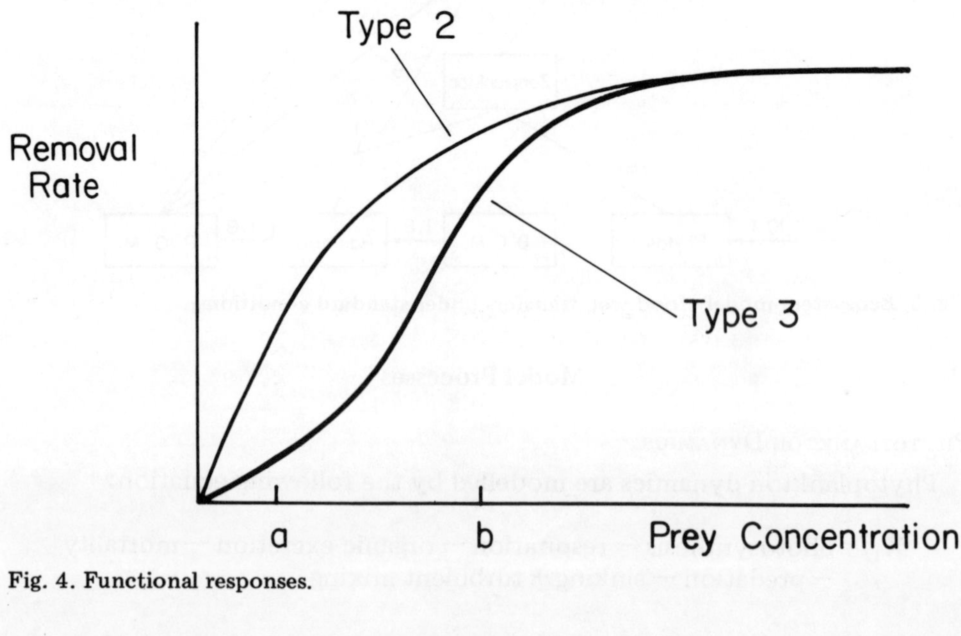

Fig. 4. Functional responses.

Laboratory studies of the functional responses of predators to a single prey species have resulted in Holling type 2 curves (Fig. 4) (e.g., for seashore snails (Murdoch 1969), starfish (Landenberger 1968) and fish (Ivlev 1961)). However, as Murdoch (1973) has pointed out, although a type 2 response to a species may occur when that species alone is present, a great range of responses may be possible when two or more species are present.

A general predation model has been developed and employed for all predator variables ($X_3, X_4, \ldots, X_{12}$) in this system. This model is based on the following parameters:

τ_0 : the time required for a satiated predator to 'handle' its food and return to the hunt;

τ_1^i : the time required for a hunting predator to achieve satiation, while feeding on the ith prey only;

γ : the ratio: food biomass required for satiation, divided by the predator biomass.

The quantity τ_1^i is assumed to depend inversely on the prey concentration, and to increase exponentially as the number of competing predators increases:

$$\tau_1^i = \frac{a_i}{x_i^r} \exp\left[\lambda\left(x_p - x_p^0\right)\right];$$

where a_i, r, λ and x_p^0 are constants. The rate of removal of the ith prey, per unit predator biomass, then takes the form:

$$R_i = \frac{\gamma \frac{1}{a_i} x_i^r}{\exp\left[\lambda\left(x_p - x_p^0\right)\right] + \tau_0 \sum_i \frac{1}{a_i} x_i^r}$$

Thus the factor $1/a_i$ represents the combined effects of the predator's preference for that prey, and the effectiveness and efficiency of hunting for that prey. For each predator, an effective food variable $\sum_i \left(\frac{1}{a_i}\right) x_i^r$ is computed and used to determine the total feeding rate. The value r = 2 has been employed, resulting in the desired density-dependent response.

Metabolic loss

A general model has been developed to describe the relationship among feeding, metabolism, production, and the excretion of nutrient (nitrogen). This model has been employed for each of the living components of the system, other than phytoplankton.

The model is based on the balance equation:

$$F = P + D + M, \text{ (in units of organic carbon)}$$

where F = total food intake;
P = net production;

D = defecation loss: food not assimilated;
M = total metabolic loss:
— excretion of non-metabolizable material;
— by-products of chemical processes taking place in digestion, maintenance, growth, activity.

The major assumption is that to achieve any given net production, there is a corresponding required metabolic loss. That is,

$$M = M(P).$$

Further, the first-order (linear) approximation is made:

$$M = M(O) + cP,$$

where c is a constant, representing the metabolic 'cost' per unit net production, and M(O) is the 'basal' metabolic loss. The metabolic nitrogen loss is then M/R_{cn}, where R_{cn} is the carbon: nitrogen ratio, which is assumed to be the same for all biomass in the system. Note that this latter assumption could be relaxed, at the expense of increasing the complexity of the model.

A proportion of this nitrogen is excreted as dissolved nutrient which can be taken up immediately by phytoplankton. The remainder is excreted as dissolved organic nitrogen, some of which may be refractory and is thus lost from the system.

The basal metabolic loss, M(O) is assumed to vary exponentially with the average individual biomass, W (Kerr 1971)

$$\frac{M(O)}{W} = aW^b.$$

Mortality, predators and defecation

Natural mortality loss is assumed to be proportional to existing biomass. The losses due to predation are found by summing the removal rates R_i (see above) achieved by each predator. Defecation loss is proportional to the amount of food ingested.

Bacterial Action

Bacteria play an important role in the recycling of nutrient and as primary producers, both in the water column and in the benthic layer (Wood 1965; Parsons and Takahashi 1973; Hargrave 1972; Rieper 1978). Michaelis-Menten saturation relationships are employed to describe bacterial attack on particulate organic matter, and the ingestion of dissolved organic matter.

Sinking and Turbulent Resuspension

A constant sinking rate is assumed for phytoplankton and for suspended particulate matter. As suggested by T. Walker (pers. comm.), the sinking process is countered by wind-driven turbulent resuspension. This mixing effect (of any substance) is assumed to be proportional to the difference in concentration between the water column and the benthic layer. It can result in interstitial sediment nutrient becoming available to organisms in the water column.

External Inputs

These include temperature, insolation, wind strength, and inanimate matter (both organic and inorganic) arising, for example, from stream run-off.

Reproduction and Maturation

For the Three Bays model, the processes of reproduction and recruitment were included only for the benthic animals (crabs, molluscs and polychaetes), since theirs were the only larvae found in significant concentrations in the plankton. Since these processes are of particular relevance to the fishery problem, a discussion of this aspect of the extended model is given further below.

Example Output

The model has been implemented in standard FORTRAN, using single-step Euler integration to solve the differential equations, with a time step of eight hours.

The model 'works' in the sense that the behavior of each state variable is qualitatively reasonable. Insufficient data are available for strict validation. Figs. 5, 6 and 7 show three sample outputs for four of the twenty state variables, firstly under 'standard' conditions as determined under the study; secondly with temperature held constant at an intermediate level; thirdly with the effects of wind-induced turbulence omitted. These are included to demonstrate the stable behavior of the model, and its qualitative response to simple changes in external conditions.

Application to Fishery Management

Application of the basic Three Bays ecosystem model to the problems of fishery management requires extension of the model in at least three directions:

- Disaggregation of the fish species variables. Many fisheries are managed as if the populations of fishes are independent (Hobson and Lenarz 1977), even when two or more important species are present. However, the natural concern is that an understanding of the interactions among fish populations may be crucial for the effective management of multi-species fisheries.
- Incorporation of reproduction, recruitment and age structure for the fish variables. Recruitment to the population has clear significance for the viability of exploited stocks, especially in view of the extreme variability often observed in fisheries. Also, the age structure of the population may be an important determinant of the effects of age-selective predation and fishing methods.
- Representation of fishing pressure. The response of populations to

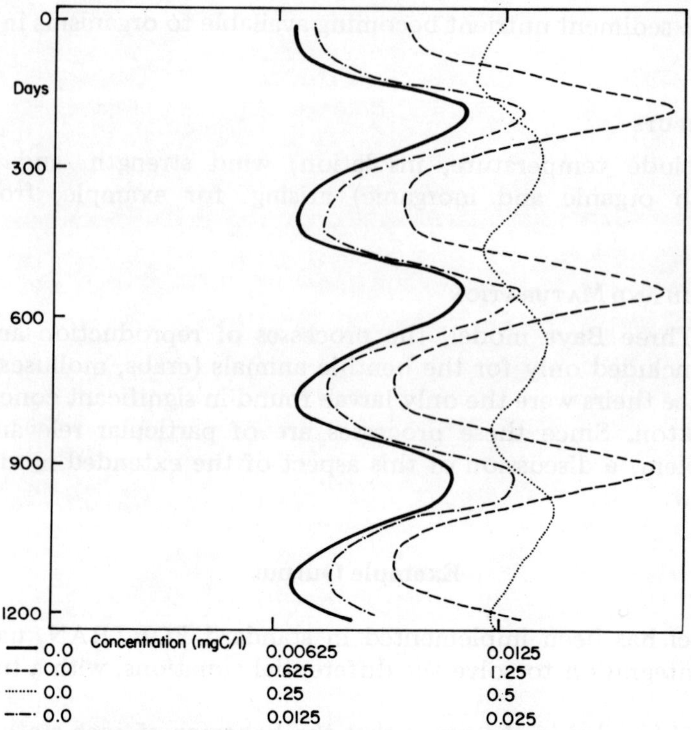

Fig. 5. Ecosystem model: standard conditions. ———: Pelagic bacteria; — — —: Benthic bacteria; · · · ·: Nekton-feeding fish; — · — ·: Plankton-feeding fish.

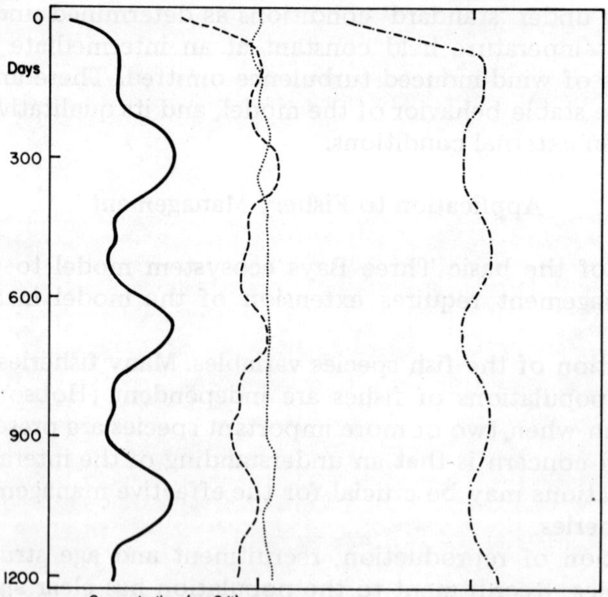

Fig. 6. Ecosystem model: constant temperature. ———: Pelagic bacteria; — — —: Benthic bacteria; · · · ·: Nekton-feeding fish; — · — ·: Plankton-feeding fish.

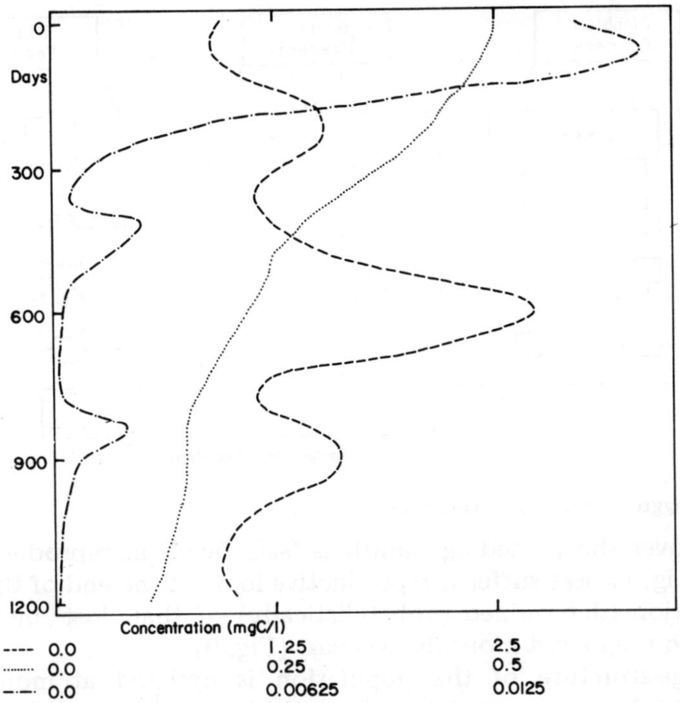

Fig. 7. Ecosystem model: zero turbulence. -----: Benthic bacteria; • • • •: Nekton-feeding fish; — • — •: Plankton-feeding fish. (Pelagic bacteria concentration dropped to zero within twenty days).

fishing pressure implies a selective induced mortality distinct from natural mortality and predation.

Disaggregation of Variables

There is no difficulty in principle in disaggregating the fish species in the model. However, there is a serious practical problem, since it would seem infeasible to model individually about 200 different species, especially when one considers the complexity of interaction among them. It is therefore necessary to identify assemblages of species that respond as units to fishing pressures and which function approximately as units in the various biological processes and interactions within the system.

Under this assumption, the proposed disaggregation is as illustrated in Fig. 8. Feeding relationships are now modelled explicitly among the different *functional* age-classes of all species.

Reproduction, Recruitment and Age Structure

As shown in Fig. 8, each species compartment is divided into a number of functional age-classes, each of which is considered to behave as a unit. These are further subdivided into monthly age-classes, with the exception of the oldest functional age-class. Reproduction by the breeding age-classes is considered at monthly intervals within each breeding season. At those times, an amount of biomass equal to some proportion of the average age-class

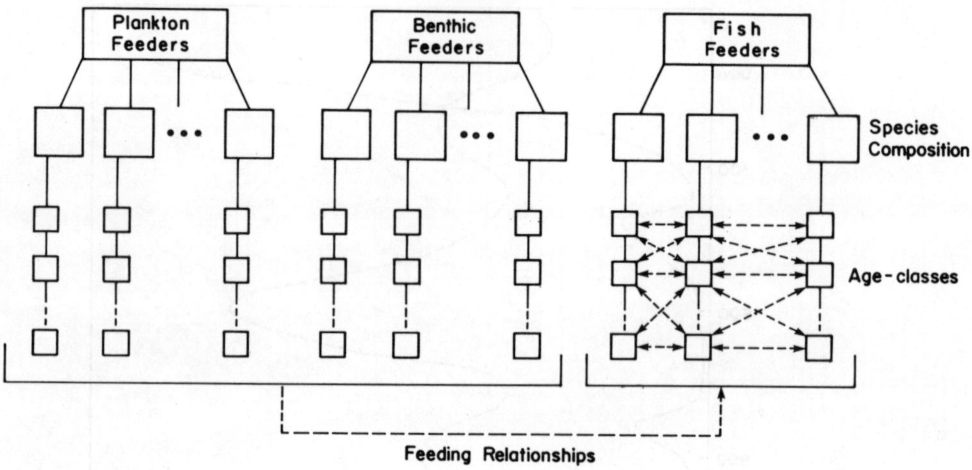

Fig. 8. Disaggregation of fish variables.

biomass over the preceding month is 'sacrificed' in reproduction: i.e., the breeding age-classes suffer a reproductive loss. At the end of the next month a proportion (determined probabilistically) of that 'lost' biomass is transferred into the lowest monthly age-class (Fig. 9).

The age-structure of the population is updated at monthly intervals by making the appropriate transfers of biomass between the monthly age-classes.

FISHING PRESSURE

The effect of fishing pressure is modelled as an induced mortality rate, acting selectively on different age-classes of different species, in accordance with variable 'catchability' under the applied fishing methods.

For a given level of fishery investment, the total yield of the fishery is computed from a saturation relationship, as a function of the 'effective' stock (Fig. 10). The effective stock is computed as the sum of the biomasses of the affected species age-classes, weighted by their 'catchabilities'. The saturation level, 'SAT', represents the maximum capacity corresponding to

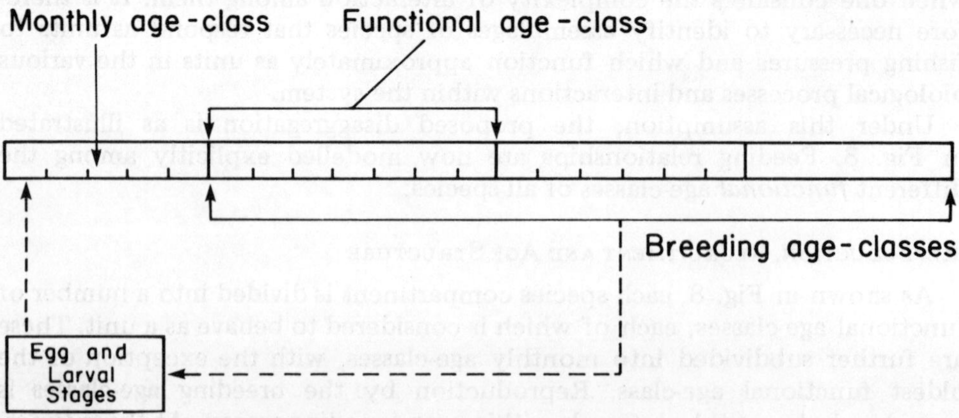

Fig. 9. Age-class structure and recruitment.

the applied fishing effort (determined in terms of the equipment used and the intensity of fishing). 'SAT' is provided as a parameter.

The yield of each age-class of each species is then determined in proportion to its contribution to the effective stock.

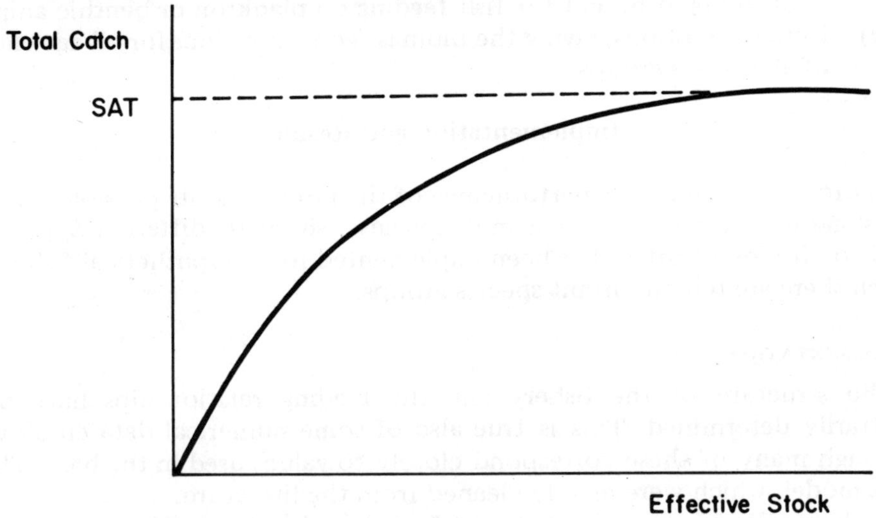

Fig. 10. Fishing saturation relationship.

Data Requirements

Data requirements of the fishery component of the model may be listed as follows:

(i) General:
- Total number of species groups
- Number of species groups defined as plankton feeders
- Number of species groups defined as benthic feeders
- Basal metabolism parameters 'a' and 'b' (see above)
- The fishery yield saturation relationship
- Monthly average temperatures

(ii) For each species group:
- Number of functional age-classes
- Age-classes which breed
- Parameters determining sensitivity of biological processes to temperature
- Breeding rate, as a function of season

(iii) For each age-class of each species group:
- Average individual biomass
- 'Metabolic cost' parameter c (see above)
- Competition factor λ (see above)
- Satiation ratio γ (see above)
- Handling time, τ_o (see above)
- Defecation rate
- Natural mortality rate (not including predation mortality)

- Reference levels $x_p{}^o$ (see above)
- Catchability, for given fishing method

(iv) Predator-prey relationship:

- Matrix of values of the parameters a_i (see above) for fish preying on other fish, and for fish feeding on plankton or benthic animals

(v) Initial conditions, giving the biomass concentrations for all age-classes of all species groups.

Implementation and Results

In order to evaluate the performance of the model and its possible value in investigating the response of a multispecies fishery to different types and levels of fishing effort, it has been implemented for a hypothetical fishery in which there are ten important species groups.

Implementation

The structure of the fishery and the feeding relationships have been arbitrarily determined. This is true also of some numerical data employed, although many of these correspond closely to values used in the basic Three Bays model, which were mainly gleaned from the literature.

Each species group is divided into four functional age-classes: one to six months, six to twelve months, twelve to eighteen months, and greater than eighteen months. Species groups one to three are plankton feeders; four to six are benthic feeders; seven to ten are fish feeders. The predator-prey relationships among the fish are illustrated in the interaction matrix of Fig. 11, in which an entry in a cell indicates that the species age-class associated with that row preys on the species age-class associated with the column.

For each species, age-classes two, three and four are assumed to breed. The breeding season for each species occupies the first five months of each year.

The catchability matrix is illustrated in Fig. 12, which shows that the main targets of fishing effort are the fourth age-classes of species seven to

Species

		1				2				3				4				5				6				7				8				9				10			
		1	2	3	4	1	2	3	4	1	2	3	4	1	2	3	4	1	2	3	4	1	2	3	4	1	2	3	4	1	2	3	4	1	2	3	4	1	2	3	4
7	1	X				X				X				X				X				X																			
	2	X	X			X	X			X	X			X	X			X	X			X	X																		
	3		X	X			X	X			X	X			X	X			X	X			X	X			X				X				X				X		
	4			X	X			X	X			X	X			X	X			X	X			X	X			X				X				X				X	
8	1	X				X				X				X				X				X																			
	2	X	X			X	X			X	X			X	X			X	X			X	X																		
	3		X	X			X	X			X	X			X	X			X	X			X	X			X				X				X				X		
	4			X	X			X	X			X	X			X	X			X	X			X	X			X				X				X				X	
9	1	X				X				X				X				X				X																			
	2	X	X			X	X			X	X			X	X			X	X			X	X																		
	3		X	X			X	X			X	X			X	X			X	X			X	X			X				X				X				X		
	4			X	X			X	X			X	X			X	X			X	X			X	X			X				X				X				X	
10	1	X				X				X				X				X				X																			
	2	X	X			X	X			X	X			X	X			X	X			X	X																		
	3		X	X			X	X			X	X			X	X			X	X			X	X			X				X				X				X		
	4			X	X			X	X			X	X			X	X			X	X			X	X			X				X				X				X	

Fig. 11. Interaction matrix: fish populations.

ten. A number of other species groups are affected but some of these may in fact be regarded as undesirable 'trash'.

It was felt unnecessary, at this stage of the study, to implement the entire ecosystem model. Instead, the basic Three Bays model was run, and its output used as input to a separate model incorporating only the fishery component.

Species	Age-class 1	2	3	4
1	0	0	1	1
2	0	0	1	1
3	0	0	1	1
4	0	0	0	1
5	0	0	0	1
6	0	0	0	1
7	0	0	1	10
8	0	0	1	10
9	0	0	1	10
10	0	0	1	10

Fig. 12. Matrix of relative catchabilities.

For simplicity in developing the model, data were chosen to be identical for each species group within each classification (plankton, benthic and fish feeders). Consequently, the results for each of these species are very similar, the differences being due to the stochastic component incorporated in the determination of recruitment.

Results

The standard run of the model, simulating a three year period with no exploitation, results in the curves of Fig. 13. Annual cycles are evident in the system state, in response to similar cycles in the 'background' provided by the basic Three Bays model, and to the annual cycle of temperature. Fig. 14 gives the total biomass transfers which have taken place as a result of predator-

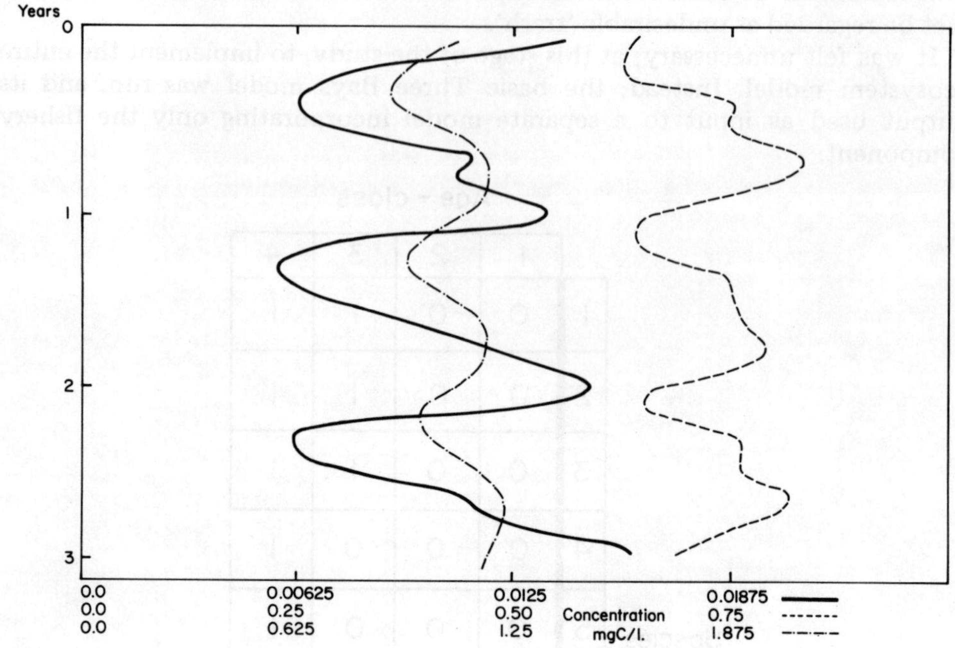

Fig. 13. Fishery component, standard run.

Prey \ Predators	7	8	9	10
1	.31	.31	.31	.31
2	.31	.31	.31	.31
3	.31	.31	.31	.31
4	3.45	3.44	3.49	3.36
5	3.40	3.40	3.45	3.32
6	3.42	3.41	3.45	3.34
7	.36	.36	.38	.34
8	.36	.36	.38	.34
9	.37	.37	.38	.35
10	.31	.32	.33	.29

Fig. 14. Standard run. Biomass transfers (mgC/l) resulting from predator-prey interactions among fish species.

prey interactions among the fish, while Fig. 15 summarizes details of feeding, defecation, respiration and natural mortality.

The model has been run for several different levels of fishing effort, measured in terms of the maximum possible fishing capacity associated with that effort. Curves showing the system population response to these levels of fishing effort are shown in Figs. 16, 17, 18 and 19. Species composition of the catch, and levels of annual sustainable yields are summarized in Tables 2 and 3. Note that species composition of the catch changes significantly with fishing effort, and that an optimal level of effort, in terms of the yield of the target species, is readily identified.

The model was also run over a period of ten years in which the level of fishing effort was gradually increased. The results, showing the changes in total landings and in species composition, are shown in Fig. 20.

Conclusion

Computer simulations performed at this stage indicate the potential value of this modelling approach as an aid in understanding the response of a multispecies fishery to various levels of fishing effort. Of particular interest is the ability of the model to simulate changes in the species composition of the catch, and to determine sustainable yields, under nominated fishing pressure. The model could readily be applied to the task of evaluating alternative control policies (closed fishing seasons, regulated fishing effort, etc.).

Evaluation of the model's full potential is not yet complete. However, a rigorous assessment of the model's validity will not be possible until it is implemented in a practical situation.

Species	Feeding	Mortality	Respiration	Defecation
1	3.5	.015	1.6	.71
2	3.5	.015	1.6	.71
3	3.5	.015	1.6	.71
4	44.0	1.1	20.1	8.8
5	43.3	1.1	19.8	8.67
6	43.7	1.1	19.9	8.74
7	12.6	1.7	6.6	2.52
8	12.6	1.7	6.6	2.52
9	12.8	1.7	6.7	2.56
10	12.	1.6	6.4	2.45

Fig. 15. Standard run. Biological transfer for fish species (mgC/l).

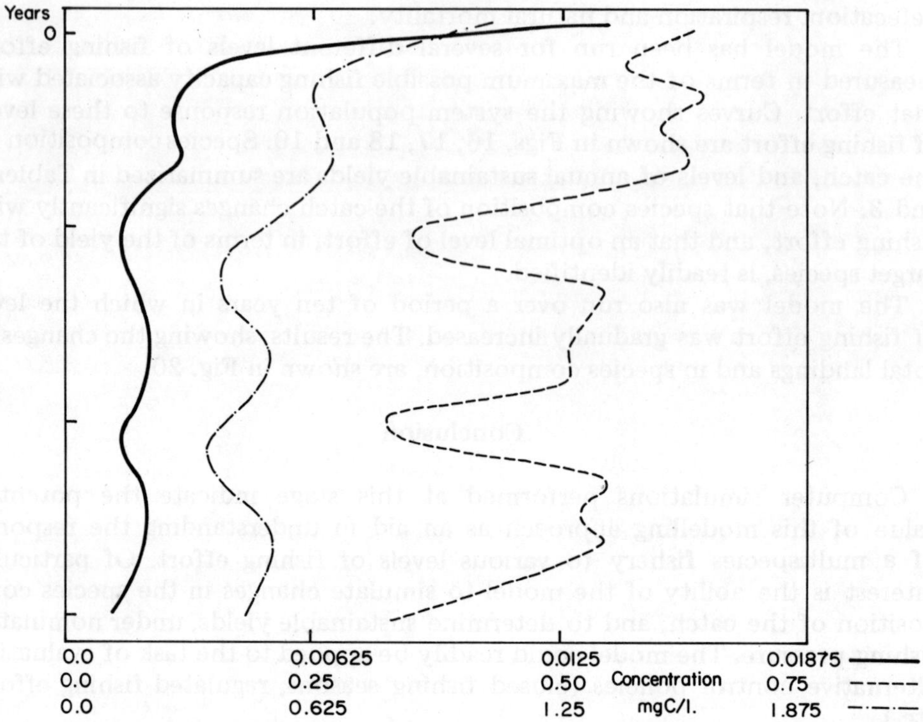

Fig. 16. Fishing capacity: .00075 mgC/l/hr. Species 1: ———; Species 4: — — —; Species 7: — • — •

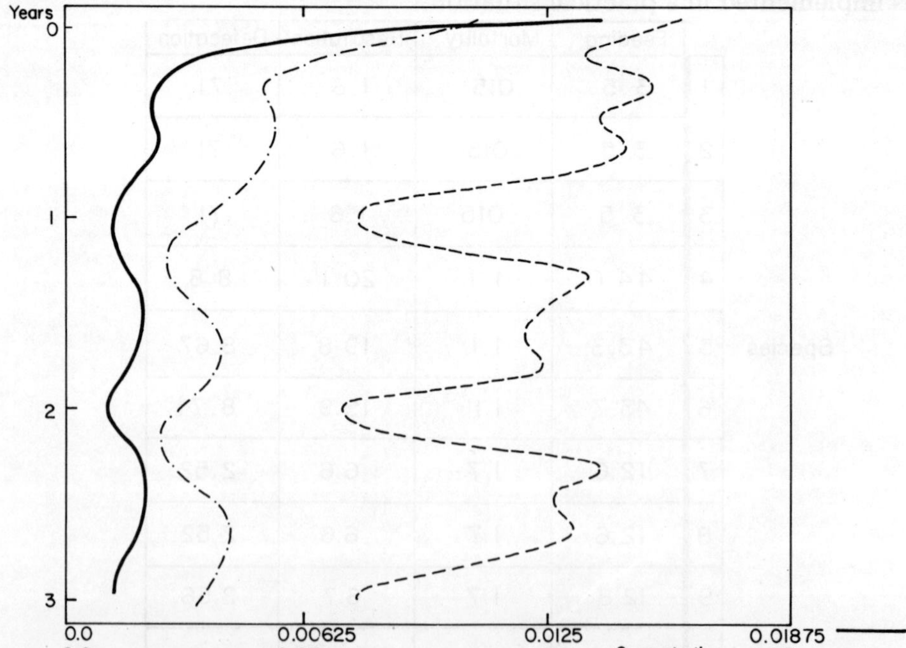

Fig. 17. Fishing capacity: .001 mgC/l/hr. Species 1: ———; Species 4: — — —; Species 7: — • — •.

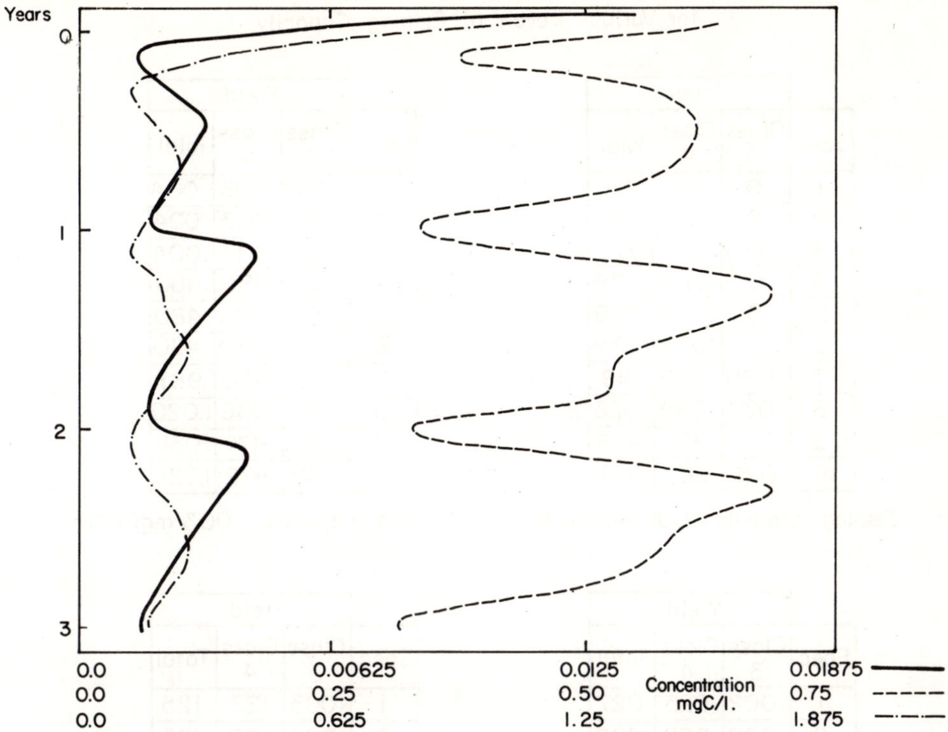

Fig. 18. Fishing capacity: .003 mgC/l/hr. Species 1: ———; Species 4: – – –; Species 7: — • — •

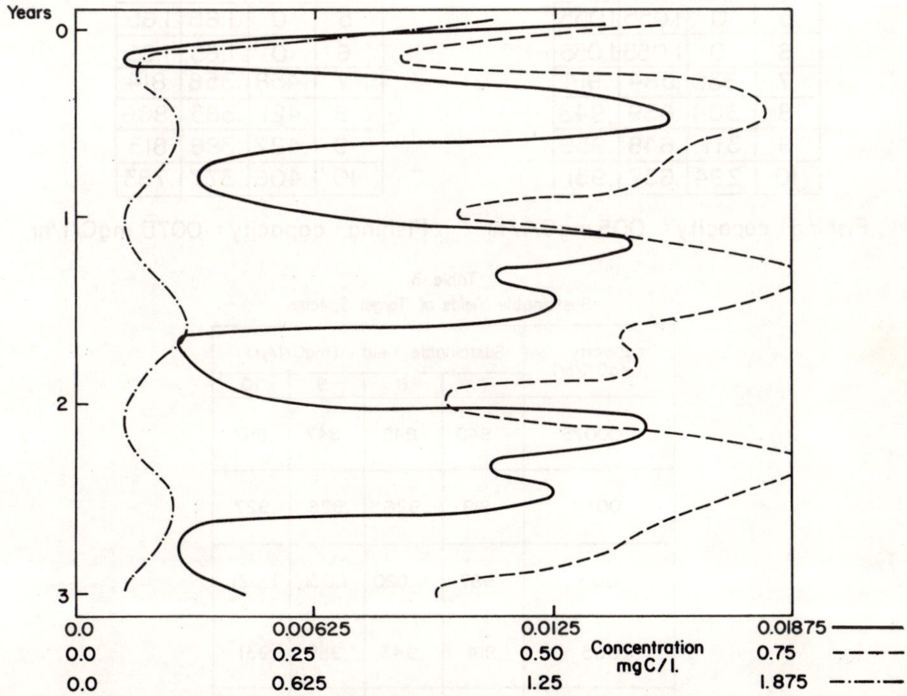

Fig. 19. Fishing capacity: .005 mgC/l/hr. Species 1: ———; Species 4: – – –; Species 7: — • — •

Table 2
Species Composition of the Catch
for Various Levels of Fishing Capacity

Spec.	Yield Class 3	Yield Class 4	Yield Total
1	0	0	0
2	0	0	0
3	0	0	0
4	0	.079	.079
5	0	.079	.079
6	0	.079	.079
7	.029	.890	.919
8	.027	.899	.926
9	.027	.901	.928
10	.026	.901	.927

Fishing capacity: .001 mgC/l/hr

Spec.	Yield Class 3	Yield Class 4	Yield Total
1	.001	.003	.004
2	.001	.003	.004
3	.001	.003	.004
4	0	.480	.480
5	0	.480	.480
6	0	.480	.480
7	.167	.820	.987
8	.154	.866	1.020
9	.157	.874	1.031
10	.149	.872	1.021

Fishing capacity: .003 mgC/l/hr

Spec.	Yield Class 3	Yield Class 4	Yield Total
1	.002	.025	.027
2	.002	.025	.027
3	.002	.025	.027
4	0	1.055	1.055
5	0	1.055	1.055
6	0	1.055	1.055
7	.330	.584	.914
8	.304	.639	.943
9	.311	.648	.959
10	.294	.637	.931

Fishing capacity: .005 mgC/l/hr

Spec.	Yield Class 3	Yield Class 4	Yield Total
1	.003	.122	.125
2	.003	.122	.125
3	.003	.122	.125
4	0	1.65	1.65
5	0	1.65	1.65
6	0	1.65	1.65
7	.458	.356	.814
8	.421	.385	.806
9	.427	.386	.813
10	.406	.377	.783

Fishing capacity: .0075 mgC/l/hr

Table 3
Sustainable Yields of Target Species

capacity (mgC/l/hr)	Sustainable Yield (mgC/l/yr) 7	8	9	10
.00075	.840	.845	.847	.847
.001	.919	.926	.928	.927
.003	.987	1.020	1.031	1.021
.005	.914	.943	.959	.931
.0075	.814	.806	.813	.783

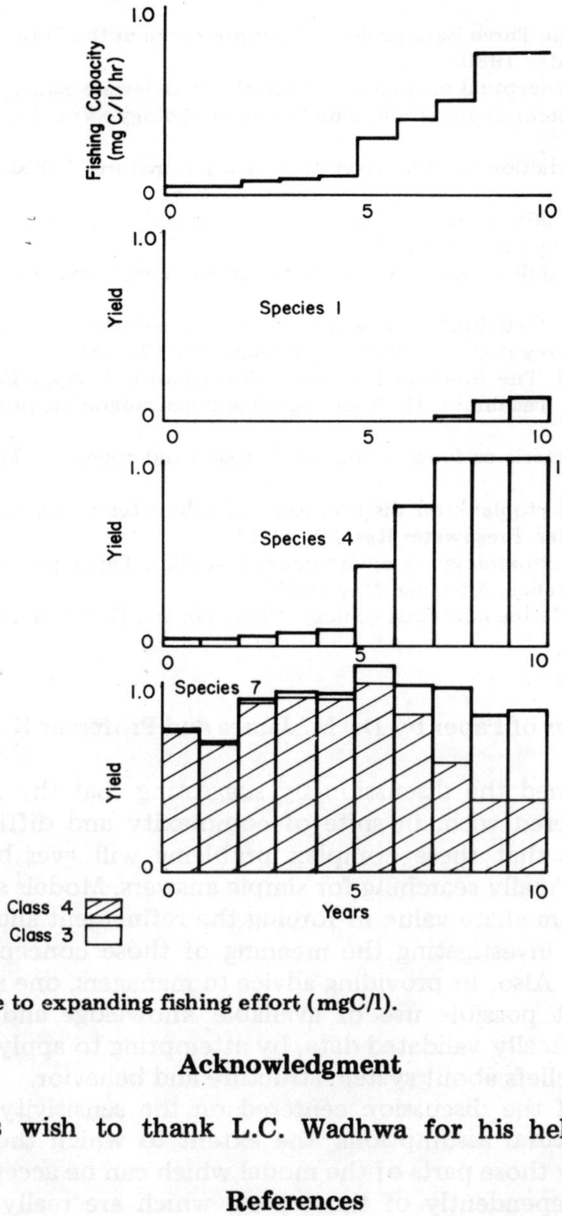

Fig. 20. Response to expanding fishing effort (mgC/l).

Acknowledgment

The authors wish to thank L.C. Wadhwa for his helpful discussions.

References

Belperio, A.P. 1978. An inner-shelf sedimentation model for the Townsville region, Great Barrier Reef Province. James Cook University, Townsville, Queensland, Australia. Ph.D. dissertation.

Hargrave, B.T. 1972. Aerobic decomposition of sediment and detritus as a function of particle surface area and organic content. Limnol. Oceanogr. 17: 583-596.

Hobson, E.S. and W.H. Lenarz. 1977. Report of a colloquium on the multispecies fisheries problem, June 1976. Mar. Fish. Rev. 39(9): 8-13.

Holling, C.S. 1959. The components of predation as revealed by a study of small-mammal predation of the European pine sawfly. Can. Entomol. 91: 293-320.

Holling, C.S. 1966. The functional response of invertebrate predators to prey density. Mem. Entomol. Soc. Can. 48: 1-86.

Ivlev, V.S. 1961. Experimental ecology of the feeding of fishes. Yale University Press, Yale.

James, M.K. 1980. The Three Bays project. Paper presented at the 50th ANZAAS Conference, Adelaide, May 1980.

Kelly, R.A. 1976. Conceptual ecological model of the Delaware estuary, p. 3-45. *In* B.C. Patten (ed.) System analysis and simulation in ecology, Vol. IV. Academic Press, N.Y.

Kerr, S.R. 1971. Prediction of fish growth efficiency in nature. J. Fish. Res. Board Can. 28: 809-814.

Landenberger, D.E. 1968. Studies on selective feeding in the starfish *Pisaster* in Southern California. Ecology 49: 1062-1075.

May, R.M. 1973. Stability and complexity in model ecosystems. Princeton University Press, Princeton.

Murdoch, W.W. 1969. Switching in general predators: experiments on predator specificity and stability of prey populations. Ecol. Monogr. 39: 335-354.

Murdoch, W.W. 1973. The functional response of predators. J. Appl. Ecol. 10: 335-342.

Parsons, T.R. and M. Takahashi. 1973. Biological and oceanographic processes. Pergamon Press, N.Y.

Rieper, M. 1978. Bacteria as food for marine harpacticoid copepods. Mar. Biol. 45: 337-345.

Scott, B.D. 1978. Phytoplankton distribution and light attenuation in Port Hacking estuary. Aust. J. Mar. Freshwater Res. 29: 31-44.

Stark, K.P. 1980. Methodology in environmental studies. Paper presented at the 50th ANZAAS Conference, Adelaide, May 1980.

Wood, E.J.F. 1965. Marine microbial ecology. Chapman and Hall, London.

Discussion of Paper by Dr. M. James and Professor K. Stark

Dr. James opened the discussion by suggesting that the modelling task should be persevered with, in spite of complexity and difficulty, because it seems unlikely that these complex problems will ever be solved satisfactorily by continually searching for simple answers. Models such as the one described have immediate value in forcing the refinement and formalization of concepts, and investigating the meaning of those concepts in terms of systems behavior. Also, in providing advice to managers, one should attempt to make the best possible use of available knowledge and experience in addition to statistically validated data, by attempting to apply models which better reflected beliefs about system structure and behavior.

A large part of the discussion centered on the sensitivity of the model behavior to structural assumptions–the extent to which the model results are determined by those parts of the model which can be accepted as "right" more or less independently of those parts which are really in an area of ignorance. The importance of the control exercised through direct interactions seems to be very important, and is achieved through assuming, on the basis of ecological theory, a type 3 functional response, in spite of laboratory tests which indicate the type 2 response. It was suggested that other possible pathways of interaction should also be considered, in an attempt to determine which pathways are the important ones. An example was cited of a modelling exercise for a small lake, in which the type 2 response was used with the result that the model was unstable. Closer experimental investigation did in fact indicate that some unknown mechanism was resulting in apparent self-regulation at the level of bacteria and protozoa, and this was sufficient to stabilize the system. Another example, in which direct interaction did not seem to be the controlling feature, was provided by the small coral reef communities described in Sale's paper (this vol.).

The need for methods of analyzing the sensitivity of the model to its structural assumptions was pointed out. It was suggested that weaknesses in the model might be identified by manipulating individual components to produce qualitative changes in model behavior that could be regarded as unrealistic. It was also observed that a modular approach should be adopted, in which model components could be separately tested during model development.

The suggestion was made that in view of current difficulties with statistical validation, some new criterion be found (e.g., based on model simplicity) for choosing between models. The view was expressed that in the present state of the art, the acceptability of a model could only be based on a belief about how the system works. A more optimistic view was that models could and should be tested on the basis of their predictions.

While the present model could be regarded as relatively simple (if compared with a number of ecosystem models which have been developed), it was pointed out that managers of multispecies systems would regard it as complex, rather than simple. However, it was suggested that in dealing with a complex system, it was necessary to accept some minimum effective level of complexity: an over-emphasis on simplicity could lead to trivial or misleading results.

A model could be viewed simply as an interpretation of current beliefs about the system, and therefore could present reality only to the extent that those concepts reflected reality. However, for models to be of any value in management, they must relate to the realities of the fishery. The Icelandic Cod Fishery was cited as an example of a fishery which seems to be successfully managed on the basis of a very simple modelling approach.

Applications of Ecological Simulation Models to Management of Tropical Multispecies Fisheries

P. A. LARKIN AND W. GAZEY

Institute of Animal Resource Ecology
University of British Columbia
Vancouver, Canada V6T 1W5

LARKIN, P.A. and W. GAZEY. 1982. Applications of ecological simulation models to management of tropical multispecies fisheries, p. 123-140. *In* Pauly, D. and G.I. Murphy (eds.) Theory and management of tropical fisheries. ICLARM Conference Proceedings 9, 360 p. International Center for Living Aquatic Resources Management, Manila, Philippines and Division of Fisheries Research, Commonwealth Scientific and Industrial Research Organisation, Cronulla, Australia.

Abstract

Simulations can be constructed of whatever may be conceptualized, and are useful in gaining understanding of complex systems. The ideal sequel of a simulation is the compression of its essential properties into a relatively small number of equations. Prediction is the ultimate goal and test of a model. Parameterization of models is a major problem that is substantially assisted by independent measurement of parameters, as well as by observing the effects of planned perturbations. It is desirable, for any particular situation, to have several models, and to design procedures to test which may be most suitable. The crucial issue is what to conceptualize, and there are questions concerning the level of aggregation, time scale, geographic scale, and inter-connected systems. Models should be seen as part of a larger context of research activity.

Complex ecological simulation models of multispecies fish communities such as those of Laevestu and Favorite, and of Andersen and Ursin are characterized by a large number of parameters, some of which can be estimated independently, but many of which must be treated as tuning variables. The models are thus unproven. Modelling tropical associations by such methods is at present impractical, and it is suggested that simpler models are appropriate.

A simple model is presented on species group interactions in the Gulf of Thailand. The model was briefly tested at the workshop, and was in part invalidated. Information on variability of natural mortality rates would be useful for more rigorous testing.

Introduction

Simulation models enable projection of the consequences of conceptualization that are not mathematically tractable or, if tractable, so profound as to be difficult to comprehend.

Anything you can imagine can be the basis for a simulation. If you can imagine a population growing at a rate of 10% per year, you can use a simulation to find out how many years it will take for the population to double in size; but with only a small amount of mathematics, the same result can be obtained directly. Similarly, a Beverton-Holt yield equation summarizes the result that would be obtained by simulating over a series of discrete time periods the increments in growth, the natural mortalities, and the fishing mortalities in an exploited population for which one wished to find the yield for various sizes of first capture of a fish population. The paper of Walters (1969) is a perfect example of just this kind of procedure, incorporating a stock-recruit relation. Even if a conceptualization is mathematically tractable, a simulation may be useful for gaining better understanding, particularly of the sensitivity of the system to variation of some of its parameters.

Conversely, the ideal sequel to construction of a model for which there is no analytical solution is the compression of its essential properties into a relatively small number of equations. Thus, one might imagine that, having built a simulation of population growth with linear density dependent birth and/or death processes, one might conceive of the logistic equation as a convenient way of summarizing the system. In like fashion, the eventual objective of a highly complex simulation model should be the development of a model that is analytically tractable and relatively simple, preserving the characteristics of the simulation and the natural circumstances it describes. There may be half-way houses to this objective in which portions of the simulation are compressed into simple analytical form, while other parts, and accordingly the whole, remain clumsy and obscure. The great value of compression is not elegance for its own sake, but rather that it is the best means of ensuring that the model is truly comprehended, is in a form suitable for communication to others, and may be applied as a generic solution to a class of problems.

Regardless of the relative degree of generality and elegance, prediction is the ultimate goal and the real test for any model. There is thus little satisfaction or scientific value in a simulation model which can only predict, with hindsight, the circumstances and the data that led to its construction. For example, it may appear from a series of observations that the abundance of one species goes up as another goes down. This can easily be imitated by postulating any of a wide variety of mechanisms. If the parameters for the model are estimated from the data that inspired the model, there should be little wonder when the model turns out to fit the data. Prediction of what will happen next year, or of some corollary circumstances, or application to another set of data which ostensibly involve the same process are the only ways in which a particular simulation may be tested and gain wider and lasting acceptance.

Except for their ostensible verisimilitude, untested simulation models may indeed be no better, or even less adequate, than straightforward statistical models that are premised only on the assumptions that the future will be like the past, and that the multitude of minor details of complex systems can be represented by a simple probability distribution. (The same kinds of things may be said more generally of simulations that are structured in dimensions other than that of time. For the fisheries manager though, temporal dynamics is all important, and successful prediction of next year is paradise.)

These kinds of considerations lead naturally into mention of the problems of parameterization. In applying a model to an independent set of data, it may be extraordinarily difficult to estimate the parameters in the circumstances that they may be correlated causally as well as statistically. Thus, an increase in biomass may have its origin in faster growth, lower natural mortality or greater recruitment. At the same time, faster growth may imply lower mortality and changes in recruitment. The modeller who is engaged in a fitting exercise commonly finds him- or herself eventually ends up in a valley or a crater of a multidimensional topography despite the sophistication of the built-in non-linear algorithmic pathfinders.

The obvious way to reduce some of the uncertainties that lead to these multi-faceted dilemmas is to estimate some of the parameters independently. The first to come to mind are growth parameters, for they may be evaluated with relative precision, relatively inexpensively, in relatively natural circumstances. The same may not be true of many other parameters which can only be independently evaluated by large-scale observation and experiment.

Another route to greater certainty is the planned perturbation of the system to assess its natural resilience. This is much easier said than done, for it implies the potential for creating substantial social and economic disturbance in the short term in the interest of gaining understandings that will perhaps reduce the risk of social and economic disturbance in the long term. One would like to have considerable confidence in a model before disrupting a fishery for testing purposes, but on the other hand, the consequences of management with an untested model can be unrewarding at least and disastrous at worst (see Larkin 1972).

It is therefore valuable to have a wide range of models for any particular situation, and to view their various properties comparatively. To the extent that they are all plausible and would all gain from a particular perturbation, all can be tested simultaneously for verisimilitude.

From a practical point of view, then, it is highly useful if a simulation model can be constructed fairly quickly and easily, and its properties explored with convenience and dispatch. The attachment of an investigator to a particular model is often directly proportional to the time spent on its construction. Thus, to the extent that the modelling process can be accelerated, alternative theories can more quickly be evaluated (and abandoned). These kinds of considerations have led, in a variety of different disciplines, to the development of software packages that facilitate assembly of a model (for example, GPSS—General Purpose System Simulator). When combined with interactive capabilities, these packages enable extraordinarily rapid appraisal of system properties. Thus, for example, using a system such as SIMCON, it is possible to build, analyze, and discard a fairly complex model in a matter of hours, during which many variations on the theme can be explored in minutes. The advantages over individual programs and batch processing are enormous and obvious.

Technical matters aside, the crucial issue in modelling of any kind is what to conceptualize in the first place. For fisheries, for example, at one extreme, all species of harvestable organisms may be lumped together as a component in a larger system; at the opposite extreme, each individual of each species may be seen as the basic system unit. Neither extreme is likely to be particularly attractive, the first yielding a relatively trite prediction that, given enough species, an ecosystem will maintain some kind of structure for producing biomass at upper trophic levels, the other requiring such an enormity of data to test as to be hopelessly impractical. Between the extremes, opinions may vary as to the best intermediate path to pursue, and this is a generic problem of simulation modelling. There is also the question of geographical constraints to consider. A pond or a small lake may indeed be considered as a microcosm for a time period long enough to exhibit dynamics independent of externalities. But for larger lakes, and more especially the world's oceans, the question of scope is difficult. Fishes migrate and oceanographic events are linked on a range of scales extending to global patterns. There are thus three immediate questions to be addressed before modelling begins: what level of aggregation of subsystems, what time scale, and what geographic scale.

Related to such questions is the matter of which other systems are interacting with the fish community. For example, while it is commonly remembered that fish production rests on a pyramid of production at lower trophic levels, it is often forgotten that fishing vessels are extremely mobile and adaptive; that fish products are marketed on local and global scales; and that fisheries are embedded in larger social contexts both nationally and internationally. Is all of this to be included in one system, or will it be viewed as a series of connected systems?

The practical answer to all of these kinds of questions is to treat one part of the whole interconnected assemblage in sufficient detail as to be revealing, and to treat the remainder as either constant or with relatively simplistic approximation. Those with greater pretention will more often fail; those with least pretention will more often be irrelevant.

It is a logical consequence of these introductory remarks to observe that simulation models should not be seen as research enterprises that stand alone, but rather should be seen as a part of a larger system of research activity. A model should be developed, tested and revised repeatedly in the light of new information and selective experiment. It is invariably desirable to predict what will happen, suffer the consequences, and go back to the drawing board.

Multispecies Systems

From the foregoing, it is clear that simulation techniques may thus be viewed as an indispensable tool for exploring the properties of multispecies systems that defy analysis, or whose complexity challenges comprehension.

The simultaneous harvesting of several species of interrelated organisms is a common phenomenon of world fisheries, frequently complicated by the catching of more than one species in the same gear. While it is relatively easy to imagine what might be happening in such circumstances, it is extremely difficult to put what is imagined into analytical mathematical form. It is accordingly attractive to construct simulations and to use them in the hope of empirically discovering the properties of multispecies systems.

There have no doubt been many unpublished efforts to examine multi-species problems using crude simulation techniques. To our knowledge, there are none published on tropical fisheries. Accordingly, what follows briefly reviews some of the better known attempts at modelling temperate situations. Because we do *not* believe these approaches will be useful in the next two decades of tropical fisheries management, they are presented by way of prelude to a discussion of what kinds of models may be useful as a next step in understanding.

The simulation of the eastern Bering Sea ecosystem by Laevestu and Favorite is described fully in their report of 1977, and is outlined summarily in Granfeldt (1979). The first stage is diagnostic of the initiating equilibrium biomasses that characterize the largely unharvested ecosystem (PROBUB = Prognostic Bulk Biomass Model). It is structured with time difference equations describing transitions by species (or groups of species) of biomasses which increase by growth and recruitment (a single variable) and decrease by natural mortalities which are substantially composed of predation. The predation coefficients linking the species are age, temperature, and area dependent, their relative magnitude reflecting the different rates of consumption by the various predators of the various prey. The model considers simultaneously the community of five species (or groups) of demersal fishes, four semi-demersal, three pelagic, four "others", all embedded in an ecosystem that represents, at a more crude level of conception, the lower trophic levels that set the gross limits on production, and the upper trophic levels (birds, mammals and fish harvests).

The PROBUB model "searches for a unique solution with a given set of inputs [comprising] one known biomass and prescribed food requirements and composition of [diet]." Thus, to the degree that these inputs are valid, the solution also may be valid.

The DYNUMES model (Dynamic Numerical Marine Ecosystem Model) is initiated with the PROBUB equilibrium state, and has a spatial resolution which permits simulation of "seasonal migrations and their effects, [and] spatial and temporal variation of composition of [diet]."

Bearing in mind that there is a shortage of historical data and that many of the coefficients are at best only rough estimates, the authors have surprising confidence that the model adequately depicts the natural system. They remark, for example, that the "limits of accuracy of the computed biomasses" are estimated to be "± 30%, varying from species to species." It is even concluded "in general, . . . the PROBUB model results are more realistic and accurate than [results of trawling surveys]."

These conclusions are probably unjustified. Because the model is so large, its inner workings are still somewhat mystifying and its sensitivity to various alternative sets of inputs is not clear. Moreover, the model has as yet an insufficient record of successful prediction to be taken seriously as a model for others to follow. Nevertheless, its holism is admirable and it has already prompted broader perceptions of how to manage the interacting community of harvestable organisms, and these contributions are useful.

The model system of Andersen and Ursin (1977) is basically similar to that of Laevestu and Favorite, but differs in not providing for spatial resolution, dealing in numbers rather than biomass (and therefore using Beverton and Holt equations for each species); and perhaps most importantly, varying the diet by reference to search equations rather than assuming that predators take prey in proportion to their abundance. As in the Bering Sea model, there are subsystems concerned with phytoplankton and the dominant groups of zooplankton, but the detail of the model is greatest in its representation of the fish community and in the circulation of phosphorus in the ecosystem.

The model was inspired by North Sea circumstances, in particular the large scale switch from demersal to pelagic fish production in the late 1960s and early 1970s. In consequence, the trick in setting it up to be ostensibly realistic is to adjust its parameters (such as the coefficients of the search equations) so as to generate stability over a relatively wide range of fishing effort, but instability such as to produce the switch that was observed when a large fishery effort of a particular kind is applied.

The authors are candid in stating that the model has required considerable readjustment so that it will behave in a manner consistent with historical data or reasonable surmise. With each adjustment, it might be expected that the model would progressively become more reliable, and indeed, the relatively well-documented history of North Sea fisheries imposes formidable constraints on the range of acceptable results. Nevertheless, the model remains "unproven". There is still enough guesswork concerning a sufficient number of the several hundred parameters that a wide range of predictions is possible. The independent assessment of many of the parameters is accordingly a matter of priority. In particular, Ursin (in press) has drawn attention to the need for data on feeding behavior in situations where there is choice, underlining the relevance of much contemporary work on optimal foraging strategies.

As in the case of the Bering Sea simulation, it seems reasonable to conclude that the North Sea model has little utility for multispecies fisheries management in tropical waters. To even construct a model at a comparable level of detail for tropical fish communities is at present impractical. Even if the data were available for estimating the several thousand parameters involved, it would take many years of observation and experiment to verify the utility of the model. It is also difficult to visualize what kind of experimental management might be adopted to test the validity of such a model.

The moral, it seems, is to simplify, and in the next section we give some specific suggestions for a simplified model. Before doing so, it seems appropriate to comment briefly on large-scale simulation models in the related field of primary production. The comprehensive modelling of growth of phytoplankton biomass and its dependence on nutrient circulation and zooplankton grazing is virtually another whole field of scientific activity. DiToro et al. (1977) present a comprehensive model based on much that has gone before, and which is aimed specifically at providing a basis for management of estuarine waters to achieve standards of water quality control. The basis of the model is the principle of conservation of mass which requires accounting of the masses of the various components of the system whether generated internally or introduced. The variables are depicted in their kinetic relationship in Fig. 1, and the equations describing these relationships follow now familiar forms.

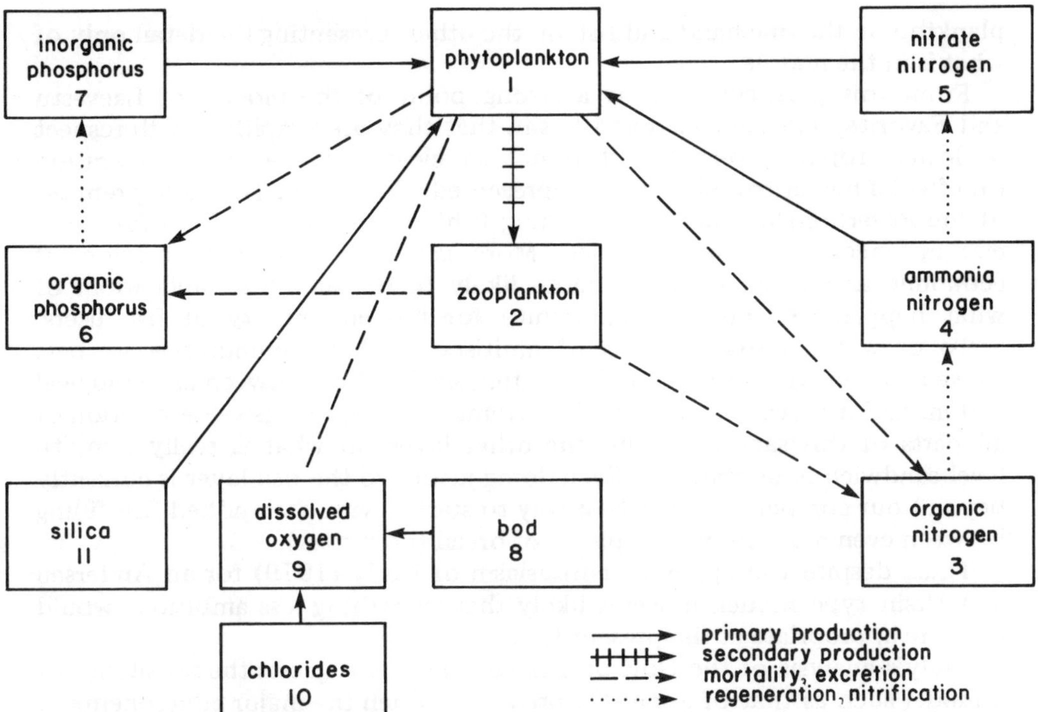

Fig. 1. Graphical representation of a model of phytoplankton dynamics proposed by DiToro et al. (1977) (see text).

The most significant feature of this model is the extensive verification process to which it has been subjected. Data are available for several years, and since the processes concerned are seasonal, one year's data can be used to test the validity of estimates based on data for other years. In addition, *in situ* and laboratory experiments may provide independent measures of many of the variables.

In brief, the modelling of such processes is much more akin to the practices of the physical sciences than is characteristic of models at the higher trophic levels. Estimates of quantities and the rates at which they change are much more readily obtained than, say, for fishes, although there are nevertheless some formidable sampling problems. Additionally, the repeatability of events on an annual basis because the organisms concerned are essentially annuals, allows for quicker evaluation of model concepts. It is noteworthy that models of this kind tend to be weakest in their depictions of the zooplankton component. As the authors say, "The formulation of the zooplankton death rate presents somewhat of a problem, because in addition to their endogenous respiration rate, the zooplankton are being preyed upon by the upper levels of the food chain. In order to simplify the model framework, it is necessary to introduce this effect empirically as an additional death rate constant."

Thus, just as fisheries models simplify the lower trophic levels, phytoplankton models simplify the upper levels. Little wonder that the model of Steele and Frost (1977) for zooplankton communities simplifies phytoplankton on the one hand and fish on the other, presenting the detail only of what is in the middle.

From this perspective, it is a strong point of the models of Laevestu and Favorite, and Andersen and Ursin, that they are simplified with respect to lower trophic levels; but it is not so clear that the fishing operation (in effect a higher trophic level) is represented adequately. The basic premises of the Beverton-Holt model concerning fishing mortality are doubtful in the circumstances of many fisheries. More generally, the incorporation of economic and social factors is more likely to translate the implications of what happens to the fish community for the community of fish users.

To draw the moral, models of multispecies fish communities are best viewed as depictions of a complex system sandwiched between an ecological system and a harvesting system. To attempt a comprehensive description of all parts of this sandwich, plus the other layers in what is really a multi-level sandwich is impractical. Even doing justice to the fish layer is currently beyond our competence. The best way to start is with a simplified fish filling between even more simplified pieces of bread (no butter).

Thus, despite the apparent enthusiasm of Pauly (1979) for an Andersen and Ursin type model, it seems likely that something less ambitious would be more appropriate at the present time.

Pauly's analysis of the Gulf of Thailand fishery suggests the possibility of a model such as that of Jones (in press), in which the major components of the ecosystems are depicted *en masse* in their functional relationship to each other (see Fig. 2). Comparable data, especially those pertaining to phyto-plankton and zooplankton abundance, may not be readily available. However, even the six groups of fishes identified by Pauly might be a sufficient basis

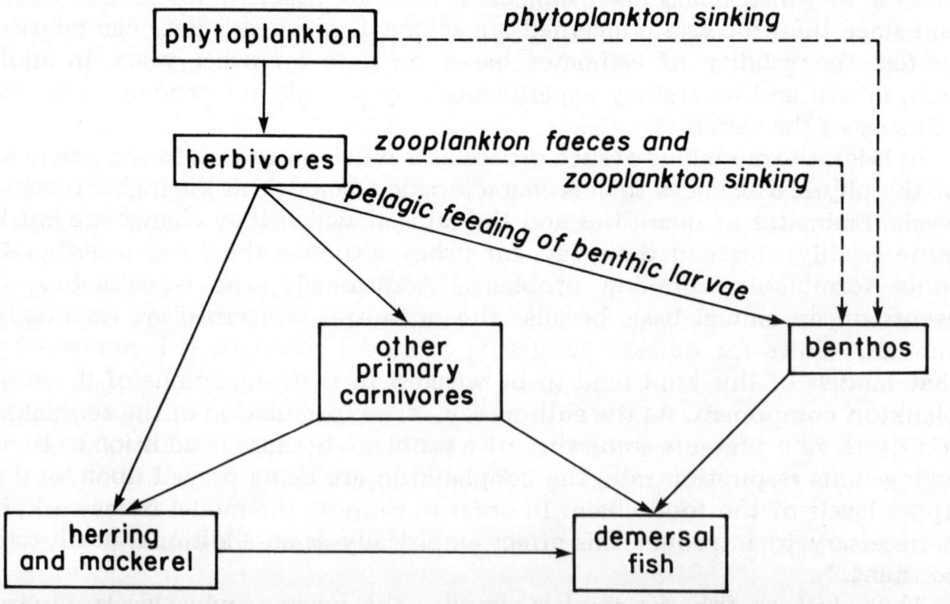

Fig. 2. Relations between major components of the North Sea fisheries ecosystem (from Jones 1981).

for modelling the gross dynamics of the community of larger organisms, making appropriate assumptions about the base of production that supports them. Just as Jones' model is a plausible explanation for the switch that occurred in the North Sea with potential application to other north temperate zone fisheries, a Gulf of Thailand model might suggest some plausible premises for management of tropical fisheries.

A Simple Model

Pauly (1979) presented a preliminary analysis on the Gulf of Thailand trawl fishery in which he identified six groups of taxa within which the species exhibited similar rates of decline or increase observed over a nine-year period (1963 to 1972). The groups are as follows:

(1) large feeders on zoobenthos characterized by their large size and longevity;
(2) small, demersal prey species characterized by high productivity;
(3) intermediate predators whose primary prey are the second group;
(4) large predators;
(5) pelagic fishes;
(6) trash fish which are small in size yet highly productive (this group includes the shrimps for the purpose of the present discussion)

Under exploitation, groups (1) to (5) decreased in biomass (group (1) showed the most decrease while group (5) showed the least), and group (6) increased in biomass. There are two notable exceptions to this scheme. Conger eel *(Muraenesox)*, which is included in large predators, and squid *(Loligo)*, which is included in pelagic fishes, both significantly increased as the total catch decreased. While these exceptions may arise from competitive interactions within a group, squid at least might be considered as a seventh group on taxonomic grounds which imply a substantially different ecology. In addition, for modelling purposes, two more groups are required to represent the lower trophic level, namely,

(8) zoobenthos which has a production of P_B; and
(9) zooplankton which has a production of P_P.

These nine groups, then, represent the level of aggregation of the subsystems. Other factors which must be considered before modelling can begin are scale of time and geography. The time scale is obviously over the period of exploitation and, since summary data are available for yearly intervals, the time step should be one year. The geographic scale is the entire Gulf of Thailand.

Next, it is necessary to consider how the groups are linked and what quantities are to be tracked through the system. Since the catches are reported in biomass and relative species composition, tracking the flow of biomass is the simplest procedure. Fig. 3 depicts the flow, implied by Pauly, through the components.

The preceding outlines a simple model which may, through explicit assumptions, be translated into mathematical expressions. First, we investigate the relationship for group interactions, then the less detailed representations of the harvest systems and plankton production.

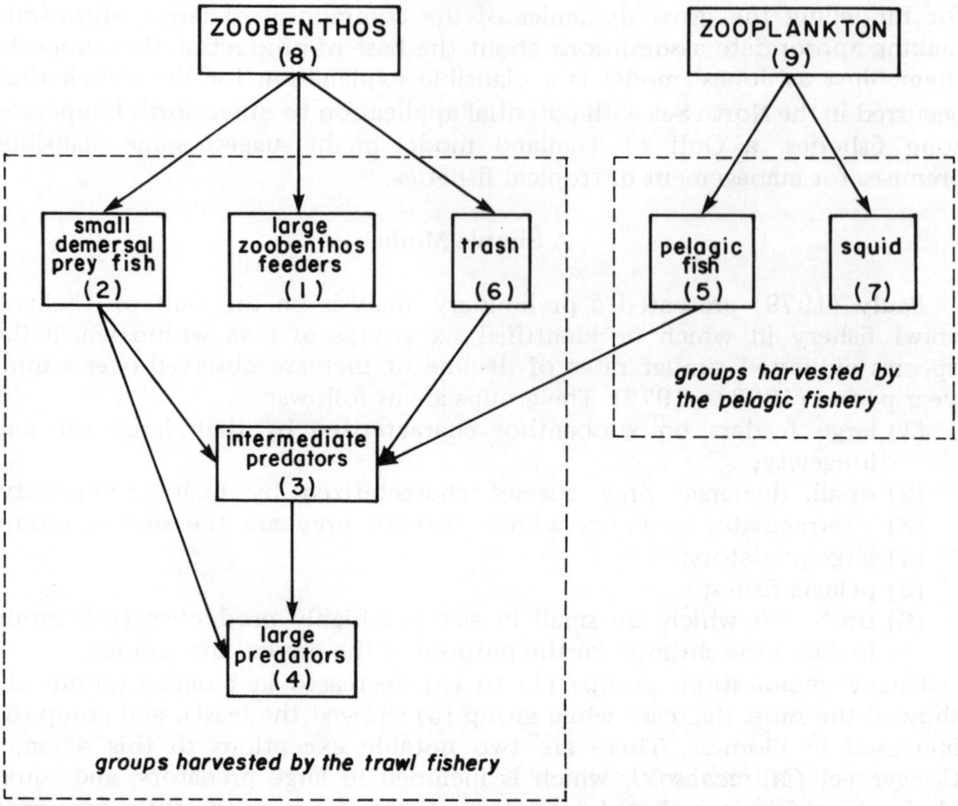

Fig. 3. Schematic relations among groups of fishes, squid, zoobenthos and zooplankton in the Gulf of Thailand (adapted from Pauly 1979).

Holling (1959) first realized that the number of prey items ingested by a predator, over any time interval, is limited by the handling time of the prey items. Assuming that the predator attacks at random and that the handling time remains constant, then the number of attacks is given by

$$\alpha' N/(1 + \beta N) \qquad (1)$$

where N is the prey biomass, α', the effective rate of successful search, and β the rate of handling prey items. Murdoch (1973) made an extension to allow a predator the choice of more than one prey species. That is, the number of attacks made by the j'th predator on the i'th prey species is

$$\frac{\alpha'_{ij} N_i}{1 + \sum_{i+1}^{n} \beta_{ij} N_i}$$

where n is the number of prey species.

If prey biomass is converted to predator biomass by a species-specific constant, C_{ij}, then material exchanged from the i'th prey to the j'th predator is

$$\frac{\alpha_{ij} N_i P_j}{1 + \sum_{i+1}^{n} \beta_{ij} N_i}$$

where $\alpha_{ij} = \alpha'_{ij} C_{ij}$ and P_j is the predator biomass. With the assumption that a group of taxa behaves as a single species, expression (2) becomes the rule for mass flow from one group to another. In addition to the flow of biomass, losses from a particular group will occur from predation by species not represented and from natural mortality other than predation. A first approximation is to treat the loss as a constant rate, m_i, for the i'th group.

Another loss to the system is the harvesting system which can be represented, in a crude fashion, by a constant rate f_T for the five groups involved in the trawl fishery, and f_P for the two groups in the pelagic fishery (see Fig. 3).

The remaining undefined component is the production of plankton. If it is argued that production is limited by essential nutrients (e.g., phosphorous), and only a very small proportion of the total nutrients in the marine ecosystem is removed by fishing, the production may be treated as constant (P_B for benthos and P_P for plankton), in relation to any fluctuation in biomass of the nine groups.

Since all inputs, linkages and losses to the system are now mathematically characterized, we can write the balance equations which constitute the model. The change in biomass in any group must be the difference between the sum of all inputs and the sum of all outputs of biomass. A single generalized equation can be written for the i'th group which in theory could prey upon all other groups and in turn be subjected to predation by all other groups. Thus,

$$\frac{dG_i}{dt} = P_i + \frac{\sum_{j=1}^{n} \alpha_{ji} G_i G_j}{1 + \sum_{j=1}^{n} \beta_{ji} G_j} - \sum_{j=1}^{n} \frac{\alpha_{ij} G_i G_j}{1 + \sum_{k=1}^{n} \beta_{kj} G_k} + G_i (m_i + f_i)$$

where n groups are identified.

Including initial conditions, the number of parameters is $4n + 2n^2$. For our structure there would be 198 parameters but because all groups do not feed on all other groups, there are only 42 non-zero parameters. The equivalent, less eloquent, but more direct, method is simply to write the specific equations for the system in question. Thus,

$$\frac{dG_1}{dt} = a_{81}/(1 + b_{81}) - G_1\,(m_1 + f_T)$$

$$\frac{dG_2}{dt} = a_{82}/(1 + b_{82}) - \{a_{23}/(1 + b_{23} + b_{63} + b_{53}) + a_{24}/(1 + b_{24} + b_{34}) + G_2\,(m_2 + f_T)\}$$

$$\frac{dG_3}{dt} = (a_{23} + a_{63} + a_{53})/(1 + b_{23} + b_{63} + b_{53}) - \{a_{34}/(1 + b_{34} + b_{24}) + G_3\,(m_3 + f_T)\}$$

$$\frac{dG_4}{dt} = (a_{24} + a_{34})/(1 + b_{24} + b_{34}) - G_4\,(m_4 + f_T)$$

$$\frac{dG_5}{dt} = a_{95}/(1 + b_{95}) - \{a_{53}/(1 + b_{53} + b_{23} + b_{63}) + G_5\,(m_6 + f_P)\}$$

$$\frac{dG_6}{dt} = a_{86}/(1 + b_{86}) - \{a_{63}/(1 + b_{63} + b_{23} + b_{53}) + G_6\,(m_6 + f_T)\}$$

$$\frac{dG_7}{dt} = a_{97}/(1 + b_{97}) - G_7\,(m_7 + f_P)$$

$$\frac{dG_8}{dt} = P_B - \{a_{82}/(1 + b_{82}) + a_{81}/(1 + b_{81}) + a_{86}/(1 + b_{86}) + G_8 m_8\}$$

$$\frac{dG_9}{dt} = P_P - \{a_{95}/(1 + b_{95}) + a_{97}/(1 + b_{97}) + G_9 m_9\} \qquad (3)$$

where

$$a_{ij} = \alpha_{ij} G_i G_j$$

$$b_{ij} = \beta_{ij} G_i$$

The system of equations (3) does not have an explicit solution. It is also to be noted that the independent variable (time) does not enter into any of the right hand side expressions, and numerical solutions will therefore be easy to program; at least up to second order methods [e.g., Euler or improved Euler methods should be used since the system is "stiff" (Gear 1971)].

Now the model has been built, the 42 parameters must be assigned values. Estimates of virgin biomass (i.e., the initial conditions) are listed for the first seven groups, along with the estimated biomasses after exploitation in Table 1.

Unfortunately, the balance of the parameters cannot be estimated readily; indeed, it is unlikely that the detailed information on functional response and conversion efficiencies required to parameterize the α_{ij}'s and β_{ij}'s can be collected. Therefore, further assumptions must be made to collapse them into a smaller set of parameters.

We start by noting that expression (1) may also be described as a Michaelis-Menten relationship with the Michaelis or half-saturation constant defined as $1/\beta$ and the maximum processing rate defined as α/β. Crowley (1975) argues that natural selection operating on a biological process with Michaelis-Menten kinetics adjusts the Michaelis constant relative to ambient substrate concentration. While Crowley's approach is from the perspective of enzyme kinetics, he does conclude that "the diversity of supportive evidence cited suggests a broad applicability of this argument across taxonomic groups and levels of biological organization". Therefore, if we assume that the virgin biomass in Table 1 is representative of mean biomass over evolutionary time, the β_{ij} of any predator group is the reciprocal of the virgin biomass of their prey. A further associated assumption is that the system was at an equilibrium at virgin biomass. Since there are nine groups, nine parameters may be collapsed into other constants. The choice should be dictated by the ease of parameterization from the available data and intuition. We proceeded as follows:

First, three new parameters were introduced,

$$\alpha_{24}/\alpha_{34} = g_1 \ ; \alpha_{23}/\alpha_{63} = g_2 \ ; \alpha_{23}/\alpha_{53} = g_3 \ ;$$

which are the relative effective rates of successful search of large and intermediate predators upon the prey they share. Given the initial equilibrium biomass (S_i) of each group and the associated natural mortality (exclusive of stated predation), the remaining parameters may be obtained from:

$$\begin{aligned}
\beta_{ij} &= 1/S_i \text{ for all i and j} \\
\alpha_{34} &= 3S_4/(g_1 S_2 + S_3) \\
\alpha_{24} &= g_1 \alpha_{34} \\
\alpha_{63} &= 4/3\,(\alpha_{34} S_4 + 3m_3)\,(g_2 S_2 + S_6 + g_2 S_5/g_3)^{-1} \\
\alpha_{23} &= g_2 \alpha_{63} \\
\alpha_{53} &= \alpha_{23}/g_3 \\
\alpha_{97} &= 2m_7/S_9 \\
\alpha_{95} &= (\alpha_{23} S_3 + 4m_5)/2S_9 \\
\alpha_{81} &= 2m_1/S_8 \\
\alpha_{82} &= (\alpha_{23} S_3/2 + 2\alpha_{24} S_4/3 + 2m_2)/S_8 \\
\alpha_{86} &= (\alpha_{63} S_3/2 + 2m_6)/S_8 \\
P_B &= S_8\,(\alpha_{82} S_2 + \alpha_{81} S_1 + \alpha_{86} S_6 + 2m_8)/2 \\
P_P &= S_9\,(\alpha_{95} S_5 + \alpha_{97} S_7 + 2m_9)/2
\end{aligned}$$

As condensed, the model requires 23 inputs which fall into four types:
(1) nine initial or equilibrium biomass levels;
(2) nine natural mortalities;
(3) three ratios of effective search rates;
(4) two fishing mortalities (trawl and pelagic).

Table 1. Initial biomass and biomass after exploitation of seven groups of "fishes" in the Gulf of Thailand (from Pauly 1979, Table 13).

Group		Initial biomass ($\times 10^3$ t)	Exploited biomass ($\times 10^3$ t)
1	Large zoobenthos feeders	150	1
2	Small demersal prey fish	800	4
3	Intermediate predators	600	60
4	Large predators	20	5
5	Pelagic fishes	15	7
6	Trash	15	30
7	Squid	20	70

The primary objective of the model is to test if a system described in Fig. 3 can demonstrate the observed dynamics reported in Table 1. One method, given the starting biomass levels, is to state the general conditions for which groups 1 to 5 decrease in biomass and groups 6 and 7 increase in biomass while holding their rank order. After repeated simulations, the conditions are:

(1) Large zoobenthos feeders have a low natural mortality in relation to small demersal prey and trash fish.

(2) The intermediate predators have a strong preference for small demersal prey over both trash and pelagic fish.

(3) Large predators do not have a strong preference for intermediate predators.

(4) Squid have a higher natural mortality than the combined predation and natural mortality of pelagic fish (if intermediate predators prefer trash fish to pelagic fish, this difference need not be large).

Since these conditions are characteristic of the biology of the groups involved, the model is capable of exhibiting the observed dynamics. Many adjustments and additions can be incorporated into the basic model framework to the extent that exact agreement between the observed data and model output could be achieved. For example, besides fine tuning the parameters, there could be the addition of time lags, lower bounds (groups have refuges), upper bounds (while a group has an ample food resource it may be habitat limited) and competitive interference. However, validation of the model will only be accomplished through blind prediction, not by agreement with existing data. Indeed, a model based on entirely different concepts than those here presented could as readily depict the observed dynamics.

For example, one paradigm in predation theory, in apparent contradiction to our model, is that a predator only takes those prey that would have died from other causes (see Jones, this vol.). The proponents of this paradigm believe that the dynamics of the system are controlled at the early life history stages when mortality rates are rarely measurable. Therefore, the measured mortality rate at older ages (besides fishing) will remain constant and independent of biomass fluctuations. One could conceptualize the system as a stock-recruitment relationship for each group modified by interacting groups. For example, Pauly (1979) explains the increase of trash fish by increased recruitment due to decreased predation on early life history stages from small demersal prey fish, a linkage not considered in our model.

Clearly, many approaches (i.e., models) will have verisimilitude but will generate widely different predictions with extrapolation from the observed events (e.g., a cessation of fishing). However, the present exercise has revealed two important impacts upon management.

First, the small demersal prey fish have been identified as a key factor. Control (i.e., management) of the system must involve the ability to selectively harvest this group. If this is impossible, another possible control is pulse fishing in space and/or time. The emphasis in modelling experimentation and data collection would switch to the dynamics of recovery (e.g., can the system "flip" across a boundary into a totally different mode of behavior?)

Second, our model predicts a large fluctuation in natural mortality as a result of changes in exploitation. Major advances in the determination of mortality rates, initiated by Doubleday (1976) with the most recent contribution from limited data sets by Paloheimo (1980), make a test possible. The minimum data requirements are age composition of total catches and effective effort for a series of years by group.

Regardless of the model scheme chosen, the management moral is, of course, the most attractive feature of any simulation. For the model here presented it is very simple to assess the consequences of various management scenarios.

Postscript

At the workshop an attempt was made to invalidate the foregoing model for the Gulf of Thailand. In addition, the model of Pope (1979) was examined for its utility for the Gulf of Thailand situation.

Concerning the Larkin and Gazey model, Pauly observed that, while the relative rates of production and fishing mortalities seemed correct, the absolute rates that occur are substantially greater than those used in the simulations. Accordingly, simulations were attempted with greater and more realistic values. The results were similar to previous simulations at lower rates with the exception that the pelagic groups (pelagic fish and squid) had reversed their behavior (i.e., an eventual increase in pelagic fish and an exponential decrease in squid). Only an extraordinarily high rate of turnover of squid (e.g., an instantaneous mortality rate of 15 at virgin conditions) could reproduce the observed dynamics. Therefore, the concept (for modelling purposes) that squid dynamics are controlled by competition with pelagic fish for the resource base is not valid. Indeed, only significant predation by at least one other group can restore the model to verisimilitude.

As for the Pope paper, the group noted that his model is equivalent to the Lotka-Volterra class of equations, the utility and flexibility of which was thoroughly discussed by Kirkwood (this vol.). At the workshop, the interpretation of Pauly concerning species groups and their interactions was used as a basis for constructing a Pope-type model. Subsequent simulations demonstrated that the biomass estimates could not be duplicated with reasonable parameter values. For example, both an extraordinarily high interaction term between intermediate predators and trash fish, and an extremely low catchability coefficient for trash fish are required to increase trash biomass under exploitation.

References

Andersen, K.P. and E. Ursin. 1977. A multispecies extension to the Beverton and Holt theory of fishing, with accounts of phosphorus circulation and primary production. Meddr. Dan. Fisk. Havunders. (New Ser.) 7: 319-435.

Crowley, P.H. 1975. Natural selection and the Michaelis constant. J. Theor. Biol. 50: 461-475.

DiToro, D.M., R.U. Thomann, D.J. O'Connor and J.L. Mancini. 1977. Estuarine phytoplankton biomass models—verification analyses and preliminary applications, p. 969-1020. *In* E.D. Goldberg et al. (eds.) The sea: ideas and observations on progress in the study of the seas. Vol. 6, Marine modelling. Wiley Interscience, New York.

Doubleday, W.G. 1976. A least squares approach to analysing catch at age data. Int. Comm. Northwest Atl. Fish. Res. Bull. 12: 69-81.

Gear, W.C. 1971. Numerical initial value problems in ordinary differential equations. Prentice-Hall Inc., New Jersey.

Granfeldt, E. 1979. Marine ecosystem simulation for fisheries management. U.S. Natl. Mar. Fish. Serv., Northwest and Alaska Fish. Cent., Seattle. 21 p. (Mimeo)

Holling, C.S. 1959. The components of predation as revealed by a study of small-mammal predation of the European pine sawfly. Can. Entomol. 91: 293-320.

Jones, R. Species interactions in the North Sea. Can. J. Fish. Aquat. Sci. (In press)

Laevestu, T. and F. Favorite. 1977. Preliminary report on dynamical numerical marine ecosystem model (DYNUMES II) for eastern Bering Sea. U.S. Natl. Mar. Fish. Serv., Northwest and Alaska Fish. Cent., Seattle. 81 p. (Mimeo)

Larkin, P.A. 1972. A confidential memorandum on fisheries science, p. 189-197. *In* B.J. Rothschild (ed.) World fisheries policy: multidisciplinary views. Univ. Washington Press, Seattle.

Murdoch, W.W. 1973. The functional response of predators. J. Appl. Ecol. 14: 335-341.

Paloheimo, J.E. 1980. Estimation of mortality rates in fish populations. Trans. Amer. Fish. Soc. 109: 378-386.

Pauly, D. 1979. Theory and management of tropical multispecies stocks: a review, with emphasis on the Southeast Asian demersal fisheries. ICLARM Studies and Reviews No. 1. 35 p. International Center for Living Aquatic Resources Management, Manila.

Steele, J.H. and B.W. Frost. 1977. The structure of plankton communities. Philos. Trans. R. Soc. London 280: 485-534.

Ursin, E. A multispecies fish stock assessment model. Can. J. Fish. Aquat. Sci. (In press)

Walters, C.J. 1969. A generalized computer simulation model for fish population studies. Trans. Amer. Fish. Soc. 98(3): 505-512.

Discussion on Paper by Dr. Larkin and Mr. Gazey

Discussion centered on the relative merits of large complex simulation models as contrasted with relatively simple simulation models of tropical multispecies fish communities.

The advantages of large models lie in their holism. They help to build awareness, especially of the interactions in systems, they help to explore the consequences of certain sets of assumption and thus help "explain" what is going on. The disadvantages are related to the need to invent functional relationships and parameter values in order to get the model to run. The major advantage of simplified models is that they are more readily constructed, tested and rejected and lead to the development of complex models that may be seen as progressively better approximations for predictive purposes.

Dr. Marten referred specifically to his experience that with greater food chain length, complex ecosystem models showed more variable behavior. He also referred to the IBP modelling experience in the United States which did some highly detailed modelling which, in its parts, was fairly impressive though not convincing. When the parts are assembled, the total model didn't come close to predicting experimental plot results; yet relatively simple models could be used for this purpose quite successfully. Dr. Marten also observed that at the time of the IBP, each group said it was in the initial stages of formulating a model but hadn't really got to that point where they would be able to publish it, but each knew about some other group somewhere else that had a fantastic model that worked really well. Of all of these models, very few ever saw the light of publication. The phenomenon of non-publication of complex models is apparently widespread.

On the other side of the ledger, Dr. Pauly stressed that the awareness created by complex modelling is not to be underestimated as a driving force for change. The entrenched paradigms of single-species assessment, particularly in Northern Europe, had led to unjustified ridicule of the multispecies approach of Andersen and Ursin. Much of this reluctance to change was based on failure to understand the complex models. More recently, there has been wider appreciation of the contribution the model has made to revising views about the kinds of research that are now needed. The value of complex multispecies models for tropical fisheries should not be underrated. It may be necessary to start with simplified versions, but we should make haste to develop complex models as a vehicle both for building awareness and in the long run, as a basis for management.

Dr. Larkin responded to these comments by stressing the need for having many models and cautioned against prolonged attachment to any model. Mr. Jones stressed the importance of alternating modelling with observation and experiment.

The question of what level of aggregation to choose for a complex model was mentioned as a particularly troublesome matter. When an attempt is made to generate conclusions at the ecosystem level from what is observed at the level of populations there are usually major difficulties. The experience of such models has led to the "sandwich" approach, in which at one trophic level there is a detailed model, but at levels above and below, there is a compression to simulate the major properties, and virtually no detail.

The discussion was in general positive on the subject of complex ecological simulation models but stressed the need for critical approaches.

The discussion on the proposed model of the Gulf of Thailand mainly focused upon its authors' contention that large fluctuations in natural mortality under exploitation, arising from predation interactions, must be present for the model to be valid. However, some participants asserted that the derived biomass data could be accommodated assuming no effect of predation.

It was pointed out that the calculation by cohort analysis of natural mortality rates, which are required to test the model, yields wide confidence limits upon the estimated values; thus, the detection of a change may be difficult. However, since the predicted fluctuations are large (an order of magnitude in some groups), reasonable catch-at-age data should be sufficient.

As an alternative, Dr. Sainsbury suggested that experimental studies utilizing fished and unfished areas, reasonably close to one another, might produce definite answers.

Further discussion pointed out that the assumption of constant production of zoobenthos and zooplankton may be over-simplistic since many studies have shown that production may vary enormously in response to feeding pressure.

Recruitment of Tropical Fishes

GARTH I. MURPHY

CSIRO, Division of Fisheries Research
P.O. Box 21, Cronulla
N.S.W. 2230, Australia

MURPHY, G.I. 1982. Recruitment of tropical fishes, p. 141-148. *In* Pauly, D. and G.I. Murphy (eds.) Theory and management of tropical fisheries. ICLARM Conference Proceedings 9, 360 p. International Center for Living Aquatic Resources Management, Manila, Philippines and Division of Fisheries Research, Commonwealth Scientific and Industrial Research Organisation, Cronulla, Australia.

Abstract

The limited case histories available on recruitment in tropical fishes suggest the situation is not much different from the temperate situation. Spawning is nearly always periodic (as opposed to continuous), often with tight seasonal control. The cases available suggest a spectrum of density control of reproduction from very tight to weak. With respect to the latter, inclusion of environmental terms in the Ricker model has been possible in two cases, greatly reducing the residual variation. It does not appear necessary to invoke interspecific interactions as the driving force in controlling reproduction.

Introduction

Biological or recruitment overfishing was largely ignored until several well studied stocks collapsed apparently owing to reducing the spawning stock to a level at which a respectable year-class could not be generated. The rude awakening was particularly evident in the North Sea where prevailing dogma held that recruitment overfishing was quite unlikely and even less likely if attention was paid to maximizing the yield per recruit.

These events led to a landmark symposium convened by ICES, ICNAF, and FAO at the University of Aarhus, Denmark from 7 to 10 July 1970 (Parrish 1973). There has not been a great deal of theoretical development since then, although several writers have modified existing models e.g., Ricker's, to allow for environmental effects. Perhaps the most elegant of these is the framework developed by Shepherd and Cushing (1980) based on the Michaelis-Menten equation.

It is not the purpose of this discussion to develop or even discuss theoretical models. Rather I wish to look more directly at stocks and recruits in the tropics. One of the more important questions that must be considered and discussed is whether there is any evidence that the stock-recruitment problem is different between tropical and temperate ecosystems. I think not. Many temperate fish spawn over several months. Spawning of the Pacific Sardine

(Sardinops caerulea) has been noted during every month of the year with the more intense activity spread over three or four months. Similar comments apply to the California Anchovy. Even such presumably seasonal spawners as sea-run rainbow trout *(Salmo gairdneri)* have been observed spawning during every month of the year in Waddell Creek, California (Shapovalov and Taft 1954). On the other hand, as will be demonstrated later, the protracted spawning (and recruitment) once held as an article of faith for tropical fishes has not withstood careful scrutiny.

Estimation

In order to examine the problem it is necessary to estimate the spawning stock and the resulting recruits year by year. This is not a simple matter, especially in the absence of detailed catch records reinforced by a sampling and aging program. Even then, changes in the fishing patterns and or the availability of the sought species will make accurate real time estimation difficult. As a substitute for sampling the catches themselves, research vessels have been employed to sample the stock annually, most notably in the Gulf of Thailand.

The key assumption in this technique is that the research vessel's catches and catching ability are some linear function of the performance of the fleet itself. In the absence of catch statistics, there is no way to test this assumption —Catch 22; obviously, in the absence of such comparisons any unusual results from the research vessel data must be regarded as tentative until verified.

Over the years a number of clever shortcut methods have been devised to extract the wanted parameters from, for example, catches that have not been aged—only measured or weighed. Many of these originated in or were inspired by Beverton and Holt (1956). Some of these methods have been extended and new ones developed and applied by Pauly (1979a, 1979b, 1980). All of these methods rest on relationships and assumptions that are difficult to verify in particular cases, and in many cases the assumptions, e.g., of steady state, are patently false. For example, estimating recruitment by dividing yield by yield per recruit clearly demands a steady state. Nevertheless, even if recognized to be in error, the methods do provide useful first estimates and insights into what is taking place.

They can be misleading in that relaxing the 'rules' enough to employ the methods can permit a variety of equally legitimate interpretations of sets of data not refined enough to support the more classical types of analyses. For instance, Pauly (1980) has analyzed Gulf of Thailand data for *Psettodes erumei* (Pauly 1980, Fig. 6). The poor fit to any reasonable model led him to conclude there was no stock-recruitment relationship. This led him to test *P. erumei* against the estimated total standing stock (Pauly 1980, Fig. 7), suggesting that recruitment of this species was under community control. Both arguments are almost totally dependent on estimated production of recruits by the virgin stock and the 1963 data set, both of which are suspect, particularly with respect to the *P. erumei* recruits.

Using a slightly different approach which, among other things, disallows the suspect two points (virgin and '63) a somewhat different picture compatible with existing models emerges (Fig. 1). There is little to be gained by further discussion. The point is that gratuitous degrees of freedom can lead to interesting results, which, however, must be regarded as preliminary and requiring additional tests.

Some Examples

Southern Bluefin Tuna (Thunnus maccoyii):

While this species is not in all respects a tropical species, it does spawn in the Indian Ocean not far south of Java and its early life is in tropical waters (Murphy and Sutherland, in press). Fig. 2 shows the spawning stocks and resultant recruits, with both axes derived from cohort analysis. For the purposes of this workshop I want to make three points from the southern bluefin tuna data all of which suggest the tropical situation is little different from the temperate. They are: (1) spawning is restricted to the period October 1 to March 31 peaking around January 1; (2) there appears to be an

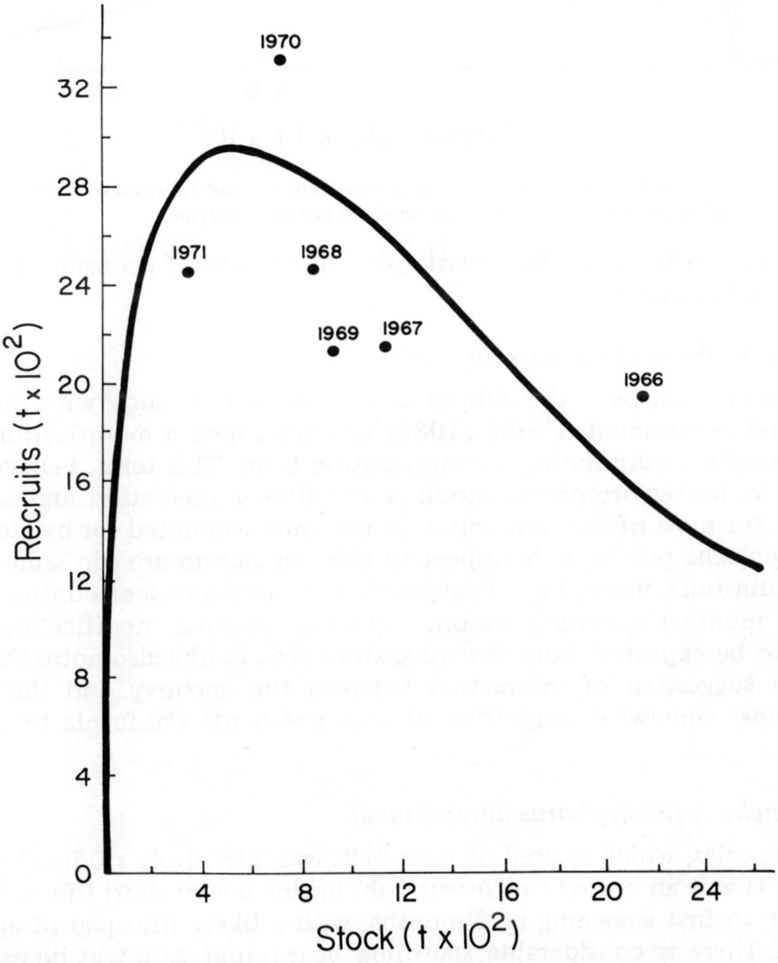

Fig. 1. Stock-recruitment relationship for the flatfish *Psettodes erumei* from the Gulf of Thailand. Curve fitted by eye. The year is the spawning year with recruits one year later.

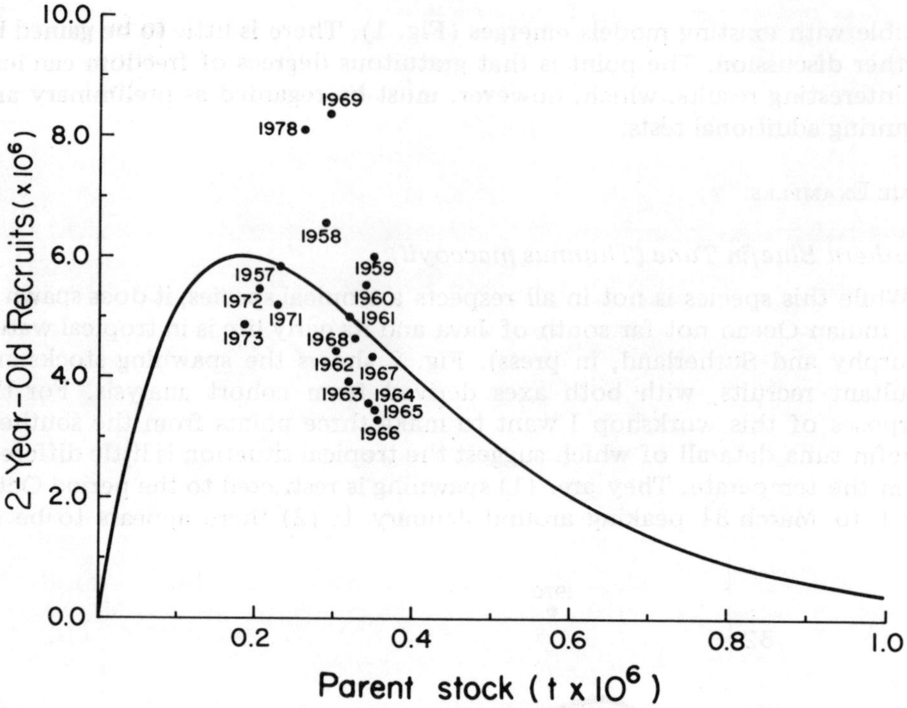

Fig. 2. Ricker curve fitted to the spawning stock biomass and recruits of southern bluefin tuna. Stock and recruit estimates were derived by cohort analysis.

underlying density dependent relationship overlaid by (3) considerable "environmental" variability.

Peruvian Anchovy (Engraulis ringens):

This stock has been the subject of a great deal of study with respect to stock and recruitment. Csirke (1980) has developed a modification of the Ricker model incorporating a concentration term. This term, believed to be related to the environment, modifies the Ricker prediction and seems to account for most of the variation in recruits not accounted for by stock size. The important points with respect to this population are the same as with the bluefin tuna above, i.e., considerable density-dependent control; a short (2 to 3 months) spawning season; and environmental modification of the results to be expected from spawning-stock size. Csirke also notes that there is some suggestion of interaction between the anchovy and the sardine *(Sardinops)* somewhat suggestive of that noted off California by Murphy (1966).

Palau Anchovy (Stolephorus heterolobus):

This species, which is used as tuna bait, was intensively studied by Muller (1976). This is an entirely different fish, having a very short life cycle going from egg to first spawning in 3 months, with a likely life span of about six months. There is considerable spawning year-round, as might be expected. Nevertheless, sharp peaks in egg deposition were observed which might be analogous to the seasonal peaks in longer-lived organisms.

Muller found only weak density dependence when the simple Ricker model was used (r^2, $\log_e$ R/S vs S = .257, P < .05). When he incorporated a rainfall term to the Ricker model, r^2 rose to .56. The *a priori* justification of this is that precipitation dissolves the phosphate rock forming most of the Palau Islands, resulting in a phytoplankton bloom. The addition of a zooplankton standing stock term raised r^2 to .606, a result which is not surprising. Even these unusual little anchovies seem to have stock recruitment characteristics similar to those of related fish in temperate waters (Cushing and Harris 1973).

Yellow Sea Bream (Taius tumifrons):

Stock-recruitment data from trawl fisheries in the East China Sea on the yellow sea bream have been reported and analyzed by Shindo (1960). Relevant material from his paper is shown in Fig. 3. With the exception of the 1949 point, there is a reasonable resemblance to a Ricker curve from 1950 to the end of the data series. The right-hand limb has a gentle slope with considerable scatter, suggesting that density-dependent control is not strong. Recruitment appears to be still increasing with stock around 1/3 of the post-World War 2-level.

The author, however, feels that the true trend is one that might be fitted to the 0, 0, and the '50, '51, and '49 points with the improved recruitment after 1952 due to environmental changes. He bases this on the observation

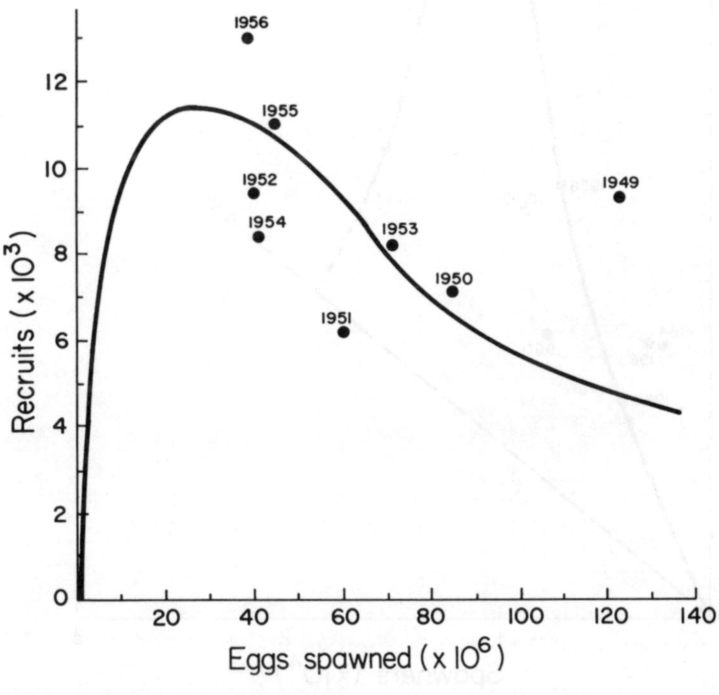

Fig. 3. Stock-recruitment relationship for the yellow sea bream *(Taius tumifrons)* of the South China Sea. Curve fitted by eye.

that during the 1920s the number of recruits was large and independent of stock size, all of which suggests weak density control.

Greater Lizardfish (Saurida tumbil):

This is another species of the East China Sea harvested by trawling, investigated by Saishu and Ikemoto (1970). This stock's density-dependent control appears to be very tight (Fig. 4), with little or no environmental control if allowance is made for measurement errors. Spawning is from June to August, a relatively short period, with recruitment from October to March of the following year.

Coral Reef Fishes:

Weak evidence in the form of some observed spawning around the year had led workers to assume that spawning and recruitment in tropical reef

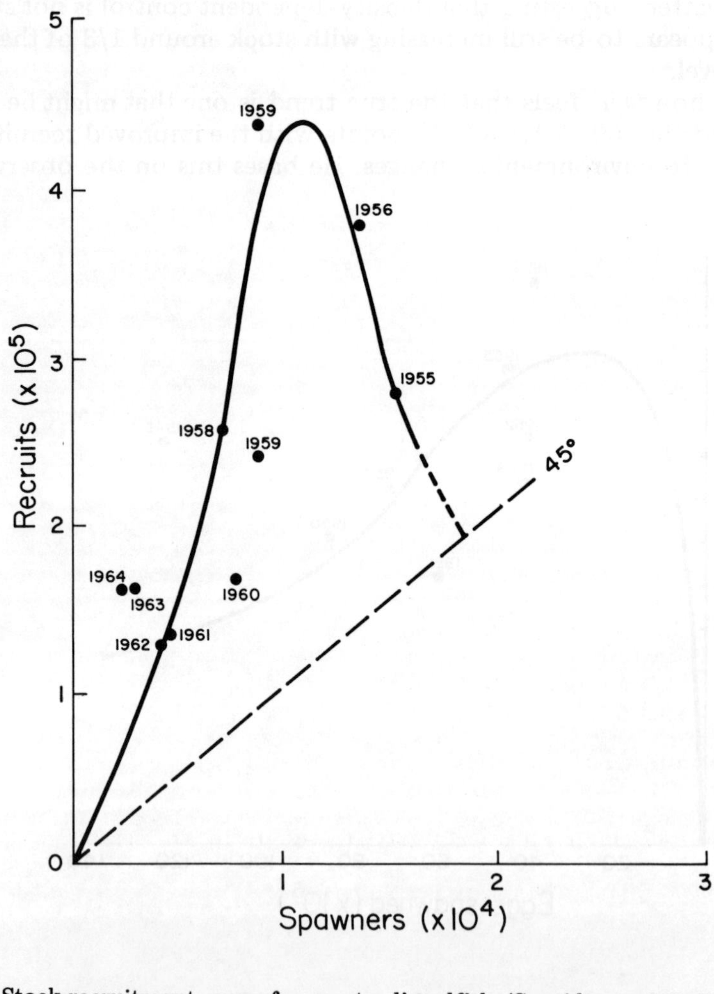

Fig. 4. Stock-recruitment curve for greater lizardfish *(Saurida tumbil)* (from Saishu and Ikemoto 1970).

fishes was a constant process. Accumulating field studies, e.g., Russel et al. (1977) and Luckhurst and Luckhurst (1977), are clearly showing that while some reproduction may take place year-round, there are peak periods similar to those in temperate waters. Recruitment to the reef populations, while usually periodic, is not necessarily at the same period from year to year.

Again, while there are no data on the stock-recruitment relation *per se*, the general pattern seems similar to temperate fishes, with the caveat that the territoriality of many tropical reef fishes imposes another dimension on the recruitment pattern not found in "demersal" and pelagic fishes whether temperate or tropical (Sale 1979).

Discussion

While the number of case histories is limited, they tend to suggest that the reproductive dynamics of most tropical species are similar to those of temperate species. Most are characterized by periodicity in spawning and recruitment, and most seem to have density-dependent reproductive curves. Some authors have improved the fit by incorporating terms thought to represent the effect of the environment. The reproductive curves that appear to apply seem to be of the Ricker type rather than the asymptotic Michaelis-Menten type, in which recruits are independent of stock size over a wide range of stock sizes.

These are important observations, if true, because they suggest that interspecific interactions might be the exception rather than the rule, vastly simplifying the construction of a framework for predicting the effect of intense fishing on the complex assemblage of fishes supporting tropical fisheries. Caution must be exercised however, because in the two best studied pelagic situations in the world, the anchovy-sardine populations off Peru (Csirke 1980) and California (Murphy 1966), there is good direct and indirect evidence of negative interactions between the species.

References

Beverton, R.J.H. and S.J. Holt. 1956. A review of methods for estimating mortality rates in exploited fish populations, with special reference to sources of bias in catch sampling. Rapp. P.-v. Réun. Cons. Int. Explor. Mer. 140: 67-83.

Csirke, J. 1980. Recruitment in the Peruvian anchovy and its dependence on the adult population. Rapp. P.-v. Réun. Cons. Int. Explor. Mer. 177: 307-313.

Cushing, D.H. and J.G.K. Harris. 1973. Stock and recruitment and the problem of density-dependence. Rapp. P.-v. Réun. Cons. Int. Explor. Mer. 164: 142-155.

Luckhurst, B.E. and K. Luckhurst. 1977. Recruitment patterns of coral reef fishes on the fringing reef of Curaçao, Netherlands Antilles. Can. J. Zool. 55: 681-686.

Muller, R.S. 1976. Population biology of *Stolephorus heterolubus* (Pisces: Engraulidae) in Palau, Western Carolines. Univ. of Hawaii, Honolulu. 174 p. Ph.D. dissertation.

Murphy, G.I. 1966. Population biology of the Pacific sardine *(Sardinops caerulea)*. Proc. Calif. Acad. Sci. 34(1): 1-84.

Murphy, G.I. and H. Sutherland. The fishery for southern bluefin tuna (*Thunnus maccoyii* (Castelnau)): catch, mortality rates, stock size, recruitment, and yield per recruit. Aust. J. Mar. Freshwater Res. (In press)

Parrish, B.B., Editor. 1973. Fish stocks and recruitment. Rapp. P.-v. Réun. Cons. Int. Explor. Mer. Vol. 164. 372 p.

Pauly, D. 1979a. Gill size and temperature as governing factors in fish growth: a generalization of the von Bertalanffy growth formula. Ber. Inst. Meereskunde Univ. Kiel. No. 63. 156 p.

Pauly, D. 1979b. Theory and management of tropical multispecies stocks: a review, with emphasis on the Southeast Asian demersal fisheries. ICLARM Studies and Reviews No. 1. 35 p. International Center for Living Aquatic Resources Management, Manila.

Pauly, D. 1980. A new methodology for rapidly acquiring basic information on tropical fish stocks: growth, mortality and stock recruitment relationships, p. 154-172. *In* S.B. Saila and P.M. Roedel (eds.) Stock assessment for tropical small-scale fisheries. International Center for Marine Resource Development, Univ. of Rhode Island, Kingston.

Russell, B.C., G.R.V. Anderson and F.H. Talbot. 1977. Seasonality and recruitment of coral reef fishes. Aust. J. Mar. Freshwater Res. 28: 521-528.

Saishu, K. and R. Ikemoto. 1970. Reproductive curve of the lizard-fish, *Saurida tumbil* (Bloch), of the East China Sea. Group. Bull. Seikai Reg. Fish. Res. Lab. 38: 41-60.

Sale, P.F. 1979. Recruitment, loss and co-existence in a guild of territorial reef fishes. Oecologia 42: 159-178.

Shapovalov, L. and A.C. Taft. 1954. The life histories of the steelhead rainbow trout *(Salmo gairdneri gairdneri)* and silver salmon *(Oncorhynchus kisutch)* with special reference to Waddell Creek, California, and recommendations regarding their management. Fish. Bulletin No. 98. 375 p. Dept. of Fish and Game, Sacramento.

Shepherd, J.G. and D.H. Cushing. 1980. A mechanism for density-dependent survival of larval fish as the basis of a stock-recruitment relationship. J. Cons. Int. Explor. Mer. 39(2): 160-167.

Shindo, S. 1960. Studies on the stock of yellow sea bream in the East China Sea. Bull. Seikai Reg. Fish. Res. Lab. 20: 1-198.

Discussion of Dr. Murphy's Paper

There was general agreement that stock-recruitment relationships in tropical fishes may be, on the evidence presented here and otherwise available, similar to those reported from temperate waters. Until shown otherwise, the safest operating hypothesis is that this is so.

The evidence of reasonably tight density dependence in the tropics suggests that competitive interactions among species may not be exercising much control, at least most of the time. It was argued that competition is wasteful of scarce resources and that evolutionary pressure should act to minimize competition.

Usefulness and Applicability of Sensitivity Analysis in a Multispecies Approach to Fisheries Management

JACEK MAJKOWSKI
Division of Fisheries Research
CSIRO Marine Laboratories
P.O. Box 21, Cronulla
N.S.W. 2230, Australia

MAJKOWSKI, J. 1982. Usefulness and applicability of sensitivity analysis in a multispecies approach to fisheries management, p. 149-165. *In* Pauly, D. and G.I. Murphy (eds.) Theory and management of tropical fisheries. ICLARM Conference Proceedings 9, 360 p. International Center for Living Aquatic Resources Management, Manila, Philippines and Division of Fisheries Research, Commonwealth Scientific and Industrial Research Organisation, Cronulla, Australia.

Abstract

This paper demonstrates how sensitivity analysis and its extensions can be helpful in overcoming, at least partly, the difficulties arising in the parameter determination and validation of complex multispecies fisheries models or procedures. Sensitivity analysis can be used for: (i) the so-called internal model validation (i.e., determination whether the levels of uncertainties in the estimated input parameters are acceptable for modelling purposes or not), (ii) estimating the relative contribution of uncertainty in each input parameter to the model output uncertainty, and (iii) determining the levels of input parameter uncertainties which would lead to acceptable model results. Implications of the information derived from (ii) and (iii) for designing a research program are discussed. It is outlined how unnecessary mechanisms included in a model can be identified with the aid of sensitivity analysis.

The theory and applicability of the following procedures are described: (i) ordinary sensitivity analysis (dealing with non-simultaneous changes in input parameters), (ii) extended deterministic sensitivity analysis (dealing with simultaneous changes in input parameters), and (iii) stochastic extensions to sensitivity analysis: (a) additive sensitivity analysis (errors in input parameters assumed to have normal distributions), (b) multiplicative sensitivity analysis (errors in input parameters assumed to have lognormal distributions), and (c) other extensions based exclusively on the Monte-Carlo simulations. These procedures are illustrated by application examples from the field of biological sciences. The potential applicability and usefulness of sensitivity analysis in a multispecies approach to management of tropical fish stocks are discussed. A simple example of application of sensitivity analysis to a method used frequently for estimating the instantaneous rate of total mortality for tropical stocks is made.

Introduction

Mathematical models and procedures used in multispecies approaches to management of fish stocks are usually more complex than those used in single-species approaches (for examples of models used in multispecies approaches see Parrish 1975; Mandecki 1976; Andersen and Ursin 1977; Majkowski 1977; Lett 1978; for an example of procedure see Majkowski 1981). The above-mentioned generalization is valid if multi- and single-species models which aim to describe fish stocks at approximately the same level of complexity are compared. The greater complexity of multispecies models in comparison to the single-species models is a consequence of accounting for interactions among species and results in a large number of input parameters for these models.

The large number of input parameters required for multispecies models causes difficulties in parameter determination and validation of these models becomes extremely difficult. This is the reason why the multispecies approach is rarely applied in practice (Larkin 1978) and its outcome rarely considered seriously. The objective of this paper is to demonstrate how sensitivity analysis and its extensions can be helpful in eliminating, at least partly, these difficulties.

Development of a simulation model of any system should be an iterative process consisting of the following two steps, both of which should be repeated as many times as necessary (Majkowski et al. 1981):

— formulation or modification of the model, and
— validation.

Validation involves the use of procedures which enable us to determine whether improvement in the model is necessary or not.

The formulation of the model can be subject to several sources of weakness (Majkowski et al. 1981):

(i) omitted "compartments (components) of the system" under consideration,
(ii) unsuitable state variables used in defining the system,
(iii) omitted mechanisms (missing pathways or interactions),
(iv) unnecessary mechanisms included,
(v) incorrect functional form of relationships used in the model, and
(vi) incorrect parameter values (due to measurement error or natural variability).

This paper concentrates on item (vi) although the first five items can also be important. It shows how sensitivity analysis and its extensions can be used for (i) the so-called internal model validation (i.e., determining whether the levels of uncertainties in the estimated input parameters are acceptable for the modelling purpose or not), (ii) estimating the relative contribution of uncertainty in each input parameter to the total uncertainty in the model output, and (iii) determining the levels of uncertainties in the input parameters which would lead to acceptable model results. Implications of the information derived in (ii) and (iii) for designing a research program are discussed. It is demonstrated that identification of the input parameters to which changes the model is most sensitive leads not only to a better understanding of the system but also eliminates unnecessary empirical and theoretical investigations concerning non-critical parameters. Additionally, it is outlined how sensitivity analysis can be helpful in identification of mechanisms unnecessarily included in a model.

This paper aims to emphasize the benefits from the use of sensitivity analysis in the multispecies approach to fisheries management, but this technique is applicable to any quantitative approach in biological sciences. The Appendix provides an example of the application of sensitivity analysis to a single-species method used frequently for estimating the instantaneous rate of total mortality in tropical fish stocks.

Methodology

Consider any deterministic simulation model or any mathematical deterministic procedure, the operation of which is specified by a number of input parameters (constants), p_i ($i = 1, \ldots, m$). These include all initial and boundary conditions, operating characteristics, etc. The output values of the model or procedure which are of particular interest are designated by X_j ($j = 1, \ldots, n$). It is assumed that the number of these values is finite (i.e., $n < \infty$). If the model or procedure provides an output in the form of continuous functions, for example, functions of time ($x_i(t)$, $i = 1, \ldots, k$), the values of X_j can be constructed, for instance, as

$$X_j = x_i\,(t_\mu) \qquad\qquad j = 1, \ldots, \tilde{\mu} \times K$$

where t_μ ($\mu = 1, \ldots, \tilde{\mu}$) are time points of particular interest. Alternatively, the values of X_j can be defined as

$$X_j = \int_{t = t_o}^{t = t_e} x_j\,(t)\,dt \qquad\qquad j = 1, \ldots, k$$

where $[t_o, t_e]$ is the time interval of particular interest. The choice of X_j in each specific case depends on the objectives of modelling and the structure of the model. It is assumed that the values of X_j are completely determined by the values of p_i and the mathematical structure of the model.

The best estimates of p_i are denoted by p_i^o and the corresponding model predictions by X_j^o. These values will be referred to as the reference case.

Ordinary Sensitivity Analysis

The objective of this analysis is the identification of input parameters which, when changed by a fixed percentage, produce either a strong or a weak effect on the model output. Only one parameter is changed in a single implementation of the model. The relative change in p_i is denoted by u_i.

The effects of such changes on the model performance can be determined in various ways. As one possibility, a scalar or vector function, D, might be chosen which measures differences between the perturbed (i.e., when an input parameter has been changed) and unperturbed (i.e., associated with the best estimates of the input parameters) results from the model (Miller 1974a, 1974b; Miller and Bramall 1976; Miller et al. 1976; Majkowski and Bramall 1980; Majkowski and Waiwood 1980, 1981; Majkowski et al. 1981). The values of this function should correlate well with intuitive estimates of the changes in the model performance. Examples of D-measures which are simple and convenient to interpret are presented in Table 1.

It is good practice to perform ordinary sensitivity analysis for various values of u_i from a certain u-interval $[-\beta, \beta]$ where the value of β should exceed the magnitude of relative errors in the most uncertain input parameter (see Majkowski and Bramall 1980; Majkowski and Waiwood 1980, 1981). It is difficult to reach conclusions from ordinary sensitivity analysis if the D-measure is strongly non-linear in respect to u_i $(i = 1, \ldots, m)$. However, such a situation does not occur often if relative estimation errors are reasonably small. The non-linearity can be overcome in some cases by a change in the definition of the D-measure.

Usually, all input parameters have errors which may reinforce or counteract one another. Therefore, ordinary sensitivity analysis provides only elementary information regarding the influence of parameter errors on the model predictions. Additional information can be obtained by simultaneous changes in the input parameters.

Extended Deterministic Sensitivity Analysis

Input parameters, p_i, are simultaneously changed by various relative magnitudes u_i. If all u_i are small, it happens frequently that the response of the model output to these changes is approximately linear. This happens even if the model output is not a linear function in respect to the input parameters.

The response of the model output to changes in the input parameters within the linearity region can be predicted with the aid of the following formulae (Tomovic 1963; Waide and Webster 1976):

$$X_j \cong X_j (u_i \underset{i}{=} 0) + \sum_{i=1}^{m} R'_{ji} u_i \tag{1}$$

where

$$R'_{ji} = \left. \frac{\partial X_j}{\partial u_i} \right|_{u_i \underset{i}{=} 0} \tag{2}$$

or

$$D_j \cong D_j (u_i \underset{i}{=} 0) + \sum_{i=1}^{m} R_{ji} u_i \tag{3}$$

where

$$R_{ji} = \left. \frac{\partial D_j}{\partial u_i} \right|_{u_i \underset{i}{=} 0}$$

It should be noted that each of these formulae can be valid for a different u-interval.

The values of R'_{ji} and R_{ji} can be determined in a numerical way even when the analytical relationships between, for example, X_j and u_j are unknown. Formulae (1) and (3) are particular cases of the well-known first-order Taylor expansion formula. These formulae are satisfactory for the most practical cases, but higher-order Taylor expansions can be used if necessary.

If the model response to changes in the input parameters of interest is linear, it is much easier from the computational point of view to examine this response on the basis of formulae (1) and (3) or any higher-order Taylor expansion than in the process of direct model implementations.

Table 1. Examples of D-measures.

$D = \sum_{j=1}^{n} (X_j - X_j^o)$	$D_j = X_j - X_j^o$	$j = 1, \ldots, n$
$D = \sum_{j=1}^{n} \frac{X_j - X_j^o}{X_j^o}$	$D_j = \frac{X_j - X_j}{X_j^o}$	$j = 1, \ldots, n$
$D = \frac{\sum_{j=1}^{n} X_j - \sum_{j=1}^{n} X_j^o}{\sum_{j=1}^{n} X_j^o}$		
$D = \prod_{j=1}^{n} \frac{X_j}{X_j^o}$	$D_j = \frac{X_j}{X_j^o}$	$j = 1, \ldots, n$

Formulae (1) and (3) also provide some insight into the interpretation of the results of ordinary sensitivity analysis. If we are interested in the model sensitivity to changes in the input parameters and the model response is linear within this perturbation interval, the coefficients R_{ji} or R'_{ji} fully describe the model responses to the changes in the input parameters. These coefficients, referred to frequently as sensitivity coefficients, can be calculated on the basis of ordinary sensitivity analysis. For example, if β is greater than 0.01, these coefficients can be expressed as:

$$R'_{ji} = 10^2 \times [X_j (u_i = 0.01, u_{k \neq i} = 0) - X_j (u_k \underset{\overline{k}}{=} 0)]$$

and

$$R_{ji} = 10^2 \times [D_j(u_i = 0.01, u_{k \neq i} = 0) - D_j (u_k \underset{\overline{k}}{=} 0)].$$

The input parameters which have small associated sensitivity coefficients in comparison to those for the remaining input parameters are interpreted as being not crucial; reduction in their uncertainties is less important. However, this statement needs qualification because the levels of acceptable uncertainties in these input parameters were not specified. The matter will be considered further in the section devoted to a more powerful procedure, extended stochastic sensitivity analysis.

If the response of the model to changes in the input parameters is not linear, the interpretation of the results of ordinary sensitivity analysis becomes more complex. For example, let us assume that the model response can be described by the second-order Taylor expansion:

$$D_j \cong D_j\,(u_i \underset{i}{=} 0) + \sum_{i=1}^{m} R_{ji}u_i + 0.5 \times \sum_{i=1}^{m} R_{jik}\,u_i u_k$$

where

$$R_{jik} = \left.\frac{\partial^2 D_j}{\partial u_i\,\partial u_k}\right|_{\substack{u_i \underset{i}{=} 0 \\ u_k \underset{k}{=} 0}}$$

and R_{ji} are the same as defined in formula (2). It can be noticed that even if R_{ji} (j = 1, . . . , n) are extremely small in comparison to the remaining first-order sensitivity coefficients, it does not mean that the i-th input parameter is not crucial in the sense described above.

Extended Stochastic Sensitivity Analysis

The deterministic values of parameter errors are never known, so extended deterministic sensitivity analysis does not provide much more useful information than does ordinary sensitivity analysis. The best obtainable information on the errors in the input parameters is of a stochastic nature.

The probability distributions of errors in the input parameters are rarely well known for complex population or ecosystem models. However, the ranges of input parameters possible from the biological point of view can usually be found from the scientific literature or by performing appropriate laboratory or field studies. In such a case, one way of assigning a probability distribution to the input parameter is to apply a triangular distribution for this parameter, using the values accepted as the minimum and maximum possible limits for this parameter (Tiwari and Hobbie 1976).

Extended stochastic sensitivity analysis can be carried out with the aid of the Monte-Carlo method or first-order theories (see sections below). There is no preference from the strictly Monte-Carlo simulation point of view in the choice of the types of distributions for the input parameters. However, the Monte-Carlo simulations are usually extremely expensive, especially if the model is complex. Thus, full formulation of recommendations for further studies is nearly impossible on the basis of these simulations (see below). These difficulties can be overcome if a linear theory is applied in conjunction with the Monte-Carlo simulations. The linear theories are at present developed only for the cases where the errors in all input parameters have either normal or lognormal distributions. Therefore, one of these two types of distributions should be preferably used in extended stochastic sensitivity analysis as an approximation of the error distributions. Fortunately, these two types of distributions cover the most frequently used error specifications for large-scale biological models: additive and multiplicative.

If a certain quantity p is measured and the relative errors are small, for instance 10 to 20%, it is usually acceptable for modelling purposes to assume the errors to have a symmetrical normal distribution with a mean of zero. This is traditionally written as $p^o \pm e$ (where p_i^o is the best estimate of the quantity p and e is the standard error), meaning that the probability that the real value of p is in the interval $[p^o - e, p^o + e]$ is approximately 68%. This type of error specification is frequently referred to as additive. Extended sensitivity analysis associated with this error specification is referred to as additive sensitivity analysis.

If the relative error in p is large, it is often specified in a multiplicative way (Aitchison and Brown 1957; Meyer 1971, 1972). For instance, the value of p might be expressed as known "within a factor of two". In this case, the real value of p is expected to be within the interval $[p^o/2, 2p^o]$. The distribution of errors specified in a multiplicative way can be usually approximated in a satisfactory way by the lognormal distribution (Aitchison and Brown 1957; Meyer 1971). Extended sensitivity analysis associated with this error specification is referred to as multiplicative sensitivity analysis.

Monte-Carlo simulations

Sensitivity analysis implemented with the aid of the Monte-Carlo method consists of performing repeated simulations, each time drawing random values for the input parameters from their prespecified distributions and calculating the resulting values of D_j. When the number of model implementations (runs) is large enough, not only the mean and the variance of D_j-values can be accurately estimated but also the collection of D_j-values can be examined to get a better idea of the type of their distributions (for detailed procedures see Miller 1974a; Miller et al. 1976; O'Neill and Gardner 1979; Gardner et al. 1980a; O'Neill et al. 1980; Majkowski et al. 1981).

First-order theories

Additive sensitivity analysis. The first-order theory associated with additive sensitivity analysis is based on two assumptions: (i) first-order expansion (formula 3) is valid within the specified error intervals and (ii) all input parameters have normal distributions (this assumption corresponds to the additive error specification; see Miller 1974a, 1974b; Miller and Bramall 1976; Miller et al. 1976; Majkowski and Waiwood 1981).

Let us assume that the D_j-measure is chosen to vanish if $u_i = 0$ $(i = 1, \ldots, m)$. Thus, the variance of D_j can be calculated on the basis of the following formula (Miller et al. 1976):

$$\text{var}(D_j) \cong \sum_{i=1}^{m} R_{ji}^2 \sigma_i^2 + 2 \times \sum_{i<k} R_{ji} R_{jk} \rho_{ik} \sigma_j \sigma_i \tag{4}$$

where R_{ji} are defined in (4), σ_i is the coefficient of variation of p_i ($\sigma_i = e_i/p_i^o$, where e_i is the standard error of p_i), and ρ_{ik} is the correlation coefficient between p_i and p_k.

The validity of this first-order theory can be examined in three steps (Miller 1974a, 1974b; Miller et al. 1976):

— testing formula (3) within ordinary sensitivity analysis,
— testing formula (3) within the Monte-Carlo simulations, and
— comparing the values of mean(D_j) and var(D_j) derived with the aid of formula (4) with those obtained on the basis of the Monte-Carlo simulations. (The distribution of D-values from the Monte-Carlo simulation can be also tested for normality.)

Multiplicative sensitivity analysis. The multiplicative error specification necessitates a refinement of the additive first-order theory in which each additive formulation should be replaced by its multiplicative analogue (Majkowski et al. 1981). Thus, it is convenient to define a new set of random variables, $f_i (i = 1, \ldots, m)$, which affect the values of p^o in a multiplicative fashion:

$$p_i = p_i^o f_i.$$

The first-order expansion suitable for multiplicative sensitivity analysis takes the following form (Majkowski et al. 1981):

$$\ell n D_j (z_i) = \ell n D_j (z_i \underset{i}{=} 0) + \sum_{i=1}^{m} \overline{R}_{ji} z_i \tag{5}$$

where

$$\overline{R}_{ji} = \left. \frac{\partial \ell n D_j}{\partial z_i} \right|_{z_i \underset{i}{=} 0}$$

and

$$z_i = \ell n f_i .$$

It is recommended that, for the sake of simplicity, a D-measure be chosen which satisfies the condition:

$$D_j (z_i \underset{i}{=} 0) = 1 \qquad j = 1, \ldots, n.$$

It can be concluded (Majkowski et al. 1981) that $\ell n D_j(z)$ has, under these assumptions, a normal distribution with a mean equal to zero and

$$\text{var} (\ell n D_j) = \sum_{i=1}^{m} \overline{R}_{ji} \overline{\sigma}_i^2 + 2 \times \sum_{i<k} \overline{R}_{ji} \overline{R}_{jk} \overline{\rho}_{ik} \overline{\sigma}_i \overline{\sigma}_k \tag{6}$$

where $\overline{\sigma}_i^2$ is equal to $(\ell n \epsilon_i)^2/6$ and $\overline{\rho}_{ik}$ is the correlation coefficient between z_i and z_k. As a consequence, $D_j(z)$ has a lognormal distribution [Aitchison and Brown (1957)].

The validity of formula (5) should be examined in a way corresponding to that for additive sensitivity analysis (for details see Majkowski et al. 1981).

Internal model validation

It was demonstrated in the previous sections how the uncertainty in the model output can be determined (visual observation of the shape of D_j-distributions resultant from the Monte-Carlo simulations, calculation of the mean and variance of D_j on the basis of the Monte-Carlo simulations and a first-order theory if applicable). At this stage, the decision has to be made by the model users whether or not this uncertainty is acceptable for modelling purposes. One way of approaching this matter is the choice of an acceptable D-interval. Since the D-measure should have a simple interpretation, this does not usually produce difficulties. Implementation of random input parameter perturbations and observation of corresponding model outputs (e.g., graphs) can be helpful in determining the acceptable D-measure interval. Then, it should be investigated whether or not the values of the D-measure, preferably derived in the process of Monte-Carlo simulations, lie within the acceptable range (for detailed procedures see Miller 1974a, 1974b; Miller et al. 1976; Majkowski et al. 1981).

If the model is implemented for the prediction of time-dependence of certain output variables, it can be expected that uncertainties in these variables will change with time. It can be examined what is the time interval within which the uncertainty in the model prediction is acceptable in the sense described above (for detailed procedure see Miller 1974b).

Formulation of recommendations for further study

If a preliminary model needs to be improved, it is very helpful to identify directions in which further empirical and theoretical investigations will be most effective. The relative importance of each of the uncertainties in the input parameters can be determined on the basis of relevant first-order theory if one of the two available theories is applicable for the system under consideration. According to these theories, the variance of D_j in the additive case is expressed by formula (4) whereas in the multiplicative case the variance of ℓnD_j is determined by formula (6). By comparing the magnitudes of the components of an appropriate sum we can discover which particular input uncertainties contribute most to the total variance and thus identify the parameters which require more accurate determination. The accuracy required for these paramet :s can be estimated on the basis of formula (4) or (6) and the range of acceptability for the output uncertainty established for internal model validation (for detailed procedures see Miller et al. 1976; Majkowski et al. 1981).

Certain R_{ji} coefficients may be equal to zero (or so small as to be effectively zero). If the linear theory is valid, this means that changes in the parameters corresponding to these coefficients do not effect D_j.

If all input parameters related to a particular mechanism in the system have sensitivity coefficients which are effectively equal to zero and if the D-measure truly reflects the objectives for modelling the system, it can be presumed in some cases that it is not necessary to include that mechanism in the model. Caution must be exercised in reaching this conclusion. For example, in some cases it might be found that the functional form of some relationships in the model is actually incorrect. Re-definition of these relationships could result in non-zero values being found for the sensitivity coefficients associated with the parameters which determine the mechanism.

If neither of the two first-order theories is applicable to the system being studied and only the Monte-Carlo simulation can be performed, the acceptable levels of uncertainties in the input parameters can be determined by performing Monte-Carlo simulations with decreased levels of uncertainties in these parameters. This method can be very expensive since many possibilities usually have to be tested.

Examples of Application

Ordinary and Extended Deterministic Sensitivity Analysis

This kind of sensitivity analysis is the one most frequently applied. The frequent use of this technique follows not from its particular usefulness but from its simplicity. The reader can encounter applications of these procedures in almost all fields of mathematical modelling. Examples relevant to the field of biological sciences are listed in Table 2. The use of this technique in fisheries sciences is very limited; only one out of six papers encountered on this subject (Mandecki 1976; Orth 1979; Rivard and Doubleday 1979a, 1979b, 1980; Moreau 1980) may be of particular interest for the reader since it deals with multispecies interactions. This paper (Mandecki 1976) concerns a mathematical model of three interacting fish populations (cod,

Table 2. Examples of ordinary sensitivity analysis relevant to biological sciences.

Reference	System examined
Brylinsky (1972)	Marine ecosystem (English Channel), energy flow
O'Brien and Wroblewski (1976)	Lower marine trophic levels off the coast of West Florida
Kitchell and Stewart (1977)	Growth of yellow perch and walleye, energy flow
Mandecki (1976)	Cod, herring, and sprat populations in the Baltic Sea
Kremer and Nixon (1978)	Coastal marine ecosystem (Narragansett Bay)
Klekowski and Majkowski (1980)	Growth of the crustacean *Simocephalus vetulus*, energy flow
Orth (1979)	Largemouth bass population in the Lake Carl Blackwell
Rose and Harmsen (1978)	Forest tent caterpillar population
Rivard and Doubleday (1979a)	Fish cohort analysis
Rivard and Doubleday (1979b, 1980)	Description of computer programs associated sensitivity analysis of procedures for stock assessment
Majkowski et al. (1980)	Growth of rotifer *Brachionus rubens*
Majkowski and Bramall (1980)	Growth of an animal, energy flow
Majkowski and Waiwood (1980, 1981)	Food consumption by a cod population
Moreau (1980)	Fish cohort, yield per recruit analysis

herring and sprat) in the Baltic Sea. In order to account for the proper diet composition for these species, the food biomass other than above-mentioned fish was introduced to the model. Sensitivity analysis for this model was limited to the input parameters considered as highly uncertain. Most of these parameter values were found by adjusting them to obtain an agreement between the catches obtained on the basis of the model and those observed. It was found that the results of the model (the catches of three fish species were examined) were extremely sensitive to changes in the two parameters determining size preferences in cod predation, especially in the parameter determining the optimal size of prey from the cod point of view. The second of these important parameters specified the range of prey sizes acceptable by cod. Parameters defining species preferences in cod predation (resultant from different distributions of species, ability to escape, etc.) do not seem to be crucial with the exception of the one determining the preference for sprat. The catch of each species is also sensitive to changes in the initial level of recruitment of the species (parental stock-independent recruitment, constant for cod and increasing linearly with time due to improvement in feeding conditions and then constant for herring and sprat was assumed). The results of the model were insensitive to changes in the parameters characterizing the time-dependence of the number of particles of "the other food" (this dependence had the same form as those for herring and sprat recruitment).

Mandecki (1976) generalized these results, concluding that predator populations should be more carefully studied than prey populations for any predator-prey fisheries system. Caution should be exercised in following this conclusion because it is based on the examination of only one system of such a kind and a theoretical support for it is not evident. It should be realized that such generalizations can be valid sometimes, but a number of systems of similar nature should be examined before their formulation and the conclusion should be supported by a general theoretical justification. An example of such a generalization of sensitivity analysis results, for bioenergetic growth models of animals, can be found in Majkowski and Bramall (1980). If such a generalization is valid, it can be of a great importance for designing a research program.

Extended Stochastic Sensitivity Analysis

This kind of powerful analysis is rarely used in modelling approaches, especially in biological sciences. The reason for this follows from the impression of relatively high degree of complexity of the method and its heavy use of computer time. The irrationality of this impression will be discussed in the next section.

Examples of sensitivity analysis limited to the Monte-Carlo simulations are listed in Table 3. In these papers, the errors in the input parameters are assumed to have triangular or normal distributions.

The papers listed in Table 4 provide examples of additive sensitivity analysis relevant to biological sciences. Implementation by Mohn (1979) of this analysis is particularly interesting since it concerns a model of the exploitation of a harp seal population. This sensitivity analysis is limited to the implementation of the first-order theory associated with additive sensitivity analysis. The author estimated standard error for the predicted size

Table 3. Examples of extended stochastic sensitivity analysis limited to the Monte-Carlo simulations relevant to biological sciences.

Reference	System examined
Gardner et al. (1980b)	Prey-predators population interactions
O'Neill et al. (1980)	Herbivores-carnivores population interactions
Gardner and Mankin (In press)	Woodland ecosystem
Garten et al. (1978)	Forest ecosystem, plutonium ecosystem
Gardner et al. (1980a)	Carbon flow through an ecosystem

Table 4. Examples of additive sensitivity analysis relevant to biological sciences.

Reference	System examined
Miller (1974a, b) Miller and Bramall (1976)	Mosquito population
Miller (1974b) Miller and Bramall (1976) Miller et al. 1976	Pollutant (mercury) transport
Miller and Bramall (1976)	Radioactive particles in a lung; Iodine-131 transfer to man
Furness (1978)	Seabird communities
Mohn (1979)	Seal population
Majkowski and Waiwood (1981)	Food consumption by a cod population

of the total and breeding stock to be equal to 81% and 47%, respectively, for the standard errors in the input parameters (10 parameters) in the range 5 to 20%. Unfortunately, the relative contributions of the input uncertainties to the output uncertainty were not evaluated. Examples of the full implementation of additive sensitivity can be found in Miller (1974a) and Miller et al. (1976). In both cases it is demonstrated that only a few input parameters contribute significantly to the total output variance. Other examples of additive sensitivity analysis support this. Thus, in all these cases, the results of additive sensitivity analysis are extremely useful for the proper allocation of research effort.

An example of application of multiplicative sensitivity analysis to a model of global transport of mercury is presented in Majkowski et al. (1981).

Potential Applicability for Fisheries Multispecies Models

Sensitivity analysis and its extensions can be regarded as a technique still in an early stage of development. This is evident from the dates of publications of the papers concerning this subject. Nevertheless, this technique is very powerful and potentially useful when applied to the design of a multispecies program and to validating its outcome; it aims to overcome critical difficulties arising in practical applications of the multispecies approach.

It may be argued that application of this technique, especially in the case of complex multispecies systems, can be expensive from the computational point of view. If it is believed that the multispecies approach is essential, the cost of implementing extended stochastic sensitivity analysis, even if high, is negligible in comparison with the benefits gained from the application of this technique. This could become clear if the reduction of the expenses due to proper allocation of effort in collecting field data were evaluated, especially, if this allocation is made at an early stage of the research program. Usually the parameter values are not accurately known at this stage but multiplicative sensitivity analysis can handle many such cases. Simulation modelling with the aid of this procedure can often provide very useful information, even in the preliminary stage of a research program.

It should be emphasized that the implementation of the first-order theory associated with relevant extended stochastic sensitivity analysis (without performing the Monte-Carlo simulations) can provide information expected to be valid if the preliminary condition (formulae (3) or (5)) of the validity of this theory is tested within ordinary sensitivity analysis (for examples of such an analysis see Mohn (1979) and Majkowski and Waiwood (1981)). Thus, extended stochastic sensitivity analysis can be carried out for complex models in some cases, even if computer time is very limited.

The majority of applications of extended stochastic sensitivity analysis were carried out assuming that the input parameters are not cross-correlated. Correlations in the input parameters can be caused by environmental effects and estimation of a few parameters on the basis of the same set of empirical data. Since modellers usually try to take important environmental effects into account in the structure of models and each input parameter is frequently estimated on the basis of independent data, it can possibly be argued (Miller 1974a; Miller et al. 1976) that such correlations are not likely to be a particularly serious problem in many cases. However, it should be emphasized that extended stochastic sensitivity analysis can handle these correlations if they are known.

Following a request by Workshop participants, a simple example of application of sensitivity analysis to a method used frequently for estimating the instantaneous rate of total mortality of tropical stocks was made. The application, given in the Appendix, shows that the total mortality method in question can produce highly uncertain results.

Acknowledgments

The author appreciates valuable comments by Dr. Garth I. Murphy (Division of Fisheries Research, CSIRO Marine Laboratories) and Mr. Dennis Reid (Division of Mathematics and Statistics, CSIRO) on a draft of this paper.

Dr. Daniel Pauly (ICLARM) suggested illustrating the implementation of sensitivity analysis using formula (7) in the Appendix, as an example of a relationship frequently used by biologists assessing tropical stocks. The associated calculations were made by Mr. John Hampton (Division of Fisheries Research, CSIRO Marine Laboratories) to whom the author is sincerely grateful.

References

Aitchison, J. and J.A.C. Brown. 1957. The lognormal distribution with special reference to its uses in economics. Cambridge University Press, Cambridge.

Andersen, K.P. and E. Ursin. 1977. A multispecies extension to the Beverton and Holt theory of fishing with accounts of phosphorus circulation and primary production. Meddr. Danm. Fish.-og Havunders. N.S. 7: 319-435.

Brylinsky, W. 1972. Steady-state sensitivity analysis of energy flow in a marine ecosystem, p. 81-102. *In* B.C. Patten (ed.) System analysis and simulation in ecology. Vol. II. Academic Press, New York.

Furness, R.W. 1978. Energy requirements of seabird communities: a bioenergetics model. J. Anim. Ecol. 47: 39-53.

Gardner, R.M. and J.B. Mankin. Analysis of compartment models as representations of woodland ecosystems. *In* D.E. Reichle (ed.) Dynamic properties of forest ecosystems. Cambridge University Press, Cambridge. (In press)

Gardner, R.M., R.V. O'Neill, J.B. Mankin and W.R. Emanuel. 1980a. A comparison of three carbon models. Ecol. Modelling 8: 313-332.

Gardner, R.M., R.V. O'Neill, J.B. Mankin and D. Kumar. 1980b. Comparative error analysis of six predator-prey models. Ecol. 61: 323-332.

Garten, C.T., Jr., R.M. Gardner and R.C. Dahlman. 1978. A compartment model of plutonium dynamics in a deciduous forest ecosystem. Health Phys. 34: 611-619.

Kitchell, J.F. and D.J. Stewart. 1977. Applications of a bioenergetics model to yellow perch *(Perca flavescens)* and walleye *(Stizostedion vitreum vitreum)*. J. Fish. Res. Board Can. 34: 1922-1935.

Klekowski, R.Z. and J. Majkowski. 1980. Sensitivity of the model of energy flow through a female *Simocephalus vetulus* (O.F. Müller) to changes in the parameters of energy budget. Pol. Ecol. Stud. 6: 55-60.

Kremer, J.N. and S.W. Nixon. 1978. A coastal marine ecosystem: simulation and analysis. Springer-Verlag, Berlin.

Larkin, P.A. 1978. Fisheries management—an essay for ecologists. Annu. Rev. Ecol. Syst. 9: 57-73.

Lett, P.F. 1978. A comparative study of the recruitment mechanisms of cod, and mackerel, their interaction, and its implication for dual stock management. Dalhousie University, Halifax, Nova Scotia, Canada. Ph.D. thesis.

Majkowski, J. 1977. Prognosis for catches of cod, herring and sprat in the Baltic Sea and their optimization with the aid of a mathematical model. Pol. Arch. Hydrobiol. 24: 361-382.

Majkowski, J. 1981. Application of a multi-species approach for assessing the population abundance and the age-structure of fish stocks. Can. J. Fish. Aquat. Sci. 38: 421-431.

Majkowski, J. and L. Bramall. 1980. Sensitivity of bioenergetic growth models of animals to changes in the energy balance parameters. J. Theor. Biol. 85: 643-655.

Majkowski, J. and K.G. Waiwood. 1980. Eco-physiological approach for estimating the food consumption of cod and the maximum possible contribution of various fish species in the diet of cod inhabiting the southern Gulf of St. Lawrence. Canadian Atlantic Fisheries Advisory Committee, Research Document 80/23.

Majkowski, J. and K.G. Waiwood. 1981. A procedure for evaluating the food biomass consumed by a fish population. Can. J. Fish. Aquat. Sci. 38: 1199-1208.

Majkowski, J., J. Pilarska and R.Z. Klekowski. 1980. Simulations of energy flow through the amictic female rotifer *Brachious rubens* Ehrbg. Can. J. Fish. Aquat. Sci. 37: 97-110.

Majkowski, J., J.M. Ridgeway and D.R. Miller. 1981. Multiplicative sensitivity analysis and its role in development of simulation models. Ecol. Modelling 12: 191-208.

Mandecki, W. 1976. A computer simulation model for interacting fish population studies, applied to the Baltic. Pol. Arch. Hydrobiol. 23: 281-307.

Meyer, C.F. 1971. Using experimental models to guide data gathering. J. Hydrobiol. Div. Am. Soc. Civil Engrs. HY10: 1681-1691.

Meyer, C.F. 1972. Surrogate modelling. Water Resour. Res. 8: 212-216.

Miller, D.R. 1974a. Sensitivity analysis and validation of simulation models. J. Theor. Biol. 48: 345-360.

Miller, D.R. 1974b. Model validation through sensitivity analysis, p. 911-913. *In* Proceedings of Summer Computer and Simulation Conference, July 1974, Houston, Texas.

Miller, D.R. and L. Bramall. 1976. Validation of ecological models, p. 911-913. *In* Proceedings of Computer Simulation Conference, July 1976, Washington, D.C.

Miller, D.R., G. Butler and L. Bramall. 1976. Validation of ecological system models. J. Environ. Manage. 4: 383-401.

Mohn, R.K. 1979. Sensitivity analysis of two harp seal *(Pagophilus groenlandicus)* population models. J. Fish. Res. Board Can. 36: 404-410.

Moreau, J. 1980. Influence des divers parameters sur l'estimation du rendement par recrue; application aux pêches continentales tropicales. Cybium, 3e série 8: 67-75.

O'Brien, J.J. and J.S. Wroblewski. 1976. A simulation of the mesoscale distribution of the lower marine trophic levels off West Florida, p. 64-103. *In* B.C. Patten (ed.) Systems analysis and simulation in ecology, Vol. IV. Academic Press, New York.

O'Neill, R.V. and R.M. Gardner. 1979. Sources of uncertainty in ecological models, p. 447-463. *In* B.P. Zeigler et al. (eds). Methodology in system modelling and simulation. North Holland Publishing Company, Amsterdam.

O'Neill, R.V., R.M. Gardner and J.B. Mankin. 1980. Analysis of parameter error in a nonlinear model. Ecol. Modelling 8: 297-311.

Orth, D.J. 1979. Computer simulation model of the population dynamics of large mouth bass in Lake Carl Blackwell, Oklahoma. Trans. Amer. Fish. Soc. 108: 229-240.

Parrish, J.D. 1975. Marine trophic interactions by dynamic simulation of fish species. Fish. Bull. 73: 695-717.

Pauly, D. 1980. A new methodology for rapidly acquiring basic information on tropical stocks: growth, mortality, and stock-recruitment relationships, p. 154-172. *In* P.M. Roedel and S.B. Saila (eds.) Stock assessment for tropical small-scale fisheries. Proceedings of an international workshop held at the University of Rhode Island, Kingston, R.I., September 19-21, 1979. International Center for Marine Resource Development, University of Rhode Island, Kingston. 198 p.

Rivard, D. and W.G. Doubleday. 1979a. Response of cohort analysis to input parameters. Canadian Atlantic Fisheries Scientific Advisory Committee, Research Document 79/45.

Rivard, D. and W.G. Doubleday. 1979b. APL programs for stock assessment including a sensitivity analysis with respect to input parameters. Technical Report of Fisheries and Aquatic Sciences No. 853.

Rivard, D. and W.G. Doubleday. 1980. APL programs for stock assessment. Canadian Technical Report of Fisheries and Aquatic Sciences No. 953.

Rose, M.R. and R. Harmsen. 1978. Using sensitivity analysis to simplify ecosystem models: a case study. Simulation 31: 15-26.

Tiwari, J.L. and J.E. Hobbie. 1976. Random differential equations as models of ecosystems: Monte Carlo simulation approach. Math. Biosci. 28: 25-44.

Tomovic, R. 1963. Sensitivity analysis of dynamic systems. McGraw-Hill, Inc., New York.

Waide, J.B. and J.R. Webster. 1976. Engineering systems analysis applicability to ecosystems, p. 330-372. *In* B.C. Patten (ed.) System analysis and simulation, Vol. IV. Academic Press, New York.

Appendix

Example of application of sensitivity analysis to a method frequently used for estimating the instantaneous rate of total mortality for tropical stocks

As discussed by Munro (this vol.), the instantaneous rate of total mortality, Z, is often evaluated for tropical stocks on the basis of the following formula:

$$Z = \frac{K(L_\infty - \overline{L})}{\overline{L} - L'} \qquad (7)$$

where K and L_∞ are positive constants of the von Bertalanffy length-age relationship, $\overline{L}$ is the mean length in the catch computed from L′ upward, and L′ is the smallest length that is fully represented in the catch (i.e., L′ is selected in such a way that $\overline{L}$ calculated as described above should be equal to the mean length of fish, in the population, characterized by lengths over L′).

Using the values of K (= 0.46), L_∞ (= 28.9), $\overline{L}$(= 16.4), and L′ (= 12) for *Nemipterus peronii* in 1966 in the Gulf of Thailand, (Pauly 1980), Z can be estimated to be equal to 1.31. Let us choose the following D-measure:

$$D = \frac{Z - Z^o}{Z^o} \times 100\% \qquad (8)$$

where superscript zero denotes the unperturbed value of Z (Z^o = 1.31). It should be noted that the value of D-measure is equal to the relative (in respect to the best estimate of Z) change (expressed in percentage) in Z.

Ordinary sensitivity analysis of expression (7) is implemented by changing the above-mentioned input parameters by various percentages, U. It is emphasized that only a single input parameter is disturbed in each calculation of Z. The resultant values of D are presented in Table 5. Several facts are evident:

— The D-measure is a linear function of U when one of the two parameters K and L_∞ is perturbed. This linearity is held for any range of perturbations.

— The D-measure remains an approximately linear function of the perturbation, U, in one (L′) of the two additional input parameters, $\overline{L}$ and L′, only when the perturbation is small (U < 5%).

— The D-measure is very sensitive to changes in all input parameters. K is the input parameter whose value affects the D-measure least although this effect is not negligible (the change by U in K results in the value of D-measure equal to U).

— The results reflect a high degree of sensitivity of the D-measure to changes in L, L′ and L_∞.

Table 5. Values of the D-measure (formula 8) for various perturbations in the input parameters. The perturbed parameter is indicated in the first column of the table and magnitude of the perturbation, U, in the first row of the table.

U	−40%	−20%	−10%	−5%	−1%	1%	5%	10%	20%	40%
K	−40.00	−20.00	−10.00	−5.00	−1.00	1.00	5.00	10.00	20.00	40.00
L_∞	−92.48	−46.24	−23.12	−11.56	−2.31	2.31	11.56	23.12	46.24	92.48
$\overline{L}$	−410.61	395.94	80.34	30.97	5.23	−4.86	−21.24	−36.71	−57.74	−80.92
L′	−52.17	−35.29	−21.43	−12.00	−2.65	2.80	15.79	37.50	120.00	−1,200.00

The values of the D-measure in Table 5 suggest that equation (7) is extremely sensitive especially to changes in the value of $\overline{L}$; it appears imperative to ensure that this parameter is estimated as reliably as possible, by using samples as large as possible to offset sampling variability and by ensuring that these samples are spread evenly over time to offset as much as possible the effect of fluctuation in recruitment (see Munro, this vol.). Additionally, full compliance with the assumption concerning L′ is necessary.

This example demonstrates that even ordinary sensitivity analysis, a very simple technique, can be helpful for identifying the uncertainty associated with a quantitative method. Obviously, extended stochastic sensitivity analysis usually provides much more useful information than that derived from ordinary sensitivity analysis. However, stochastic sensitivity analysis requires quantitative information about the nature of uncertainties in the input parameters, which was not available for the example considered.

Discussion of Dr. Majkowski's Paper

The discussion concentrated on the usefulness of sensitivity analysis and its extensions to formulate recommendations for appropriate effort allocation in a research program. Also, comments concerning the methodology itself were made.

The participants of the Workshop emphasized that it is absolutely essential to carry out sensitivity tests for complex multispecies fisheries models. They also strongly recommended the use of this technique in quantitative single-species approaches. It was noted that sensitivity analysis and its extensions should be applied particularly in all cases when the input data are uncertain to a high degree; this would allow assessment of the output uncertainty due to input uncertainties.

It was stressed that the data are frequently collected without any real appreciation of what role their quality is eventually going to play in the process of their use; the desired accuracy is usually evaluated on the basis of some sort of tradition or "gut" feelings. It was pointed out that application of sensitivity analysis to the standard quantitative procedures and models used in the single-species studies may lead to redesigning of routine data collections. Participants of the discussion argued that sensitivity analysis can be particularly useful in developing countries where the resources available for research are very limited and have to be distributed effectively among various research activities.

One of the participants asked Dr. Majkowski for details on "model improvement." Essentially, "model improvement" in the context of the paper presented is related to the reduction of uncertainties in the appropriate input parameters. This definition is very limiting since a model can have weaknesses other than incorrect parameter values; these, however–with the exception of unnecessary mechanisms included to the model–cannot be detected by sensitivity analysis. It can be argued that the mathematical formulation of mechanisms which have large sensitivity coefficients associated with their input parameters should be validated in a particularly careful way. However, sensitivity analysis cannot detect whether the mathematical structure of these mechanisms is correct or not.

The discussion then led to two separate problems which occur when considering the sensitivity of any model to its input parameters:

— the sensitivity of the output to changes in the input parameter, and
— the uncertainty in the input parameter.

Also, two extreme cases can be conceived: (i) the output is very sensitive to changes in the input parameter, but the parameter is very accurately determined, (ii) the sensitivity of the output to changes in the parameter is low but the parameter is inaccurately known. Clearly, the uncertainty of the output can be judged only when both factors are simultaneously taken into account. Dr. Majkowski stressed that this was precisely the reason why he recommended in his paper the use of extended stochastic sensitivity analysis, which alone can deal with such problems.

A participant suggested that the conclusions derived from sensitivity analysis can be erroneous if the best estimates of input parameters are very different from the real values. Dr. Majkowski agreed with this, although generally, the results of sensitivity analysis should not be very sensitive to changes in the best estimates of input parameters.

The advantages of the use of first-order theories, associated with extended stochastic sensitivity analysis in conjunction with the Monte-Carlo simulations, were emphasized by one of the participants.

The Ecological Basis of Tropical Fisheries Management

K. J. SAINSBURY

Division of Fisheries Research
CSIRO Marine Laboratories
P.O. Box 21, Cronulla
N.S.W. 2230, Australia

SAINSBURY, K.J. 1982. The ecological basis of tropical fisheries management, p. 167-188. *In* Pauly, D. and G.I. Murphy (eds.) Theory and management of tropical fisheries. ICLARM Conference Proceedings 9, 360 p. International Center for Living Aquatic Resources Management, Manila, Philippines and Division of Fisheries Research, Commonwealth Scientific and Industrial Research Organisation, Cronulla, Australia.

Abstract

Confident multispecies fisheries management implies an understanding of community dynamics, from which the ecological and fishery implications of alternative exploitation strategies may be predicted.

Two approaches toward understanding community structure and dynamics are reviewed, the cybernetic approach and the ecological niche approach. Their implications and relationships to models used in fishery management are then explored, and parameter estimation is examined.

It is concluded that neither approach provides an adequate model of community dynamics, and that parameter estimation even for these over-simplistic models is prohibitively difficult. An empirical approach to fishery management is suggested, which could also provide suitable situations for examination of the basic assumptions of the available models.

Introduction

Confident management of a fishery, or any other form or renewable resource harvesting, requires that reliable predictions can be made of the consequences of alternative exploitation strategies, from which the strategy deemed "best" may be selected. Amongst other things, this progression implies a sound knowledge of (1) how the community in question operates and (2) how this operation is influenced by exploitation.

This paper examines the current level of ecological understanding in relation to these two questions, particularly as they apply to the biological management of tropical marine fisheries.

Both questions impinge upon the very basis of ecological thought, by definition toward understanding the structure and function of ecosystems (Odum 1962), and both have generated a voluminous literature. Rather than provide an exhaustive review of the many points of view on community structure and function, I have concentrated on two subject headings, the 'cybernetic approach' and the 'ecological niche approach'. The cybernetic approach, developed mainly by Margalef, has had an important influence upon the approaches taken to studying groups of organisms and the concept of a community. Niche theory forms the basis of most recent ecological work directed toward understanding communities, and its underlying mathematical model has also been applied in a fisheries context. These headings encompass most (but not all) of the currently held views and approaches. In particular, the bio-energetics models are not discussed, as they are examined in a separate review (Jones, this vol.). The two headings are not entirely distinct and research under one has clearly had an influence on development of the other. They do, however, form a convenient point of division of an otherwise massive subject.

The ecological theory of these topics is reviewed and the implications and relationships to models used in fishery management are examined. The likely fidelity of the ecological and fishery models in describing the dynamics of communities is discussed, and parameter estimation is examined. In conclusion, some possible research approaches are suggested which might lead toward better management of trawl fish resources and understanding of community dynamics.

The Cybernetic Approach

This holistic approach is due mainly to Margalef (1957, 1963, 1968, 1969) who combined earlier ecological work, particularly that of MacArthur (1955) and Hutchinson (1959), with elements of systems control and communication theory. Margalef considered that biological communities evolve to reduce the degree of outside influence on their composition by developing the ability to track and damp predictable environmental fluctuations. Unpredictable changes are not damped, and remain as perturbations to the biological system. The ability of a community to achieve this environmental tracking was determined by the quantity of information contained within the community. This followed from Margalef's concept of a community operating as a communication channel, influenced by the environment, relating its present state to its state at some future time (Fig. 1). The capacity of this communication channel is measured approximately by the well-known Shannon-Weaver equation,

$$H' = \sum_{i=1}^{s} p_i \log_e p_i$$

where p_i is the proportion of species i in a community composed of s species. H' has an established history in information theory (Shannon and Weaver 1949), and in some circumstances may also be considered an index of diversity (Pielou 1966). H' was introduced to ecology by MacArthur (1955) as a measure of food web diversity (p_i being the proportion of energy flowing down pathway i), but is used in the context of Margalef's model to simultaneously measure species diversity and information content.

Margalef (1968) then reasoned that communities evolve to increase their

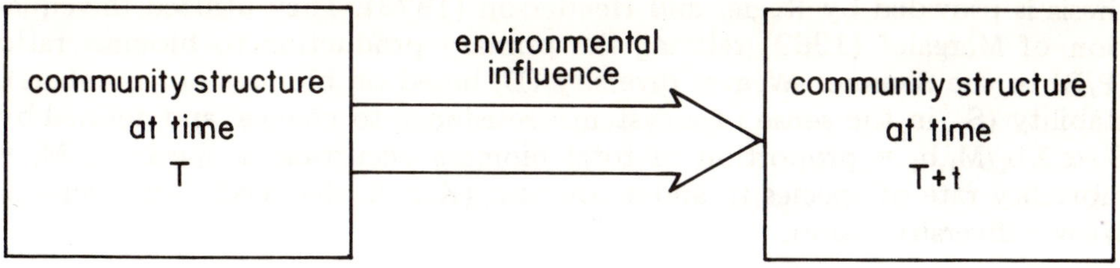

Fig. 1. Margalef's conceptual view of the regulation of community structure. From Caswell (1976). Copyright 1976, the Ecological Society of America.

information content (and hence diversity) until ultimately reaching a limit at which all the predictable elements of environmental variation are accommodated, leaving only the unpredictable component. If a large fraction of the environmental fluctuation is unpredictable, this limit will be reached at a lower species diversity than could be achieved if more of the environmental influence was predictable. The model thus provided a link between species diversity, environmental predictability and the degree of biological accommodation within communities.

Margalef (1968, Chapter 2) re-interpreted a body of empirical information on correlates of ecological succession, and indicated that they also applied to the evolutionary development of communities. It was suggested that during succession total community biomass increased, primary production increased, the primary production/biomass ratio decreased, diversity increased, food chains lengthened and organisms became more specialized. A mature community was seen as being characterized by a low flow rate of energy, implying a low rate of turnover, and giving rise to the prediction that any factor which accelerated energy flow through an ecosystem reduces its maturity. Margalef regarded harvesting as one such factor; others include artificial eutrophication and elevated temperatures.

Margalef (1968, Chapter 1) suggested a connection between energy flow and diversity, based on the concept that individuals of a complex, stable community would require less maintenance energy than those of a more simple, variable community. This was later (Margalef 1969) extended by empirical considerations to give a particular function relating diversity, stability, primary production and biomass. A qualitatively similar link between these variables was also suggested by Connell and Orias (1964) on different grounds.

There have been few cases of quantitative application of Margalef's hypothesis, due in part to uncertainty regarding the definition of stability (Levin 1975). The hypothesis has, however, had an important bearing on qualitative thought, particularly with respect to harvesting complex tropical ecosystems, and usually leading to the suggestion that highly diverse communities are fragile to exploitation. It is argued that harvesting is a form of population control external to the community (similar to the harsh physical environment of high latitudes) to which the species or species group is not evolutionarily adapted, and that the community may abruptly collapse under the influence of this control. This view is common in the 'conservation literature' (e.g., Wallace 1974), but is also encountered in the fisheries literature (e.g., Regier and Henderson 1973; Pauly 1979).

One of the few attempts at quantitative application of Margalef's hypothesis is provided by Regier and Henderson (1973). They utilized the equation of Margalef (1969) relating the primary production to biomass ratio (P/B) to the Shannon-Weaver diversity (D, based on biomass not numbers), stability (S′, in the sense of a systems 'resistance to change' and defined by $S' \alpha \Sigma b_i/M_i$, b_i = proportion of total biomass occurring as species i, M_i = mortality rate of species i), and a constant ($K \cong 5$, the maximum Shannon-Weaver diversity value),

$$\frac{P}{B} = e^{K-(D-1/S')} - 1$$

Regier and Henderson (1973) argue that both stability (S′) and diversity (D) decrease under exploitation. That S′ could decrease may be seen from its definition and noting that 'totally opportunistic' exploitation would increase the mortality rate of all species. The behavior of b_i under fishing also influences S′, and solely determines D. Regier and Henderson (1973) unfortunately equated b_i with the total biomass of species i, rather than proportionate biomass, as intended by Margalef (1969). Regier and Henderson (1973) then argue that both S′ and D decrease with fishing since optimal exploitation of each stock generally implies reduction of each species biomass by 1/3-2/3. The use of biomass rather than proportionate biomass in the equation for S′ is correct if discussion concerns a constant biomass system. Total community biomass may then be simply incorporated into the constant of proportionality in the equation for S′. However, it is not valid when discussing exploitation, in which community biomass is variable. For example exploitation which reduced all populations by the same proportion would decrease S′ but not effect D, while both D and S′ could increase if exploitation was highly selective on an initially abundant species with a large M_i. The implications to b_i and diversity of other harvesting regimes depends upon both the selectivity of the fishery and the response of the community. Consideration of these influences by Regier and Henderson (1973) is via discussion of a series of suggested empirical relationships between various biotic and abiotic variables based on 'considerations not fully elaborated', which while extremely interesting are inconclusive.

The one quantitative relationship presented was obtained by considering the primary production to community biomass ratio to be directly proportional to the ratio of equilibrium fish catch (Y) to fish biomass (B_F). Substituting this into Margalef's equation gives

$$Y \alpha B_F (e^{K-(D-1/S')} - 1),$$

a monotonic decreasing yield with increasing diversity (Fig. 2). In support of this expression, Regier and Henderson note that if (D − 1/S′) is considered proportional to the logarithm of the mean depth, the equation has the same form as part of the empirically determined relationship between catch and mean depth of some North American lakes (e.g., Ryder 1970). Margalef (1969) regarded (D − 1/S′) an expression of the 'degree of organisation of an ecosystem' which he suggested should increase with increasing environmental stability. To the extent that deeper lakes provide a more stable physical environment, these empirical findings may be judged consistent with theory.

Evaluation of the validity of Regier and Henderson's equation as a mechanistic representation of a fish community requires examination of the underlying hypothesis. Margalef's hypothesis, as part of the 'diversity-stability' hypothesis, found widespread acceptance in ecology as it provided a consistent explanation for many ecological observations. These observations were mostly the results of correlative studies of diversity and environment (Dobzhansky 1950; Williams 1964; Klopfer 1959; Fischer 1960; Southwood 1961; Pianka 1966; Odum 1969) and the latitudinal gradient of diversity in particular (e.g., Sanders 1969; Slobodkin and Sanders 1969).

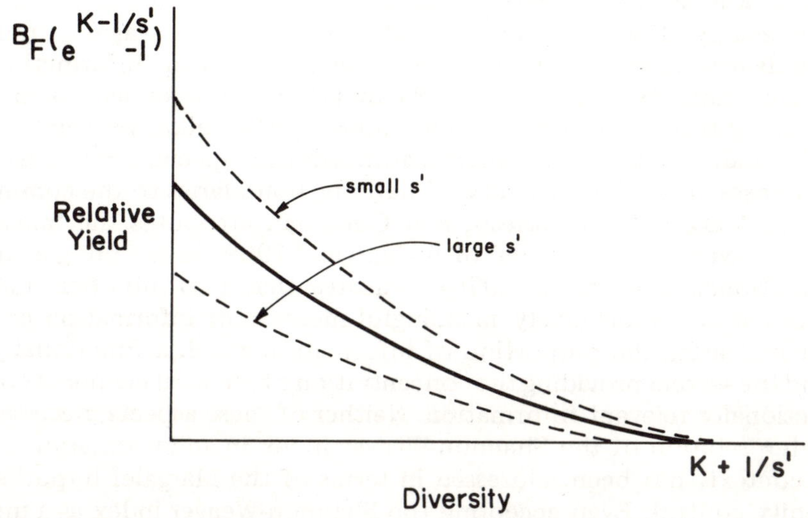

Fig. 2. The relationship between fish yield and diversity utilized by Regier and Henderson (1973). A small value of S' implies that a large proportion of the fish communities biomass consists of short lived species, while a large S' results from a community dominated by long lived species. B_F is fish biomass and K is a constant.

Consistency with observed patterns of community diversity however does not constitute proof of the mechanism used to derive the explanation, and there were also many empirical studies which did not fully agree with the hypothesis (Goodman 1975). Indeed, the correlative studies of community diversity and environment are marked by the wide range of mechanisms which have been invoked to explain the observations. Other than the stability-diversity hypothesis, explanations of the high diversity of the tropics have included as arguments high levels of predation, and 'predator switching' in particular (Paine 1966; Glasser 1979; Teranoto et al. 1979), intense biological competition (Dobzhansky 1950) and the form this competition takes (Roughgarden 1974; Glasser 1978, 1979), a high productivity (Connell and Orias 1964), the length of time available for the evolution of complex communities (Simpson 1964; Fischer 1960; Nursall 1977) spatial heterogenity (e.g., Whittaker and Levin 1977) and a faster rate of evolution (Fischer 1960).

Critical evaluation of the stability-diversity hypothesis has followed two main approaches. The first is a thorough examination of the logic invoked during erection of the hypothesis (Goodman 1974, 1975; Johnson 1970; Pielou 1966) and the second is construction of a null or neutral hypothesis, specifically omitting the mechanisms thought to be important, to generate expectations for comparison with the real world (Caswell 1976).

The stability-diversity hypothesis as developed by Margalef (and applied by Regier and Henderson 1973) requires that the information content of a community is measured by the Shannon-Weaver formula and that this is relevant to both stability of a community under environmental influences and energy flow through a community.

In his penetrating review of the stability-diversity hypothesis, Goodman (1975) negates each of these required elements. The Shannon-Weaver diversity index was shown to have no clear ecological interpretation in terms either of energy flow in a food web (MacArthur's 1955 usage) or numbers and abundances of species. Its value as an estimator of the information content of a community was shown by Pielou (1966) to be subject to considerable sampling bias if applied to communities with an unknown total number of species and/or contagious species distributions. A special kind of sampling problem arises from the difficulty of defining boundaries to the community relevant to Margalef's hypothesis, and Goodman attributes the 'maximum Shannon-Weaver diversity value' of 5 (Margalef 1969) to sampling inadequacies. Additionally, Johnson (1970) indicates that two modifiers must be introduced into a biologically meaningful measure of information content, the first indicating the proportion of information which is functionally relevant, and the second providing the concentration of physical elements containing functionally relevant information. Neither of these aspects, necessary for strict interpretation of the Shannon-Weaver index in an information-control theory context, has been addressed in terms of the Margalef hypothesis of community control. Even accepting the Shannon-Weaver index as a measure of relevant information, however, Goodman (1975) questions whether the small amount of information indicated by the index could contribute greatly to maintenance of a community, particularly as compared to the very much greater information content of the genetic structure of each individual. Goodman found no theoretical basis for necessarily connecting the information content of a system to energy flow through that system (except the tiny quantity of energy obtainable from the system if all contained information, as measured by the Shannon-Weaver index, were convertible to energy from its information-entropy equivalent).

Attempts to test aspects of Margalef's theory empirically have been inconclusive largely because of the uncontrolled nature of the observations and experiments (see Goodman 1975). It is also clear that many of these studies suffered from the absence of a null hypothesis, i.e., what patterns of diversity are to be expected if there is no relation between community structure and biological interactions, i.e., no accommodation of environmental variability. To examine such a null hypothesis Caswell (1976) applied a 'neutral model', here a model which specifically excludes biological interactions and which had originally been derived to explore the behavior of selectively neutral alleles in a gene pool (e.g., Ewens 1972). The model consists of two processes, the arrival of an individual of a new species (by immigration or evolution), and a subsequent linear birth and death process. The instantaneous Poisson probability of immigration is ν, while the birth and death parameters are λ and μ respectively. It is assumed that $\lambda = \mu$, implying that the population is certain to ultimately become extinct, but with an infinite expected waiting time. Ewens (1972) derived a means of estimating ν from the number of species and individuals in a sample, so that the model can be used to predict the community structure (diversity) expected for the observed number of species and individuals in the absence of biological interactions. Caswell (1976) applied this model to several published data sets illustrating ecological succession, and to the structure of communities in temperate and tropical regions. These comparisons, across what had been reported as gradients in the intensity of biological interaction, indicated a general lack of departure from the predictions of the neutral model. Further, these was a distinct tendency for diversity in 'climax communities' and in communities from stable physical environments (e.g., tropics) to be lower than that predicted by the neutral model, indicating dominance by some species. Caswell (1976) also used the neutral model to predict species-area relationships which were consistent with observations, while others (Crowder 1980; Connor and Simberloff 1979) have shown by similar techniques that many of the 'laws' of biogeography can be explained without recourse to special biological properties of groups of species.

Caswell's (1976) work shows that much of a community's structure can be explained by a very simple model without species interactions, although it cannot be inferred that the mechanism of community maintenance is as modelled. The findings are, however, quite contrary to Margalef's suggestions that biological interactions become more important in shaping and maintaining community structure in going from variable to less variable environments or from 'pioneer' to 'climax' communities. There remains, of course, the interesting problem of interpreting changes in ν with community structure.

Caswell's (1976) approach warrants further examination, particularly with regard to community dynamics. However, that the model fails to adequately represent the observed persistence of species within communities under exploitation may be shown by examining the effect of increased mortality on the time to extinction of a species.

Under a linear birth-death process (Cox and Miller 1965) with $\lambda = \mu$, the probability that extinction of a species has occurred at or before time t (measured from the time of introductions of the single founder) is

$$\frac{\lambda t}{1 + \lambda t}$$

If exploitation increases the death rate to $\mu = \delta\lambda$ ($\delta > 1$) this probability of extinction becomes

$$\frac{\delta - \delta e^{-(1-\delta)\lambda t}}{1 - \delta e^{-(1-\delta)\lambda t}}$$

From these the ratio of the time taken to reach a certain probability of extinction (α) with $\mu = \delta\lambda$ to the time taken with $\mu = \lambda$ is

$$R = \frac{T_{(\mu = \delta\lambda)}}{T_{(\mu = \lambda)}} = \frac{1-\alpha}{\delta(\delta-1)} \ell n \left(\frac{\alpha - \delta}{\alpha\delta - \delta}\right)$$

This is illustrated in Fig. 3 for α = .95 and shows the extreme sensitivity of species persistence to the assumption of $\lambda = \mu$. Exploitation which doubles the mortality rate (δ = 2, not uncommon in fisheries) would dramatically reduce species persistence. This could be compensated for by the introduction of some form of density dependence of μ or λ. A thorough examination of time series data on community structure in the light of this class of model is needed, emphasizing the dynamics of communities under changes in μ. This should first examine the question of just how much of the behavior (at both the species and community level) can be explained by such non-interaction stochastic models, and then explore the magnitude and forms of whatever interaction-dependent or regulatory processes need to be invoked to explain the remaining behavior.

It is interesting to note that if Caswell's (1976) model is considered to apply to small somewhat isolated sub-communities, rather than the large

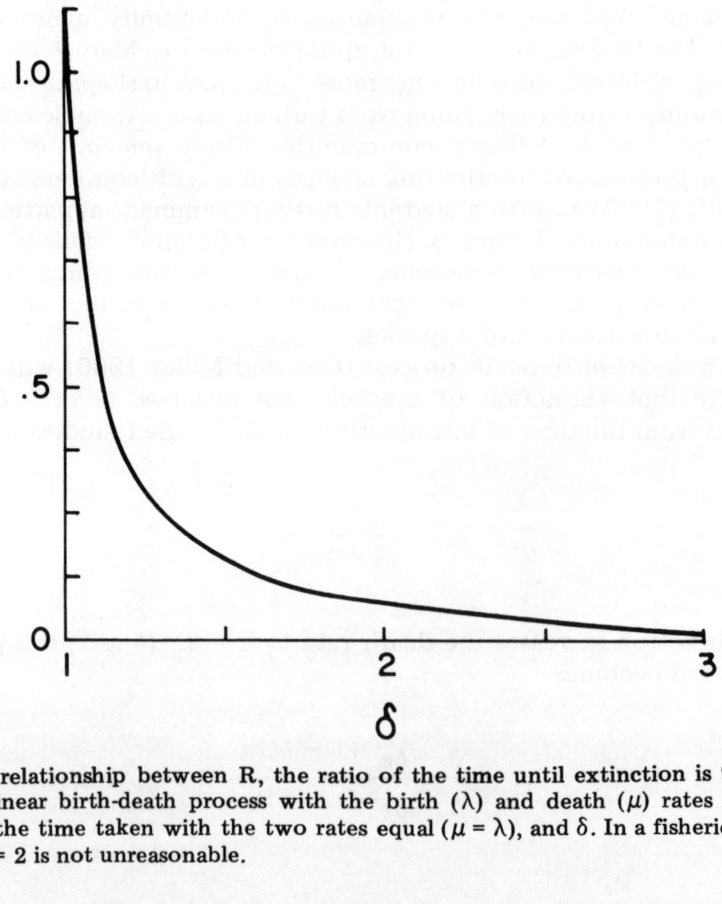

Fig. 3. The relationship between R, the ratio of the time until extinction is 95% probable for a linear birth-death process with the birth (λ) and death (μ) rates not equal ($\mu = \delta\lambda$) to the time taken with the two rates equal ($\mu = \lambda$), and δ. In a fisheries context a value of δ = 2 is not unreasonable.

assemblages usually examined in successional and biogeographic studies, it becomes remarkably similar to the spatial heterogeneity-random immigration concept of community structure and maintenance proposed by several workers (e.g., Huffaker 1958, for mites on oranges; Pimentel et al. 1963, for wasps and flies; Sale 1978; Sale and Dybdahl 1978, for coral reef fish; Whittaker and Levin 1977 and Caswell 1978, for a treatment of theory). This concept de-emphasizes the attainment of overall equilibrium conditions within communities, and hence the importance of mechanisms such as competitive abilities, food resource partitioning, etc., in generating stability.

In summary, then, it must be concluded that the mechanistic base of Margalef's reasoning, the link between information theory, thermodynamics and community structure, is not justified. This does not of course alter the substance of the many empirical observations embroidered into Margalef's hypothesis (particularly as used by Regier and Henderson 1973). It means simply that these observations stand without a connecting mechanism, and so are open to the difficulties of interpretation experienced when dealing with any correlative, empirical database. In particular the cybernetic mechanism does not provide clear reasons to expect diverse tropical communities to exhibit more or less biotic regulation than temperate ones, nor to expect them to be more or less fragile to harvesting than temperate communities. Similarly, the suggestion of Regier and Henderson (1973) that lower yields are obtainable from diverse communities than from less diverse communities has no mechanistic foundation, and if true in some instances (e.g., temperate lakes) it must remain an empirical finding of unknown generality.

The Ecological Niche Approach

The concept of the ecological niche appeared very early in the history of ecology, the first formal use apparently being by Grinnell (1917, 1924) to describe the 'position occupied by a species in both its biological and physical environment.' A somewhat different definition was introduced by Elton (1927), who regarded the niche as an organism's 'place' in the biotic environment. This was developed to include the organism's function within a community, rather than just its 'position' (e.g., Odum 1953).

Hutchinson (1958, 1965) related the ecological niche to the n-dimensional hyper-volume occupied by a species within the space defined by a set of coordinates, each relating to a variable affecting the species. As noted by Kerr (1980), there is a strong resemblance between Hutchinson's definition of the niche (or at least the coordinates of the physical variables) and Fry's (1947) quantification of fish-environment interactions. It is usually accepted that Hutchinson intended both physical and biotic variables to be included as coordinates (e.g., Odum 1971), although more recently Whittaker et al. (1973) have argued that this is neither desirable nor what Hutchinson intended. They suggest that a distinction be made between the purely physical tolerances of a species, relevant both to inter- and intracommunity and termed 'habitat' by Whittaker et al. (1973), and the position occupied by a species in its intracommunity environment, emphasizing biotic influences and termed the 'niche'. This distinction appears justified to the extent that selection processes could be perceived to operate differently under the two concepts; 'a response to gradients of environmental factors' leading to 'habitat differentiation' and β diversity in the first case, and selection on characters relating to relationships with other species, leading to 'niche differentiation' and α diversity in the second. The concept of the ecological niche is by no means clearly defined however. The distinctions made by Whittaker et al. (1973), while helpful, depend to an extent on the definition of 'community' and are not unambiguous. For example, Whittaker et al. (1973) suggest that 'when intercommunity and intracommunity variables are difficult to separate they may be grouped as "environmental variables" '.

Recent quantitative application of 'niche theory' (MacArthur 1968; Levins 1968; May 1974) has emphasized the role of the biotic environment in determining community structure and conforms reasonably to the definition of niche proposed by Whittaker et al. (1973). Studies under this approach generally seek to relate the statics and dynamics of community structure to biological variables such as the pattern of resource usage, overlap of requirements of species, competition, etc., and it is usually assumed that these variables apply most crucially to trophic activities. The underlying model used is the Lotka-Volterra equation set for interacting species (see Levins 1968; Odum 1971, p. 215)

$$\frac{dN_i}{dt} = N_i \left(r_i - \frac{r_i}{K_i} \sum_{j=1}^{j=n} a_{ij} N_j\right) \qquad a_{ii} \equiv 1 \qquad i = 1, n$$

$$= N_i f_i (N_1, N_2 \ldots N_n) = F_i (N_1, N_2 \ldots N_n)$$

where N_i is the abundance of species i, r_i its intrinsic rate of increase and K_i its carrying capacity. a_{ij} is an interaction coefficient describing the effect of species j on species i. At equilibrium

$$N_i^* = K_i - \sum^{j \neq i} a_{ij} N_j^*$$

and so a_{ij} may be considered to be the amount by which an individual of species j reduces the equilibrium population size of species i from its carrying capacity.

The equilibrium equation may also be expressed

$$0 = \underset{\sim}{K} - A\underset{\sim}{N}^* \quad \text{or} \quad \underset{\sim}{N}^* = A^{-1}\underset{\sim}{K}$$

where $\underset{\sim}{K}$ is a column vector of K_i values, $\underset{\sim}{N}^*$ is a column vector of equilibrium population sizes, N_i, and A is the community matrix (Levins 1968) with elements α_{ij}, often termed competition coefficients. If the Lotka-Volterra equations are assumed to completely describe the community $\alpha_{ij} = a_{ij}$ by definition. Different symbols have been used to emphasize that the parameters of the equations governing the communities dynamics and the elements of the community matrix are not always identical.

If the equations governing a community's behavior are not the Lotka-Volterra equations, they and the concept of the community matrix can still be applied, as an approximation, in a region near the equilibrium point. The range of population sizes for which this approximation is reasonable depends of course, on the nature of the 'true' community equation set. Several authors have suggested methods of obtaining formulations for α_{ij} from community equation sets which are not of Lotka-Volterra form. The literature is rather confused on this point, and so it is worth examining the methods and their interrelationships.

—MacArthur (1972) obtains α_{ij} by re-arranging the given community equation set into the form

$$0 = K_i' - \sum^{j} \alpha_{ij} N_j^* \qquad \alpha_{ii} = 1$$

at equilibrium. As implied by Schoener's (1974a) calculations, and explicitly stated by Levine (1977), this is equivalent to evaluating

$$\alpha_{ij} = \frac{\partial N_i}{\partial N_j}$$

with all populations except i and j held constant at their equilibrium values. Clearly $\alpha_{ij} = a_{ij}$ for the Lotka-Volterra equation.

—Abrams (1977, 1980), following Gill (1974), suggests evaluating α_{ij} from the given community equations, in the form

$$\frac{dN_i}{dt} = N_i f_i\,(N_1, N_2 \ldots N_n)$$

as

$$\alpha_{ij} = \frac{\partial f_i^*(\cdot)}{\partial N_j} \Big/ \frac{\partial f_i^*(\cdot)}{\partial N_i}$$

By the chain rule this equals $\frac{\partial N_i}{\partial N_j}$ evaluated with all other species at their equilibrium levels, and so this formulation is equivalent to that of MacArthur (1972) and Schoener (1974a). If $f_i(\cdot)$ is the Lotka-Volterra equation set,

$$\frac{\partial f_i^*(\cdot)}{\partial N_j} = \frac{-r_i}{K_i} a_{ij} \text{ (termed the intensity of competition by Abrams 1980)}$$

$$\frac{\partial f_i^*(\cdot)}{\partial N_i} = \frac{-r_i}{K_i}$$

and so $\alpha_{ij} = a_{ij}$ as before.

—With the given community model in the form $\frac{dN_i}{dt} = F_i(N_1, N_2 \ldots N_n)$, May (1974, p. 22) equates the community matrix with the Jacobian matrix of $F_i(\cdot)$ at equilibrium, and evaluates its elements as

$$\alpha_{ij}' = \frac{\partial F_i^*(\cdot)}{\partial N_j}$$

This usage is different from those above (note that for the Lotka-Volterra equation set $\alpha'_{ij} = r_i a_{ij} N_i^*/K_i$ so that $\alpha'_{ii} \neq 1$), with α'_{ij} describing the behavior of the equation set near equilibrium. Specifically, the behavior of a small perturbation, $x_{it} = N_{it} - N_i^*$, is described by

$$\frac{dx_t}{dt} = \Sigma\, \alpha'_{ij}\, x_{jt}$$

The relationship of May's (1974) α'_{ij} to α_{ij} may be seen from

$$\alpha'_{ij} = \frac{\partial F_i^*(\cdot)}{\partial N_j} = N_i^* \frac{\partial f_i^*(\cdot)}{\partial N_j}$$

so $\alpha_{ij} = \dfrac{\alpha'_{ij}}{\alpha'_{ii}}$, and α'_{ij} is equivalent to N_i^* times the 'intensity of competition' of Abrams (1980).

Application of niche theory to community structure and function has followed closely the initial work of MacArthur (1968, 1970, 1972; MacArthur and Levins 1967; May and MacArthur 1972) and what has become known as the 'MacArthur model' of a community. This model considers an array of consumers, not directly influencing one another, but competing exploitatively for an array of non-interacting resources. The consumer species, C_i are described (e.g., Schoener 1974a) by

$$\frac{dC_i}{dt} = E_i C_i \left(\sum_{k=1}^{k=m} f_{ik}\, b_{ik}\, R_k - T_i \right) \qquad i = 1, n$$

and the resource species by

$$\frac{dR_k}{dt} = R_k \left(r_k - \frac{r_k R_k}{K_k} - \sum_{j=1}^{j=n} f_{jk} N_j \right) \qquad k = 1, m$$

b_{ik} is the net energy extractable from an individual of resource k by consumer i, f_{ik} is the consumption rate on resource k by consumer i (this is the predator functional response of Holling 1959, the number of prey eaten per predator in some time interval as a function, here assumed linear, of prey abundance), T_i is the maintenance and replacement cost for consumer species i, E_i is the number of individuals of consumer i produced per extracted unit of energy, and r_k, K_k are the Lotka-Volterra parameters as before.

This model has formed the basis of most studies of the implications of resource utilization pattern to community stability and dynamics. An important outcome was the derivation of a 'limiting similarity' to resource usage; a limit to how similar species could be and yet co-exist at equilibrium (MacArthur and Levins 1967; MacArthur 1970; May and MacArthur 1972; May 1974). This is in agreement with Gause's 'law' of competitive exclusion (see Gause 1934; Odum 1971) and provided the foundation for the belief that competition, via its influence on resource partitioning and permissible overlap, is the major factor shaping community structure (e.g., May 1974; MacArthur 1970; Schoener 1974b; Roughgarden 1974, 1976; Moermond 1979).

Practical application of 'niche theory' requires measurements of the competition coefficients. The approach taken has been to utilize data on the overlap of use by the competing species of some resource or resources deemed relevant (Levins 1968; May 1975; Schoener 1974a). The general form of the determination follows from the definition of α_{ij} derived for competition between the consumers of MacArthur's model (see Schoener 1974a).

$$\alpha_{ij} = \frac{\sum^{k} f_{ik} f_{jk} \left(\frac{b_{ik} K_k}{r_k}\right)}{\sum^{k} f_{ik}^2 \left(\frac{b_{ik} K_k}{r_k}\right)}$$

The equation used by MacArthur and Levins (1967) and Levins (1968),

$$\alpha''_{ij} = \frac{\sum^{k} p_{ik} p_{jk}}{\sum^{k} p_{ik}^2}$$

where p_{ik} is the fraction resource k makes of the total resource usage of species i, was justified in the case where the limiting resources are associated with certain habitats, and competition related to the probability of species i and j jointly using these habitats and hence resources. If $b_{ik}\ K_k/r_k$ may be considered identical (or irrelevant) for all such habitats (Schoener 1974a) the the two equations become comparable. In the terminology of Levins (1968) α''_{ij} is a niche overlap measure, and its denominator is a measure of niche breadth.

A large number of studies have used this methodology to examine the interrelationships between species (e.g., Pianka 1973; Cody 1968; Preston 1973; Yeaton and Cody 1974; Cody 1974; Leviten and Kohn 1980; Wilbur 1972; Hariston 1980; Schoener 1970, 1974b; Vandermeer 1969) and have provided interesting documentation of the ecological differences between species. Some studies have also shown a predictive ability (e.g., Yeaton and Cody 1974; Cody 1968), and it would appear that some observed patterns of resource usage were not due to random divisions of the available resource spectrum (see Sale 1974; Inger and Colwell 1977; Lawlor 1980).

There has been considerable difficulty in fully applying and testing the equation set, due mainly to 'flexibility' in application of the model. The Lotka-Volterra equation set, basic to quantitative niche theory, is seen as a 'first order approximation' to the actual community equations (May 1974) and the community matrix applies in the vicinity of equilibrium (Levins 1968; Schoener 1974a). However there is no means by which a field worker can *a priori* ascertain this domain of applicability. Furthermore, the method of estimating α_{ij} given by Schoener (1974a) is strictly applicable only if the MacArthur model is correct and the system is at equilibrium, and then requires that r_i, K_i and b_{ik} be determined in addition to resource availability and usage. There do not appear to be any studies in which the necessary parameters are estimated and the assumed model and equilibrium conditions are verified. Use of the MacArthur and Levins (1967) and Levins (1968) formulation to measure α_{ij} must be regarded, in the words of Schoener (1974a), as 'simply an assumption'. To these problems is added the potential bias introduced by reliance upon the ecologists interpretation of which resources are of relevance in a community, and hence those to study. Under this combination of uncertainties falsification is difficult.

There would also appear to be a difference between usage of the model by ecologists and that implied by the mathematics. Obtained exactly as defined, $\alpha_{ij} = 0$ if N_j does not appear explicitly in the equation for dN_i/dt near equilibrium. This is the case, for example, in the MacArthur (1972) model of consumers whose only interaction is via exploitative competition for resources. An approximate measure of competition under these circumstances was obtained by MacArthur (1972) and Schoener (1974a) by first re-arranging the equilibrium equations into a form which explicitly included N_j. This was achieved by assuming that the resource species respond to any perturbation far more rapidly than consumers, so that they may be considered to rapidly transmit the influences of a perturbation to the consumers but always remain at or very near equilibrium with respect to consumer abundances themselves. The equilibrium resource abundances were then obtained as functions of consumer abundances from the resource equations alone, and substituted into the non-equilibrium consumer equations. This provided the desired equation set for the rate of change of one consumer as a function of the abundance of the other consumers. (If, as in the MacArthur model, resources are non-interacting, this function does not explicitly include resource abundances.) This procedure is probably adequate for considerations at equilibrium, but is questionable under all but extremely small perturbations, as MacArthur (1970) was fully aware.

More importantly however, the use of α_{ij} in this somewhat relaxed context has obscured the implications of the strict interpretation of α_{ij}; a measure of direct, linear interaction between species i and j with all other species held constant at their equilibrium levels. In many instances, particularly when trying to predict the effect of some non-infinitesimal perturbation, this is not what ecologists are really concerned about, and greatly decreases the utility of the community matrix model.

A more complete treatment, including indirect effects, introduces many difficulties both biologically and mathematically which have not as yet been overcome. Examinations available, however, indicate that system behavior as defined from consideration of the community matrix is substantially incomplete, even when dealing strictly with the case of exploitative competition and linear consumer functional responses. Two related factors are involved.

Firstly, more realistic descriptions of the direct effects of species upon one another's dynamics introduce higher level interaction terms into the equation set. Such interactions have been demonstrated to influence system behavior in experimental situations, causing considerable deviation in the outcome of interaction experiments from predictions made on the basis of linear interaction and neighborhood stability considerations (Neill 1974; Preston 1973; Wilbur 1972; Vandermeer 1969).

Non-linearities may enter the community equation set in a wide variety of ways. For example, utilization efficiency of ingested food is known to vary greatly under various influences such as temperature, meal size, previous feeding history (Fänge and Grove 1979), predator functional responses are in general non-linear, being influenced among other things by threshold responses, prey switching, prey refugia (Steele 1974; Rozenzweig 1977; Glasser 1979; Teranoto et al. 1979); constraints of 'handling time' (Holling 1959, 1973), and perhaps 'optimal foraging' considerations (Pyke et al. 1977); and it is expected that the age/size composition of the population has an influence on the capacity for reproduction, trophic status, competitive ability and metabolic requirements of a population.

The second factor is the potential for 'indirect mutualism' between species, brought about by the network of direct interactions (Levine 1976, 1977; Vandermeer and Boucher 1978; Lawlor 1979; Vandermeer 1980; Neill 1974). The simplest form of this arises by slightly modifying MacArthur's (1972) resource-consumer model to include negative interaction between resource populations (Fig. 4). The net interaction between consumers can then range from negative (competition) to positive (mutualism) depending, among other things, upon the strengths of the various intermediary interactions (Vandermeer 1980), whether the consumers are obligate (i.e., cannot persist in the absence of the effects of mutualistic partner) or facultative mutualists (Vandermeer and Boucher 1978), and the extent to which specialization by a consumer involves increasing the effectiveness of obtaining a particular resource versus increasing the efficiency of converting a unit of consumed resource into consumer numbers (Levine 1977; Vandermeer 1980). As a measure of these various effects, Lawlor (1979) has proposed the coefficient

$$\gamma_{ij} = -\frac{\Delta N_i}{\Delta N_j}$$

where ΔN_i is the change in equilibrium level of population i caused by altering the level of population j by ΔN_j, and then holding N_j constant for sufficient time to allow attainment of a new equilibrium. Lawlor (1979) has shown that γ_{ij} can be obtained from the community matrix. However the resulting Γ matrix is applicable only in the range of population densities over which the community matrix is valid. Thus, the Γ matrix obtained is globally applicable only if the community is fully described by the n-species Lotka-Volterra equation set. Under other circumstances, where γ_{ij} would depend both upon ΔN_j and population sizes, the measure would have to be determined empirically.

While the inclusion of indirect effects and non-linear responses into the models of community dynamics is highly desirable, there is at present no clear mechanistic and general method of doing so. Levine (1977) introduces a mathematical framework based on energy flow, and Levins (1975) provides a possible alternative from loop analysis.

The practical implications of higher level interactions to parameter estimation and interpretation of experiments are rather daunting. If such interactions are important even experimental species manipulations will be of little use in predicting the outcome of different manipulations. The problem is well put by Neill (1974), 'To interpret competition . . . . a field ecologist is there-

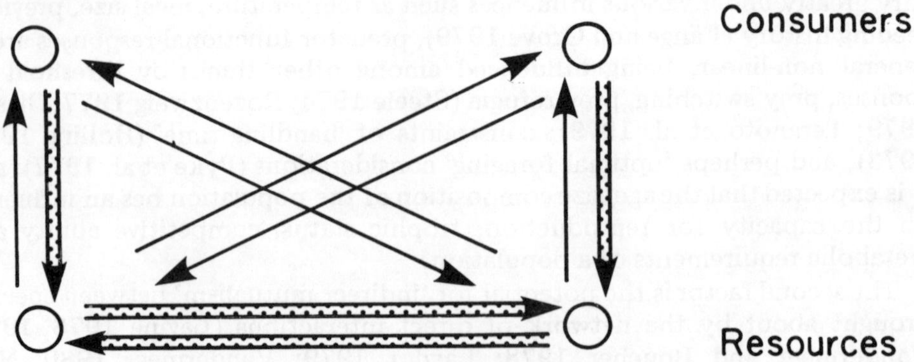

Fig. 4. Interactions between two consumers and two resource species. The arrows indicate a negative influence by the source species on the recipient species. If the thickened arrows are dominant interactions the two consumers have a net mutualistic interaction despite being competitors or independent on more direct pathways. From Vandermeer (1980), by permission of The University of Chicago Press.

fore in a difficult position: on the one hand, he cannot hope to measure the dynamics of the interactions between pairs of species without perturbing them, and on the other, results of this perturbation depend on the effects of other species on the competing pair. Properly done, the appropriate experiment becomes hopelessly difficult for all but the very simplest communities.'

These are difficulties arising from the application of the MacArthur model, or some similar model of exploitative competition. However, current dependence of the bulk of quantitative niche theory upon the mechanism of exploitative competition at equilibrium is also of concern in that it has tended to displace attention from a variety of other potentially important mechanisms. Even within the subject of competition, for example, the role of interference competition has received very little attention (e.g., Park 1962; Schoener 1973; Nixon 1980; Ebersole 1980). Still within the 'biotic environment', the importance of predation to maintenance of community structure, largely by preventing competition from reaching resource limitation levels and hence preventing competitive exclusion, has also been emphasized by some ecologists (e.g., Paine 1966, 1971; Dayton 1971; Connell 1978; Glasser 1978, 1979). There has been little attempt to integrate this with quantitative niche theory. May (1974) and Cramer and May (1971) have shown that such an effect is possible under the Lotka-Volterra equation set at equilibrium, but only for a very restricted set of parameter values. However, Caswell (1978) and others (e.g., Dayton 1971; Connell 1978; Huston 1979; Sousa 1979) have suggested that predator mediated coexistence (and some other mechanisms) should be viewed in the context of a 'non-equilibrium' model of the community, which would render equilibrium analysis and most contemporary niche theory irrelevant.

Quantitative niche theory, as developed in the last 10-20 years, has also resulted in de-emphasis of the role of the physical environment in community dynamics. The effects of uncorrelated, random fluctuations in equation parameter values have been examined (May 1974), but the major part played by the physical environment under current theory is in setting a limit to the 'allowed' similarity of the trophic niches of competitors (May and MacArthur 1972; May 1974) rather than on community dynamics *per se*. Even this effect, however, has been shown to depend critically upon model structure and the assumed small amplitude of fluctuations (Nisbet et al. 1978). This de-emphasis of the role of the physical environment is in contrast to the views of many earlier ecologists. Andrewartha and Birch (1954) regarded the physical environment as the major controlling agency of populations and communities. Others indicated the need for some form of density dependence for regulation of populations (e.g., Nicholson 1954; Chitty 1960; Wellington 1960; Klomp 1962; Solomon 1964), but still emphasized the major role played by the environment in community and population dynamics. Most importantly, they emphasized the mutual modification of biotic and abiotic factors in their effects on populations (the 'legislative' influences of Nicholson 1954). These approaches have been somewhat eclipsed by the development of quantitative niche theory, but not negated. The possibility that such mechanisms play an important role in community dynamics should not be overlooked (see Caswell 1978).

The multispecies Lotka-Volterra equation set was first applied in a fisheries context by Larkin (1963) in his examination of the exploitation of two interacting species, and later by Silliman (1975), Anon. (1978) and Pope (1979). Larkin (1963) used the equation set, as originally intended, to describe the numbers in each population, while subsequent workers applied the same equations to the biomass of each population (B_i replaces N_i). The implications of fishing the community may be examined by adding a term for fishing removals in the usual way, with q_i being the catchability of species i and E being the fishing effort. Using the same notation as before (but noting the parameters have a different meaning when describing changes in biomass), the equation set becomes

$$\frac{dN_i}{dt} = N_i\,(r_i - \frac{r_i}{k_i} \overset{j}{\Sigma}\, \alpha_{ij} N_j - q_i E)$$

and at equilibrium

$$0 = r_i - \frac{r_i}{K_i} \overset{j}{\Sigma}\, \alpha_{ij} N_j - q_i E = K_i - \overset{j}{\Sigma}\, \alpha_{ij} N_j - \frac{K_i q_i}{r_i}\, E$$

If E = 0

$$\frac{r_i}{K_i} \overset{j}{\Sigma}\, \alpha_{ij} N_j = r_i$$

or in matrix notation

$$\underset{\sim}{n}_o = \Delta^{-1} \underset{\sim}{r}$$

where Δ is a matrix of elements $\delta_{ij} = r_i \alpha_{ij}/K_i$ (these elements are equivalent to the b and c terms of Pope 1970). $\underset{\sim}{n}_o$ may alternatively be expressed as

$$\underset{\sim}{n}_o = A^{-1} K$$

For $E \neq 0$

$$\Delta \underset{\sim}{n}_E = \underset{\sim}{r} - E\underset{\sim}{q}$$

$$\underset{\sim}{n}_E = \Delta^{-1}\underset{\sim}{r} - E\Delta^{-1}\underset{\sim}{q}$$

$$= \underset{\sim}{n}_o - E\Delta^{-1}\underset{\sim}{q}$$

i.e., all populations linearly decrease with increasing effort. The yield for each species is $q_i EN_i$, and the total yield, Y, is given by

$$Y = E\underset{\sim}{q}^T \underset{\sim}{n}_E = E\underset{\sim}{q}^T \underset{\sim}{n}_o - E^2 \underset{\sim}{q}^T \Delta^{-1} \underset{\sim}{q}$$

i.e., total yield is parabolic in fishing effort, and catch per unit effort is linear in effort. The maximum yield of this parabola is

$$Y_{max} = \frac{(q^T n_o)^2}{4\,(\underset{\sim}{q}^T \Delta^{-1} \underset{\sim}{q})}$$

As Pope (1979) has shown, the maximum on this particular parabola is obtained when the total catch per unit effort of all species combined is reduced to half its initial level. It is important to note that the vector of catchabilities must be constant throughout, and that the fishery transverses a particular yield parabola because of this and the form of the underlying community model. If individual species catchabilities are under management control, the global maximum yield can be achieved at some species mix $\underset{\sim}{n}^*_{E_{max}}$, obtained by setting the derivative of Y with respect to each N_k to zero

$$Y = \sum^{i} q_i EN_i = \sum^{i} r_i N_i - \sum^{i} \frac{r_i}{K_i} \sum^{j} \alpha_{ij} N_j$$

$$\frac{\partial Y}{\partial N_k} = r_k - \sum^{j} N_j^* (\delta_{kj} + \delta_{jk}) = 0$$

$$\underset{\sim}{n}^*_{E_{max}} = (\Delta + \Delta^T)^{-1} \underset{\sim}{r} = (\Delta + \Delta^T)^{-1} \Delta \underset{\sim}{n}_o$$

Without further assumptions this cannot usefully be simplified. However if Δ is symmetric, then

$$(\Delta + \Delta^T)^{-1} = \tfrac{1}{2} \Delta^{-1}$$

giving

$$\underset{\sim}{n}^*_{E_{max}} = \tfrac{1}{2} \underset{\sim}{n}_o$$

and in this case the global maximum is obtained by reducing each species to half its original biomass. Symmetric Δ requires that

$$\frac{r_i a_{ij}}{K_i} = \frac{r_j a_{ji}}{K_j}$$

for all i and j (i = j). A sufficient condition for symmetric Δ is that all r_i/K_i values are equal and that A is symmetric.

Symmetric A requires that all species pairs be competitive equals, and specifically excludes predator-prey relationships. While a symmetric A matrix results from the application of some techniques for estimating α_{ij} (see May 1975), it would seem likely that this is an artefact rather than a true attribute of communities (May 1975; Lawlor 1980). This treatment also requires that the Lotka-Volterra equation set with constant parameters must describe the community over a wide range of species abundances (at least from virgin to half virgin abundance for each species), although it is usually justified as being a reasonable approximation at equilibrium (e.g., Anon. 1978).

The behavior of groups of species within the community may also be examined using the above equations. If a set, Φ, of n_Φ species are combined, so that the sum of their abundances, ϕ, is

$$\phi = \sum^{j\epsilon\Phi} n_j$$

then the behavior of the species group to fishing is given by

$$\phi_E = \phi_o - E\underset{\sim}{X}\underset{\sim}{q}$$

where $\underset{\sim}{X}$ is a row vector with elements

$$x_i = \sum^{j\epsilon\Phi} \delta_{ij}^{-1}$$

i.e., each element is the column sum of the elements of the inverse of Δ, the sum being taken over the rows corresponding to species in the set Φ. As before, the abundance of the group decreases linearly with fishing effort, with the gradient determined by Δ and $\underset{\sim}{q}$. This gradient can be further decomposed into a component due to the nature of the system (the elements of Δ) and a component due to the species chosen for combination by noting that

$$x_i = \frac{x_i'}{|\Delta|} \text{ giving } \phi_E = \phi_o - \frac{E\underset{\sim}{X}'\underset{\sim}{q}}{|\Delta|}$$

where | Δ | is the determinant of Δ and x_i' is the element of a row vector made up of the column sums of the adjoint of Δ (Searle 1966), the sums being taken over the species chosen for combination. The gradient for each species group is thus made up of sums of products of all elements in the matrix Δ and all catchabilities. The above equations describe the behavior of 'combined species' data, but no general properties due to particular groupings of species (e.g., all predators) are apparent.

While this model may be used to obtain an indication of overall behavior, quantitative application is restricted by the difficulty of parameter estimation. All of the earlier described problems associated with estimating the α_{ij} terms apply, and their use in a fisheries context additionally requires that q_i and r_i/K_i be known for each species. Separate and direct estimation of parameters for even this simple model appears prohibitively difficult. Some advance might be made by applying an approach similar to that employed by Yeaton and Cody (1974). Noting that Δ = BA, where B is a diagonal matrix of elements r_i/K_i,

$$\underset{\sim}{n}_E = B^{-1}A^{-1}\ (\underset{\sim}{r} - E\underset{\sim}{q}) = \underset{\sim}{n}_o - E\Delta^{-1}q = n_o - E\,B^{-1}A^{-1}\underset{\sim}{q}$$

If $\underset{\sim}{n}_o$ is known and A is estimated by one of the earlier described techniques (but noting their limitations), then the elements of B and $\underset{\sim}{q}$ might be estimated from a time series of abundance and fishing effort data. Separate estimation of q may be necessary to avoid difficulties of indetermination in the estimation of B and q, caused by their occurring only as a product. If $\underset{\sim}{n}_o$ was not known it would also have to be estimated, bringing the number of parameters estimated to 3n for n species (the full model contains n^2 + 3n parameters). The matrix Δ may then be constructed, and the yield-effort relationship and maximum yield calculated using the equations provided earlier. Such a technique would at best be very approximate, given the large number of assumptions made and the sampling difficulties involved in obtaining α_{ij} and $\underset{\sim}{n}$. An indication of sensitivity may be obtained by noting that simply estimating the matrix A to within a constant of proportionality of the true matrix (i.e., $\hat{A}$ = kA) results in the estimated maximum yield being k times the true maximum.

Alternatively the parameters may be estimated by fitting the equation set to data. This approach was applied to laboratory populations by Gause (1934), Vandermeer (1969) and Silliman (1975), and to a continental shelf fish community by Sissenwine et al. (1982). The large number of parameters involved necessitates collection of a long time series of abundance data before this estimation procedure could be applied to communities of even moderate diversity. Besides it being difficult to reliably collect such data, the assumption of constant model parameters (particularly $\underset{\sim}{q}$) becomes increasingly tenuous with the length of the time series. The possibility of confounding temporal effects, which could lead to incorrect conclusions, also increases.

The study by Sissenwine et al. (1982) provides a good example of the difficulties involved. Using a sixteen year data set of population abundances (from research vessel surveys) and fishery catches covering 29 species, they estimated 740 parameters in an equation set of Lotka-Volterra form. Only 27 (4%) of these estimates were significantly different from zero at the 5% level, which is less than the number expected on the basis of type I error* alone in this number of comparisons. It had been suggested previously, on the basis of food consumption (Edwards and Bowman 1979), that one of the main predator species in this system, the silver hake *(Merluccius bili-*

*Type I error is the rejection of a true null hypothesis (Sokal and Rohlf 1969).

neatus), 'plays a principal predatory role in regulating the ecosystem' and that 'the species composition of the ecosystem is dependent to some significant degree on the population status of the silver hake, and to a lesser extent the spiny dogfish'. Yet Sissenwine et al. (1982) found only four significant interactions involving silver hake and/or dogfish out of 116 examined, again within the limits of type I error.

There are several possible interpretations of these findings, one being that over the range of population sizes examined interspecific interactions were unimportant. This interpretation is consistent with the view that behavior of the Lotka-Volterra equation set at equilibrium is largely irrelevant (e.g., Caswell 1978; Huston 1979) and/or that the physical environment plays a major role in community dynamics (Nicholson 1954; Klomp 1962; Solomon 1964). The work of Lasker (1978a, b) on the influence of the physical environment on recruitment provides one possible mechanism for this role, and the mutual modification of biotic and environmental influences (emphasized by the early ecologists) has been stressed by Hempel (1978) in his review of stock changes in the North Sea.

Other interpretations, however, include that the considerable sampling variability in the data set masked the effects of interactions, that the time series was not sufficiently long to provide precise estimates of the large number of parameters, that confounding temporal influences involving the fishery and/or physical environment obscured the interactions, and that the fishery provided a non-orthogonal experimental design with respect to species abundances from which detection of interactions would be difficult. Changes in community composition would occur with development of a fishery on an assemblage of completely independent species if the catchability and/or population parameters of all species were not identical. The difficulty is to separate this effect from that of true interactions. Even with an extremely good data set unambiguous interpretation of fitted parameters may be impossible. The finding of significant interactions could be due to fitting the wrong model or confounding temporal changes in the physical environment and/or species catchabilities; failure to find evidence of such interactions could be due to the same influences, plus an inadequate data set or experimental design.

In part, these problems arise from the difficulties in falsifying a complex, multi-parameter model with many alternative underlying mechanisms. They ensure, however, that despite intentions the use of such data in the biological management of multispecies fisheries will remain on a strictly empirical level for some time to come.

Conclusions

Examination of ecological and fishery theory indicates that at present there is no ecologically adequate model of community dynamics and no adequate method of estimating the parameters of the models that are available. No coherent body of methods and rules which might be applied to management of tropical multispecies fisheries is apparent in the approaches reviewed.

The mechanism invoked by the holistic community model of Margalef cannot be supported, and it has been shown that mechanisms which do not involve biological interactions can also generate many of the community attributes the diversity-stability hypothesis attempted to explain. The empirically determined community properties utilized by Margalef remain however, and their existence supports the continued development of holistic concepts of community structure and function.

The ecological niche approach historically began with a broad definition of the niche, encompassing both physical and biotic factors, but its recent quantitative development and application has focused on an unrealistically narrow range of mechanisms associated with exploitative competition for food resources at equilibrium. It is alarming that parameter estimation even for this restricted class of mechanism is prohibitively difficult. Inclusion of more realistic and varied mechanisms could be expected to compound the problems of parameter estimation.

Development of equations governing the behavior of groups of species from the n-species equation set, an attempt to derive 'intermediate level' community behavior from the behavior of the individual species, produced no generalities.

The problems involved in construction of a realistic community model from reductionist approaches (i.e., from examination of the dynamics of individual species and their interactions) are great, and will not be easily overcome. It is particularly difficult because of the lack of consistently collected, controlled, time series of species abundances and catches (unencumbered by possible but unknown changes in $\underset{\sim}{q}$), from which to obtain a clear indication of just what behavior a community model must describe.

There is a need for continued development of understanding of the details of individual species behavior and the nature of interaction processes, but it is also apparent that serious attention should be given to collection of empirical data sets describing fishery catches and the fish communities response to various fishing regimes. These management experiments should be conducted in as controlled a manner as possible to help reduce the influence of confounding temporal factors, and emphasis should be placed upon determining the constancy, or otherwise, of the species catchabilities. The collection of reliable data on both the fishery withdrawals and the structure of the community (probably from research vessel surveys) is essential to this approach, implying a considerable level of commitment if it is to be pursued. Casual attempts are unlikely to succeed. Collection of these data sets would seem essential, however, if only for the crucial role they could play in hypothesis testing.

One such experiment has been suggested for the Northwest shelf of Australia (Sainsbury 1982), and a further potential opportunity is provided by the Indonesian Government's recent closure of the Java Sea to trawling (Sardjono 1980).

Empirical management advice could be formulated on the basis of the data from experimental management regimes (augmented by cautious application of existing single-species and community models) on a more immediate time scale than that imposed by ecological research. Indeed it could reasonably be claimed, on the basis of the current understanding of community dynamics, that empirically supported guidance is the only guidance which can be given with reasonable confidence.

Accumulation of data from several such experiments would help formulation of a body of empirically determined observations or 'rules' at the level of species groups, providing the much needed focus for integration of knowledge of single species and their interactions into a community model.

Development of community theory requires that mechanisms be examined in addition to collection of empirical data on system behavior. Studies must correct an alarming lack of knowledge concerning the biology of tropical continental shelf fishes. At present many cannot even be consistently identified. Basic items in need of attention include taxonomy, life history descriptions and estimation of current population parameters for at least the more abundant species. Concurrently the validity of the basic assumptions of quantitative niche theory, and the MacArthur model in particular, must be examined for tropical fish communities. Most importantly these include the assumption that the availability of food resources to post-larval animals limits population sizes and that biotic factors are more important than abiotic ones in determining population sizes.

Finally, it is apparent that current moves toward the management of fisheries on a multispecies or community basis is bringing fisheries biology and ecology closer together than has generally been the case over the past few decades. While solutions are clearly distant as yet, fisheries studies with their extensive data gathering facilities and need for prediction on perturbed systems may prove to be an excellent testing ground for some of the ecological hypotheses generated in recent years.

References

Abrams, P.A. 1977. Density-independent mortality and interspecific competition: a test of Pianka's overlap hypothesis. Amer. Nat. 111: 539-552.

Abrams, P.A. 1980. Some comments on measuring niche overlap. Ecology 61: 44-49.

Andrewartha, N.G. and L.C. Birch. 1954. The distribution and abundance of animals. Univ. Chicago Press, Chicago.

Anon. 1978. Some scientific problems of multispecies fisheries. FAO Fish. Tech. Pap. No. 181. 42 p.

Caswell, H. 1976. Community structure: a neutral model analysis. Ecol. Monog. 46: 327-354.

Caswell, H. 1978. Predator-mediate co-existence: a non-equilibrium model. Amer. Nat. 112: 127-154.

Chitty, D. 1960. Population processes in the vole and their relevance to general theory. Can. J. Zool. 38: 99-113.

Cody, M.L. 1968. On the methods of resource division in grassland bird communities. Amer. Nat. 102: 107-147.

Cody, M.L. 1974. Competition and community structure. Princeton Univ. Press, Princeton.

Connell, J.H. 1978. Diversity in typical rainforests and coral reefs. Science 199: 1302-1310.

Connell, J.H. and E. Orias. 1964. The ecological regulation of species diversity. Amer. Nat. 98: 399-414.

Connor, E.F. and D. Simberloff. 1979. The assembly of species communities: chance or competition. Ecology 60: 1132-1140.

Cox, D.R. and H.D. Miller. 1965. The theory of stochastic processes. Methuen, London.

Cramer, N.F. and R.M. May. 1971. Interspecific competition predation and species diversity: a comment. J. Theor. Biol. 34: 289-293.

Crowder, L.B. 1980. Ecological convergence of community structure: a neutral model analysis. Ecology 61: 194-204.

Dayton, P.K. 1971. Competition, disturbance, and community organisation: the provision and subsequent utilization of space in a rocky intertidal community. Ecol. Monogr. 41: 351-389.

Dobzhansky, T. 1950. Evolution in the tropics. Amer. Sci. 38: 209-221.

Ebersole, J.P. 1980. Food density and territory size: an alternative model and a test on the reef fish *Eupomacentrus leucostictus*. Amer. Nat. 115: 492-509.

Edwards, R.L. and R.E. Bowman. 1979. Food consumed by continental shelf fishes, p. 387-406. *In* H. Clepper (ed.) Predator-prey systems in fishery management. Sport Fishing Institute, Washington, D.C.

Elton, C. 1927. Animal ecology. Sidgwick and Jackson, London.

Ewens, W.J. 1972. The sampling theory of selectively neutral alleles. Theor. Popul. Biol. 3: 87-112.

Fänge, R. and D. Grove. 1979. Digestion, p. 161-260. *In* W.S. Hoar, D.J. Randall and J.R. Brett (eds.) Fish physiology, Vol. VIII. Academic Press, New York.

Fischer, A.G. 1960. Latitudinal variations in organic diversity. Evolution 14: 64-81.

Fry, F.E.J. 1947. Effects of the environment on animal activity. University of Toronto Studies Biological Series 55. Publications of the Ontario Fisheries Research Laboratory. 68: 1-62.

Gause, G.F. 1934. The struggle for existence. Hafner, New York.

Gill, D.E. 1974. Intrinsic rate of increase, saturation density, and competitive ability. II. The evolution of competitive ability. Amer. Nat. 108: 103-116.

Glasser, J.W. 1978. The effect of predation on prey resource utilization. Ecology 59: 724-732.

Glasser, J.W. 1979. The role of predation in shaping and maintaining the structure of communities. Amer. Nat. 113: 631-641.

Goodman, D. 1974. The validity of the diversity-stability hypothesis, p. 75-79. *In* Structure, functioning and management of ecosystems. Proc. First Int. Congress of Ecology, The Hague. Centre for Agricultural Publishing and Documentation, Wageningen.

Goodman, D. 1975. The theory of diversity-stability relationships in ecology. Q. Rev. Biol. 50: 237-266.

Grinnell, J. 1917. The niche-relationship of the California thrasher. Auk 34: 427-433.

Grinnell, J. 1924. Geography and evolution. Ecology 5: 225-229.

Hariston, N.G. 1980. Species packing in the salamander genus *Desmognathus*: what are the interspecific interactions involved. Amer. Nat. 115: 354-336.

Hempel, G. 1978. North Sea fisheries and fish stocks—a review of recent changes. Rapp. P.-v. Réun. Cons. Int. Explor. Mer 17: 145-167.

Holling, C.S. 1959. The components of predation as revealed by a study of small mammal predation of the European saw fly. Can. Entomol. 91: 293-320.

Holling, C.S. 1973. Resilience and stability of ecological systems. Annu. Rev. Ecol. Syst. 4: 1-23.

Huffaker, C.B. 1958. Experimental studies on predation: dispersion factors and predator-prey oscillations. Hilgardia 27: 343-383.

Huston, M. 1979. A general hypothesis of species diversity. Amer. Nat. 113: 81-101.

Hutchinson. G.E. 1958. Concluding remarks. Cold Spring Harbor Symp. Quant. Biol. 22: 415-427.

Hutchinson, G.E. 1959. Homage to Santa Rosalia or why are there so many kinds of animals. Amer. Nat. 93: 145-159.

Hutchinson, G.E. 1965. The ecological theater and the evolutionary play. Yale Univ. Press, New Haven.

Inger, R.F. and R.K. Colwell. 1977. Organization of contiguous communities of amphibians and reptiles in Thailand. Ecol. Monogr. 47: 229-253.

Johnson, H.A. 1970. Information theory in biology after 18 years. Science 168: 1545-1550.

Kerr, S.R. 1980. Niche theory in fisheries ecology. Trans. Amer. Fish. Soc. 109: 254-260.

Klomp, H. 1962. The influence of climate and weather on the mean density level, the fluctuations and the regulation of animal populations. Arch. Neerl. Zool. 15: 68-109.

Klopfer, P.H. 1959. Environmental determinants of faunal diversity. Amer. Nat. 93: 337-342.

Larkin, P.A. 1963. Interspecific competition and exploitation. J. Fish. Res. Board Can. 20: 647-648.

Lasker, R. 1978a. Ocean variability and its biological effects—regional review—northeast Pacific. Rapp. P.-v. Réun. Cons. Int. Explor. Mer 173: 168-181.

Lasker, R. 1978b. The relation between oceanography conditions and larval anchovy food in the California current: identification of factors contributing to recruitment failure. Rapp. P.-v. Réun. Cons. Int. Explor. Mer 173: 212-230.

Lawlor, L.R. 1979. Direct and indirect effects of n-species competition. Oecologia (Berlin) 43: 355-364.

Lawlor, L.R. 1980. Structure and stability in neutral and randomly constructed competitive communities. Amer. Nat. 116: 394-408.

Levin, S.H. 1975. Ecosystem analysis and prediction. Society for Industrial and Applied Mathematics, Philadelphia.

Levine, S.H. 1976. Competitive interactions in ecosystems. Amer. Nat. 110: 903-910.

Levine, S.H. 1977. Exploitation interactions and the structure of ecosystems. J. Theor. Biol. 69: 345-355.

Levins, R. 1968. Evolution in changing environments: some theoretical explorations. Princeton Univ. Press, New Jersey.

Levins, R. 1975. Evolution in communities near equilibrium, p. 16-50. *In* M.L. Cody and J.M. Diamond (eds.) Ecology and evolution of communities. Harvard Univ. Press, Cambridge.

Leviten, P.J. and A.J. Kohn. 1980. Microhabitat resource use, activity pattern, and episodic catastrophe: *Conus* on tropical intertidal reef rock benches. Ecol. Monogr. 50: 55-75.

MacArthur, R.H. 1955. Fluctuations of animal populations and a measure of community stability. Ecology 36: 533-536.

MacArthur, R.H. 1968. The theory of the niche, p. 159-176. *In* R.C. Lewontin (ed.) Population biology and evolution. Syracuse Univ. Press, Syracuse.

MacArthur, R.H. 1970. Species packing and competitive equilibrium for many species. Theor. Popul. Biol. 1: 1-11.

MacArthur, R.H. 1972. Geographical ecology: patterns in the distribution of species. Harper and Row, New York.

MacArthur, R.H. and R. Levins. 1967. The limiting similarity, convergence, and divergence of co-existing species. Amer. Nat. 101: 377-385.

Margalef, R. 1957. La teoria de la informacion en ecologia. Mem. R. Acad. Cienc. Artes Barcelona 32: 373-449. (Transl. in Gen. Syst. 3: 36-71. 1968.)

Margalef, R. 1963. On certain unifying principles in ecology. Amer. Nat. 97: 357-374.

Margalef, R. 1968. Perspectives in ecological theory. University of Chicago Press, Chicago.

Margalef, R. 1969. Diversity and stability: a practical proposal and a model of interdependence. Brookhaven Symp. Biol. 22: 25-37.

May, R.M. 1974. Stability and complexity in model ecosystems. Princeton Univ. Press, New Jersey.

May, R.M. 1975. Some notes on estimating the competition matrix, α. Ecology 56: 737-741.

May, R.M. and R.H. MacArthur. 1972. Niche overlap as a function of environmental variability. Proc. Nat. Acad. Sci. U.S.A. 69: 1109-1113.

Moermond, T.C. 1979. Resource partitioning: a dynamic competitive balance, p. 303-309. *In* H. Clepper (ed.) Predator-prey systems in fishery management. Sport Fishing Institute, Washington, D.C.

Neill, W.E. 1974. The community matrix and interdependence of the competition coefficients. Amer. Nat. 108: 399-408.

Nicholson, A.J. 1954. An outline of the dynamics of animal populations. Aust. J. Zool. 2: 9-65.

Nisbet, R.M., W.S.C. Gurney and M.A. Pettipher. 1978. Environmental fluctuations and the theory of the ecological niche. J. Theor. Biol. 75: 223-237.

Nixon, M.A. 1980. Food production and competitor density as the determinants of feeding territory size. Amer. Nat. 115: 510-530.

Nursall, J.R. 1977. Speculation concerning speciation in coral reef fishes. Mar. Res. Indones. 17: 133-139.

Odum, E.P. 1953. Fundamentals of ecology. Saunders, Philadelphia.

Odum, E.P. 1962. Relationships between structure and function in the ecosystem. Jap. J. Ecol. 12: 108-118.

Odum, E.P. 1969. The strategy of ecosystem development. Science 164: 262-270.

Odum, E.P. 1971. Fundamentals of ecology. 3rd ed. Saunders, Philadelphia.

Paine, R.T. 1966. Food web complexity and species diversity. Amer. Nat. 100: 65-75.

Paine, R.T. 1971. A short-term experimental investigation of resource partitioning in a New Zealand rocky intertidal habitat. Ecology 52: 1096-1106.

Park, T. 1962. Beetles, competition and populations. Science 138: 1369-1375.

Pauly, D. 1979. Theory and management of tropical multispecies stocks: a review, with emphasis on the South East Asian demersal fisheries. ICLARM Studies and Reviews No. 1. 35 p. International Center for Living Aquatic Resources Management, Manila.

Pianka, E.R. 1966. Latitudinal gradients in species diversity: a review on concepts. Amer. Nat. 100: 36-46.

Pianka, E.R. 1973. The structure of lizard communities. Annu. Rev. Ecol. Syst. 4: 53-74.

Pielou, E.C. 1966. Shannon's formula as a measure of specific diversity: its use and misuse. Amer. Nat. 100: 463-465.

Pimentel, D., W.P. Nagel and J.L. Madden. 1963. Space-time structure of the environment and the survival of parasite-host systems. Amer. Nat. 97: 141-167.

Pope, J.G. 1979. Stock assessment in multispecies fisheries, with special reference to the trawl fishery in the Gulf of Thailand. SCS/DEV/79/19. 106 p. South China Sea Fisheries Development and Coordinating Programme, Manila.

Preston, E.M. 1973. A computer simulation of competition among five sympatric congeneric species of Xanthid crabs. Ecology 54: 469-483.

Pyke, G.H., H.R. Pulliam and E.L. Charnov. 1977. Optimal foraging; a selective review of theory and tests. Q. Rev. Biol. 52: 137-154.

Regier, H.A. and H.F. Henderson. 1973. Towards a broad ecological model of fish communities and fisheries. Trans. Amer. Fish. Soc. 102: 57-72.

Roughgarden, J. 1974. Species packing and the competition function with illustrations from coral reef fish. Theor. Popul. Biol. 5: 163-186.

Roughgarden, J. 1976. Resource partitioning among competing species—a co-evolutionary approach. Theor. Popul. Biol. 9: 388-424.

Rozenzweig, M.L. 1977. Aspects of biological exploitation. Q. Rev. Biol. 52: 371-380.

Ryder, R.A. 1970. Major advances in fisheries management in North American glacial lakes, p. 115-127. *In* N.G. Benson (ed.) A century of fisheries in North America. Amer. Fish. Soc. Spec. Pub. 7. Washington, D.C.

Sainsbury, K.J. 1982. The biological management of Australia's multispecies tropical demersal fisheries: a review of problems and some approaches. CSIRO Marine Laboratories Report No. 147.

Sale, P.F. 1974. Overlap in resource use, and interspecific competition. Oecologia (Berlin) 17: 254-256.

Sale, P.F. 1978. Coexistence of coral reef fishes—a lottery for living space. Environ. Biol. Fish. 3: 85-102.

Sale, P.F. and R. Dybdahl. 1978. Determinants of community structure for coral reef fishes in isolated coral heads at lagoonal and reef slope sites. Oecologia (Berlin) 34: 57-74.

Sanders, H.L. 1969. Benthic marine diversity and the stability-time hypothesis. Brookhaven Symp. Biol. 22: 71-81.

Sardjono, I. 1980. Trawlers banned in Indonesia. ICLARM Newsl. 3(4): 3.

Schoener, T.W. 1970. Non-synchronous spatial overlap of lizards in patchy habitats. Ecology 51: 408-418.

Schoener, T.W. 1973. Population growth regulated by intraspecific competition for energy or time: some simple representations. Theor. Popul. Biol. 4: 56-84.

Schoener, T.W. 1974a. Some methods for calculating competition coefficients from resource utilization spectra. Amer. Nat. 108: 332-340.

Schoener, T.W. 1974b. Resource partitioning in ecological communities. Science 185: 27-39.

Searle, S.R. 1966. Matrix algebra for the biological sciences. John Wiley and Sons, New York.

Shannon, C. and W. Weaver. 1949. The mathematical theory of communication. Univ. of Illinois Press, Urbana.

Silliman, R.P. 1975. Experimental exploitation of competing fish populations. Fish. Bull. 73: 872-888.

Simpson, G.G. 1964. Species diversity of recent North American mammals. Syst. Zool. 13: 57-73.

Sissenwine, M.P., B.E. Brown, J.E. Palmer, R.J. Essig and W. Smith. 1982. An empirical examination on population interactions for the fishery resources off the northeastern U.S.A., p. 82-94. *In* M.C. Mercer (ed.) Multispecies approaches to management advice. Can. Spec. P bl Fish. Aquat. Sci. 59. 169 p.

Slobodkin, L.B. and H.L. Sanders. 1969. On the contribution of environmental predictability to species diversity. Brookhaven Symp. Biol. 22: 82-95.

Sokal, R.R. and F.J. Rohlf. 1969. Biometry. Freeman and Co., San Francisco.

Solomon, M.E. 1964. Analysis of process involved in the natural control of insects. Adv. Ecol. Res. 2: 1-54.

Sousa, W.P. 1979. Disturbance in marine intertidal boulder fields: the non-equilibrium maintenance of species diversity. Ecology 60: 1225-1239.

Southwood, T.R.E. 1961. The numbers of species of insect associated with various trees. J. Anim. Ecol. 30: 1-8.

Steele, J.H. 1974. The structure of marine ecosystems. Harvard Univ. Press, Cambridge.

Teranoto, E., K. Kawasaki and N. Shigesada. 1979. Switching effect of predation on competitive prey species. J. Theor. Biol. 79: 303-315.

Vandermeer, J.H. 1969. The competitive structure of communities: an experimental approach with protozoa. Ecology 50: 362-370.

Vandermeer, J.H. 1980. Indirect mutualism: variations on a theme by Stephen Levine. Amer. Nat. 116: 441-448.

Vandermeer, J.H. and D.H. Boucher. 1978. Varieties of mutualistic interaction in populaion models. J. Theor. Biol. 74: 549-558.

Wallace, G. 1974. Chairman. Australian Government and Government of the State of Queensland Royal Commissions into exploratory and production drilling for petroleum in the sea of the Great Barrier Reef. Australian Government Publishing Service, Canberra.

Wellington, W.G. 1960. Qualitative changes in natural populations during changes in abundance. Can. J. Zool. 38: 289-314.

Whittaker, R.H. and S.A. Levin. 1977. The role of mosaic phenomena in natural communities. Theor. Popul. Biol. 12: 117-139.

Whittaker, R.H., S.A. Levin and R.B. Root. 1973. Niche, habitat and ecotope. Amer. Nat. 107: 321-338.

Wilbur, H.M. 1972. Competition, predation, and the structure of the *Ambystoma-Rana sylvetica* community. Ecology 53: 3-21.

Williams, C.B. 1964. Patterns in the balance of nature. Academic Press, New York.

Yeaton, R.I. and M.L. Cody. 1974. Competitive release in island song sparrow populations. Theor. Popul. Biol. 5: 42-58.

Discussion of Paper by Dr. Sainsbury

There was general support for the experimental approach to management, including experiments on a restricted scale that involve fishing techniques which may at first appear undesirable. In agreeing with the conclusions drawn in the review, Dr. Sale emphasized the view that much of current community theory is based on the assumption that competition for food is the principal determinant of community structure and function, and that this assumption may not be justified.

Discussion continued on the general problem of examining and falsifying models which contain a multiplicity of hypotheses, and the need to clearly identify the hypothesis being examined during a particular experiment.

There was discussion on the interrelationships between experiments designed to answer a scientific question and those designed to empirically explore the response of a fishery to a new fishing regime (e.g., if a new gear is being considered for use, it could be tried in a restricted region and the results accessed). It was felt that management of most fisheries proceeds in this empirical manner, and that documenting the biological response under the new and old fishing regimes would aid both empirical management of the fishery and research toward better understanding of how the community functioned. It was recognized however, that not all research questions could be answered in this way.

Ecosystems, Food Chains and Fish Yields

R. JONES

Department of Agriculture and Fisheries
Marine Laboratory
P.O. Box 101, Victoria Road
Aberdeen AB9 8DB, Scotland

JONES, R. 1982. Ecosystems, food chains and fish yields, p. 195-238. *In* Pauly, D. and G.I. Murphy (eds.) Theory and management of tropical fisheries. ICLARM Conference Proceedings 9, 360 p. International Center for Living Aquatic Resources Management, Manila, Philippines and Division of Fisheries Research, Commonwealth Scientific and Industrial Research Organisation, Cronulla, Australia.

Abstract

This paper reviews the marine and freshwater literature on ecosystems, food chains and fish yields. Principal conclusions are:

- fish yields depend partly on the level of primary production and partly on the harvesting level;
- the level of primary production and the harvesting level of commercial-sized animals are not independent. As primary production is increased, there is a tendency for particle size to increase at each trophic level, and this is associated with the production of animals of a size large enough to be commercially valuable at progressively lower levels in the food chain;
- in temperate regions, fish biomass is made up of a relatively small number of species. This may be because there are relatively few strategies for making the most effective use of large seasonal variations in food availability;
- a tropical marine system tends to be nearer to a steady-state system and a more uniform predation pressure may be at least one factor responsible for the relatively large number of species. It would seem reasonable to expect exploitation to lead to changes in species composition. In general, replacement of species of commercial value is likely to be by smaller-bodied species; however, the exact nature of these changes generally remains unpredictable;
- except for bacteria, average transfer efficiencies are unlikely to be much more than 10%. Higher values can be used in calculations, but only at the expense of including additional steps. This is because higher values are associated with juvenile stages, and hence with the small-particle-sized parts of the food web. To get from a plant cell to a fish of harvestable size requires the equivalent of about one more step if high transfer efficiencies are assumed.

Introduction

Reviewed here is the literature which pertains to the general topic of aquatic ecosystems, food chains and fish yield; the review is organized such that all themes are covered which are of relevance to the other contributions in this volume. This review covers the following topics:

— Characteristics of some selected ecosystems
— Major differences between ecosystems
— Components and structure of food chains
— Transfer and ecological efficiencies
— Role of predation
— Predation pressure and species replacement
— Relationships between predation and species number in tropical waters.

An attempt has been made throughout this review to include as much information from tropical systems as possible; however, reference to temperate and other non-tropical systems was made when an important principle could not be illustrated otherwise, and in order to present material leading to the emergence of temperature or latitudinal trends which culminate in the tropics.

Characteristics of Some Selected Ecosystems

Biomass levels and rates of production can vary considerably from one ecosystem to another (Table 1). This section summarizes the main features of selected ecosystems.

North west Atlantic and North Pacific

The north Atlantic and the north Pacific are characterized by comparatively low rates of carbon fixation (about 0.2 $gC/m^2/day$). Annual production values are about 50 to 100 $gC/m^2/year$ (Table 2).

In both areas, the mixed layer (about 80 m) and euphotic zone (about 100 to 130 m) are comparatively deep. In the summer, nitrate levels in the mixed layer are often below the levels of detection, and chlorophyll levels also tend to be relatively low (Table 1).

In the Sargasso Sea there is an annual winter breakdown of the seasonal thermocline with subsequent circulation down to the permanent thermocline. Nutrients are mixed upwards and there is a spring bloom of diatoms from January through March or April. Following the formation of the seasonal thermocline, primary production declines and remains low for the remainder of the year. Nitrate appears to be the limiting nutrient (Dugdale and Goering 1967).

The central gyre of the north Pacific ocean is a trans-Pacific body of water extending approximately from 40° N to 15° N latitude and maintained by the surrounding, anti-cyclonic pattern of surface circulation. There is a thermocline which persists within the euphotic zone over a time measured at least in months if not in years. All the information suggests that an approximately steady state might pertain with respect to nutrient concentrations, plankton standing stocks, growth rates and grazing (Eppley et al. 1973).

In both areas, there is a zone of chlorophyll at the bottom of the euphotic zone, at a depth of about 100 to 120 m, in which the concentration of plant material is considerably higher than that in the mixed layer above. This zone contributes an important fraction of the total primary production of the water column (Venrick et al. 1973; Ortner et al. 1980). Presumably production in this zone is limited by light, whereas production above is limited by nutrients.

As one moves from coastal to offshore waters, the plants change from large micro-plankton (100 μ or larger) to much smaller nannoplankton of 5 to 25 μ (Ryther 1969).

Table 1. Comparison of biomass levels and rates of production in various regions.

	N.W. Atlantic and North Pacific	Northern North Sea	N.W. Africa	Peru	Lake George[o]
Carbon fixation ($g\ C/m^2/day$)	0.2	0.5[j]	2.4[a], 0.4[k]	4.8[a], 0.6[k]	—
Assimilation number (mg C/mg Chl/day)	27[p] in mixed layer	—	37[a]	90[a]	—
Productive season (days)	semi-continuous	200	200	270	365
Euphotic depth (m)	100-130[p]	40	21[a]	23[a]	< 75 cm
Mixed layer depth (m)	80[p]	40	38[a]	13[a]	
Nitrate (μg at/1)	—	—	4[a]	16[a]	Very low
(Seasonal range)	0.3-1.5[g]	0.5-7[j]	—	—	—
Chlorophyll (mean mg/m^3)	—	—	3.5[b], 2.5[d]	2.7[b], 1.5[e]	250
(Seasonal range mg/m^3)	0.1-0.5[f]	0.2-2.5			
mg/m^2	34	6-75	60	97	600
Zooplankton (mg/m^3 dry weight) (mean)	1[h]	7-70[j]	65[n], 60[l]	42[i], 32[l] 16-160[m]	490
g/m^2 dry weight	0.5	1-10	6	6	1
Zooplankton depth (m)	500	150	100	100	2.4
Fish biomass gC/m^2	—	1.5	29[c]	14[q]	1.4
Fish production $gC/m^2/yr$	—	0.3	—	—	8

[a]From Huntsman and Barber (1976).
[b]Calculated from data in Huntsman and Barber (1976) assuming 100 mg $Chl/m^2 \equiv 4.6$ mg/Chl/m^3
[c]Pelagic biomass only—from Whitledge (1978).
[d]Calculated from data in Watson (1978) assuming 1 $gC/m^2 \equiv 0.8$ mg Chl/m^3.
[e]Calculated from data in Beers et al. (1971) assuming 1 $gC/m^2 \equiv 0.8$ mg Chl/m^3.
[f]From Beers (1966).
[g]From Steele and Menzel (1962).
[h]From Beers (1966) assuming 1 mg $C/m^3 \equiv 1.67$ mg/m^3 dry weight.
[i]From Whitledge (1978).
[j]Based on annual primary production of 90g C/m^2 in Steele (1974).
[k]From Cushing (1969).
[l]From Whitledge (1978) assuming 100 m depth range for zooplankton.
[m]Calculated from Cushing (1969) assuming 1 ml/1,000 $m^3 \equiv 0.16$ mg/m^3 dry weight.
[n]Calculated from Houghton and Mensah (1978) assuming 1 ml/1,000 $m^3 \equiv 0.16$ mg/m^3 dry weight.
[o]From Greenwood (1976), Burgis et al. (1973) and Ganf (1975).
[p]In central gyre of N. Pacific. From Eppley et al. (1973).
[q]Anchoveta only—from Whitledge (1978).

The zooplankton biomass per square meter is relatively low and is distributed down to considerable depths. For example, Longhurst and Williams (1979) give details of the vertical zooplankton distribution in the north Atlantic. About 20 to 60% of the carnivores and 50 to 75% of the filter feeders above 1,000 m are located in a discrete epiplankton layer, which extends from the surface to a depth of about 300 to 400 m. Below the epiplankton, the zooplankton concentration is 15 to 30 times lower than in the epiplankton and extends down to 1,000 m or more.

The maximum concentration of filter feeders occurs at about half the depth of the chlorophyll maximum and that of carnivores is about coincident with the maximum chlorophyll concentration.

Ryther (1969), quoting Grice and Hart (1962), states that the percentage of carnivorous species in the zooplankton increased from 16 to 39% in a transect from the coastal waters of the northeastern United States to the Sargasso Sea. According to Longhurst and Williams (1979), as much as 60 to 70% of the zooplankton biomass in the north Atlantic consist of carnivores.

Table 2. Estimates of primary production from various oceans.[a]

	Area 10^6 km^2			Primary production gC/m^2/day		gC/m^2/year[c]		
Ocean	Total	Shelf	Offshore	Shelf	Offshore	Average[b]	Shelf	Offshore
Indian	73.8	2.8	71.0	0.71	0.23	89	259	83
Atlantic	92.6	8.6	83.9	0.41	0.28	105	150	101
Pacific	177.6	10.7	166.9	0.52	0.15	64	191	56
Antarctic	24-12	4.8		0.89		130		
Arctic	13.1	6.1	7.0					
Total		33.0						

[a]From Platt and Subba Rao (1975).
[b]Averages weighted by area.
[c]These data imply 365 days production/year, except for the Antarctic (146 days).

The food chain is comparatively long. This is partly because of the relatively small size of the primary producers in oceanic waters (Ryther 1969) and partly because the biomass of invertebrate carnivores appears to be large enough to require most of the zooplankton herbivore production. Small-bodied mesopelagic fish species can be regarded either as primary carnivores that depend on a relatively small proportion of the herbivore production, or as secondary carnivores that depend primarily on the production of the invertebrate carnivores. Harvesting tends to be at an even higher trophic level (tunas, squid, etc.). The number of steps from primary production to the fishery appears to be equivalent to at least four and may be as many as five (Ryther 1969).

Because of the considerable depth of water, benthic production is too far removed from the euphotic zone to contribute to commercial fish production. In the productive zone, the detritus food chain presumably leads to protozoa and micro-zooplankton and can make little contribution to fish production.

A Temperate Coastal Area (the North Sea)

The North Sea is typical of the relatively more productive coastal, continental shelf regions. Carbon fixation averages about 0.5 to 1.0 $gC/m^2/day$, with the higher levels of production occurring in the shallower water. Annual primary production ranges from about 60 gC/m^2 on the Fladen ground (Steele 1958) to 200 gC/m^2 in the English Channel (Cushing 1973a).

Because of the shallower water (mainly < 200 m), benthic production contributes to the existence of commercially exploitable, demersal fish species.

The food chain is of intermediate length, with harvestable fish species occurring mainly as primary and secondary carnivores.

Upwelling Areas

Some upwelling areas are found in the eastern boundary currents of the sub-tropical anti-cyclones where the winds tend to blow parallel to the coast towards the equator (Cushing 1969). The principle upwelling areas within this category are found off the coasts of Peru, California, Oregon, northwest and southwest Africa. There is also upwelling off Somalia, in the Arabian Sea, and in the Antarctic. In the Antarctic, upwelling occurs along the continental slope and also at the boundaries of the water masses of the Southern Ocean (Walsh 1971).

Associated with upwelling is the horizontal transport of surface water away from the point of upwelling. For example Cushing (1969), referring to upwelling associated with coastal regions, notes that the surface waters move offshore for about 100 km before the water sinks again. An important feature of an upwelling area is not simply that nutrient is brought continuously to the surface, but that it is transported horizontally for a long enough time for it to become incorporated into higher trophic levels before the water sinks again. This mechanism differs from a simple mixing process which also brings nutrient to the surface but at the expense of carrying plant cells down below the euphotic zone.

Ryther (1969), referring to the Peruvian upwelling system, states that many of the phytoplankton species are relatively large, and that some species are colonial and form large gelatinous masses or long filaments. Ryther, quoting Hart (1942) also refers to the colonial habit of all of the most abundant species of phytoplankton in the Antarctic where colonies up to several millimeters, and in some cases several centimeters in diameter occur. This could account for the relatively large size of many of the herbivores in upwelling systems (i.e., krill in the Antarctic, and sardines, anchovy and *Sardinella* in coastal upwelling systems). In the Antarctic, some of the primary carnivores also are exceptionally large (e.g., baleen whales).

The food chain, from phytoplankton to animals large enough to be of commercial value to man, tends to be relatively short.

Coastal Lagoons, Mangrove Swamps, Reefs and Weed Beds

The basic shape of a coastal lagoon with a sand barrier is rectangular, with a long axis parallel to the coast (Fig. 1). There may be an associated river valley more or less perpendicular to the shoreline. The sand-lagoon barrier

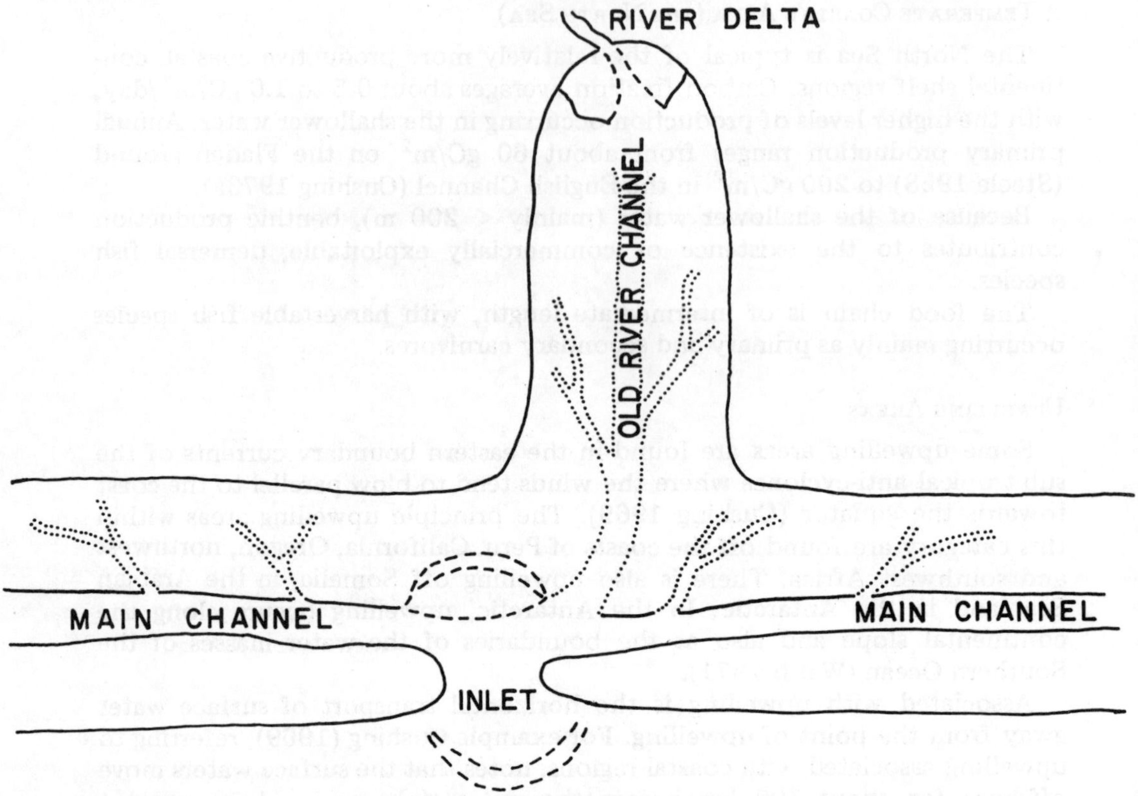

Fig. 1. Diagram illustrating some of the features of coastal lagoons (modified from Phleger 1970).

progrades seaward at a rate depending on the amount of sand available in the near-shore wave zone. The number and size of lagoon inlets are a function of the amount of water flowing through them, which in turn is a function of lagoon area, tidal range and frequency of river flow. The deepest channel in the lagoon is characteristically adjacent to the lagoon barrier (Phleger 1970).

Tidal rise and fall leads to currents with maximum velocities at approximately half-tide. In the channels, relatively high velocities produce intense turbulence and there is a regular decrease in velocity inside lagoons away from the inlet. As a result the coarsest sediments are found in and near the inlet, while the finer sediments are in the inner lagoon where the current velocity is lowest (Phleger 1970).

Qasim (1973a) estimated that the net annual production in a tropical estuary (a "backwater" near Cochin, India) was 124 gC/m^2/year and that of this, only about 30 gC/m^2/year was consumed by zooplankton herbivores.

Mangrove swamps can be regarded as a special type of coastal lagoon. These are found around the world in the tropical belt. They build up land by entangling soft muddy sediments between the roots and are commonly found near the estuaries of large rivers, often behind sand bars. Salinity may vary from almost zero to that of seawater. The water is usually turbid and the sediments rich in organic matter.

In mangrove swamps, primary production is usually high and includes production by numerous benthic algae which grow on the mud of the tidal flats and on the relatively enormous surface area offered by the stilt-roots and trunks of the mangroves, which are covered with diatoms and algae. Wium-Anderson (1979) estimated the rate of plankton primary production in a tropical mangrove bay in southwestern Thailand. Daily production varied from 560 mg to 2,410 mgC/m^2/day. Nannoplankton ($< 30\ \mu$) accounted for 35% of the production. Annual production was estimated as 468 gC/m^2.

Many shallow-water regions lend themselves particularly well to direct observation of fish behavior, and feeding habits by underwater observation (Sale 1980, this vol.). For example, Hobson (1973) lists tropical reef fish which feed during the day and rest at night, and others which feed at night and rest during the day. The feeding grounds of many are some distance from the resting grounds and migration between the two locations, which is usually at twilight, is part of a general change-over routine. As a result many of the migrations are predictable both as to time and route.

Ebeling and Bray (1976) investigated the feeding activities of kelp-bed fish off Santa Barbara, California. They found that the community followed the same basic pattern of diurnal activity as the tropical reef community but in a much looser manner.

Coral reefs are considered to be very productive, but make up only 0.65% of the aquatic environment (De Vooys 1979).

Tropical Fresh Waters, with Emphasis on Lake George, Uganda

Lowe-McConnell (1975) gives a useful, general account of fish communities in tropical fresh waters. Tropical regions are characterized by high temperatures, except where they are modified by altitude, and there is relatively little seasonal variation in temperature compared with temperate regions. The day length is practically constant at 12 hours throughout the year, and even at 10° N and 10° S annual variations in day length are less than 1 hour. Seasonal variations in most parts of the tropics are caused primarily by fluctuations in rainfall on a scale quite unknown outside the tropics. Rainfall maxima occur mainly around May-July north of the equator and November-January south of it. Large river systems may receive tributaries from both sides of the equator which flood at opposite times of the year causing a bimodal or prolonged flood season in the lower reaches. Fish respond to the rise in water level rather than to the local rains.

At flood time, water is greatly enriched with nutrient salts from the breakdown of organic matter, decaying vegetation and the droppings of animals. This leads to an explosive growth of plant and animal life, and juvenile production is generally timed to coincide with this period. As nutrients are depleted, the water level falls and de-oxygenation may occur. Many tropical fishes have respiratory adaptations to enable them to survive the dry periods. Mortality appears to be greatest when the water level is falling and during the dry season.

Lake George is a shallow (2.5 m) lake on the equator situated in Uganda. Gross primary production (including respiration) is estimated as 10 g O_2/m^2/day (Ganf 1975), which is equivalent to about 3 gC/m^2/day or 1,000 gC/

m^2/year. To determine particulate plus soluble production, an allowance must be made for respiration. According to Ganf (1974), respiration in Lake George makes up about 10 to 50% of the gross production. Annual production of particulate + soluble carbon can therefore be taken as between 500 and 900 gC/m^2/year.

Ninety-five percent of the total biomass in the open water is phytoplankton and 70 to 80% of this is blue-green algae (Ganf 1975). In spite of its extreme shallowness, the chlorophyll biomass under 1 m^2 of surface is about 18 times as great as it is under 1 m^2 of surface of the deep oceans. The chlorophyll concentration is at least 500 times as great as that in the euphotic zones of the mid-Atlantic and Pacific Ocean (Table 1).

The general, physical and chemical characteristics of the water column show marked diurnal changes. At dawn, the column is isothermal (about 25° C). During the day the column becomes thermally stratified and by mid-afternoon there may be a temperature gradient of 10° C. With the onset of nocturnal winds, the column rapidly returns to isothermy. The diurnal thought that this was mainly due to fishing (Lawrie 1978; Smith 1968). 9 and 10 (Ganf 1975).

A characteristic feature of the crustacean zooplankton is the small size of the individuals, the largest cladoceran *(Daphnia barbata)* being little more than 1 mm in length (Greenwood 1976). An important zooplankter is *Thermocyclops* and this is reported to be able to digest blue-green algae (Moriarty et al. 1973).

The fish community is dominated by the family Cichlidae of which at least one species *(Sarotherodon niloticus)* can digest blue-green algae (Moriarty et al. 1973).

According to Greenwood (1976), comparatively little of the primary production is transferred to higher trophic levels. The food chain is comparatively short, with fish being mainly herbivores and primary carnivores.

Major Differences Between Ecosystems

Differences between Ecosystems associated with Physical Features

Depth

Useful generalizations about depth appear to be that:

— benthic biomass per unit area decreases with increasing depth (Rowe et al. 1974);

— the most productive lakes appear to be very shallow (i.e., less than about 10 m). Fig. 2 shows how fish yield is related to depth in a group of African lakes;

— the highest levels of primary production are recorded from very shallow lakes. Hickman (1979) gives 1,400 gC/m^2/year for Ministik Lake, Alberta, Canada, a lake which has a mean depth of 1.8 m. Talling et al. (1973) give values of daily gross production of about 50 g O_2/m^2 for two shallow Ethiopian soda lakes. Off Long Island Sound, Smayda (1973) states that there is a seaward gradient in total annual primary production, decreasing from 470 gC/m^2/year inshore to 78 gC/m^2/year in the Sargasso Sea. Very deep bodies of water, such as the deep oceans, are generally associated with levels of primary production less than 100 gC/m^2/year.

Latitude (temperature)

Latitude and temperature tend to be correlated. Differences in ecosystems that appear to be associated with these features are benthic biomass, proportion of demersal fish in fishery landings, seasonal variability, numbers of species, body size and level of productivity.

Benthic biomass appears to be larger at high than at low latitudes (Table 3). Alton (1974), for example, gives values ranging from 55 to 905 g/m^2 wet weight for the Barents Sea. For the North Sea and English Channel, values range from less than 5 to 106 g/m^2 wet weight. For the west coast of Thailand, Petersen and Curtis (1980) give values ranging from 0 to 10 while for the waters between Iran and the Arabian peninsula, Thorson (1957) gives a number of values averaging about 40 g/m^2 wet weight. Although comparisons are made difficult by the fact that benthic biomass decreases very markedly with depth, the data indicate that benthic biomass tends to be larger at high than at low latitudes.

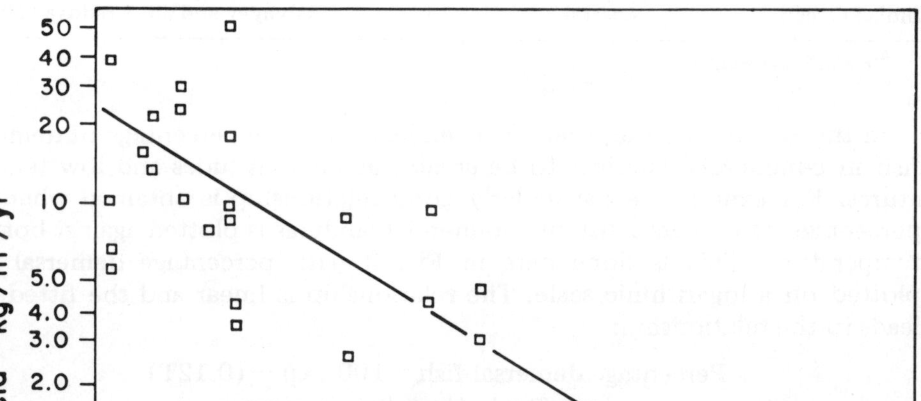

Fig. 2. Fish yields from African lakes of varying depths—after Fryer and Iles (1972) (modified from Lowe-McConnell 1975).

Table 3. Estimates of benthic biomass.

Area	Depth (m)	Biomass[a]	Source
Arctic: Scoresby Sound	10	35-360	Thorson (1957)
	100	20-200	Thorson (1957)
	150	15-110	Thorson (1957)
Bering Sea	—	55-905	Alton (1974)
Bering Sea	215	100	Zenkevitch (1963, in Crisp 1975)
Labrador	30-80	227-801	Mills (1975)
Plymouth	mud flat	13	Warwick and Price (1975)
Rame Head	50	108	Mare (1942)
Scottish East Coast	10-40	5	McIntyre (1958)
Fladen (N. Sea)	140	6	McIntyre (1961)
English Channel	80	55	Holme (1953)
Southern Bight of North Sea	50	106	Heip et al. (1979)
Waters between Iran and Arabian Peninsula	—	ca 40	Thorson (1957)
W. Coast of Thailand	—	mainly 0-10	Petersen and Curtis (1980)
Pakistan coast	shelf	10-15	Savich (1972)
Pakistan coast	slope	0.3-1.0	Savich (1972)
Sierra Leone	inshore	7	Longhurst (1957)
Sierra Leone	offshore	12	Longhurst (1957)
South China Sea	inshore	10	Mokyevsky (1960)
Indian Ocean	coastal	14	Belyayev and Vinogradova (1961)

[a]In g/m^2, wet weight.

In the North Atlantic, there is a tendency for the percentage of demersal fish in commercial landings to be greater at high latitudes and low temperatures. For example, a particularly good relationship is obtained when the percentage of demersal fish in commercial landings is plotted against bottom temperature. This is done here in Fig. 3 with percentage demersal fish plotted on a logarithmic scale. The relationship is linear and the fitted line leads to the relationship:

$$\text{Percentage demersal fish} = 100 \exp - (0.12T)$$
$$\text{where } T = \text{bottom temperature}$$

In the North Atlantic, the percentage of the landings made up of demersal fish decreases from over 90% at Labrador to about 30% in the Bay of Biscay area.

Seasonality also varies with latitude. Near the equator, production in the sea tends to be steady or semi-steady throughout the year. At high latitudes, production is seasonal and highly so at very high latitudes. For example, most of the primary production in the Bering Sea occurs in a period of only 3 to 4 weeks (Alexander and Cooney 1978).

The number of species also varies with latitude. For example, tropical fish communities are characteristically made up of very large numbers of species with none being outstandingly dominant. At high latitudes on the other hand, there tends to be a comparatively small number of species which make up a relatively large part, often as much as 80%, of the total fish biomass.

Body size at any one trophic level tends to increase with increasing latitude (e.g., some whales in the Antarctic are primary carnivores). Tropical plankton species tend to be smaller than boreal plankton (Pomeroy 1974, quoted by Sheldon and Sutcliffe 1978).

Regarding productivity, Brylinsky and Mann (1973) summarized data for 43 lakes and 12 reservoirs from the tropics to the Arctic. They concluded that the most productive were in latitudes 10 to 12° N. The least productive were at high latitudes or at high altitudes.

Different Types of Pathways

One of the ways in which ecosystems can differ is in the relative importance of the 'grazing' pathway as distinct from the detritus pathway. The grazing pathway means here the energy flow associated with grazing on living phytoplankton by herbivores. The detritus pathway refers to the breakdown of dead organic matter by bacteria and associated energy flows.

Grazing of phytoplankton within the water column

There is considerable variation in the extent to which phytoplankton is grazed within the water column by zooplankton herbivores. For example, grazing is believed to reach its maximum in the tropical oceans where values of the percentage going to the pelagic food chain range from 60% (Fenchel and Jørgensen 1977) to 100%.

In coastal regions, zooplankton grazing is usually believed to take up from about 50% to 80% of the plant production, although Riley (1956), referring to Long Island Sound, puts the figure as low as 25%.

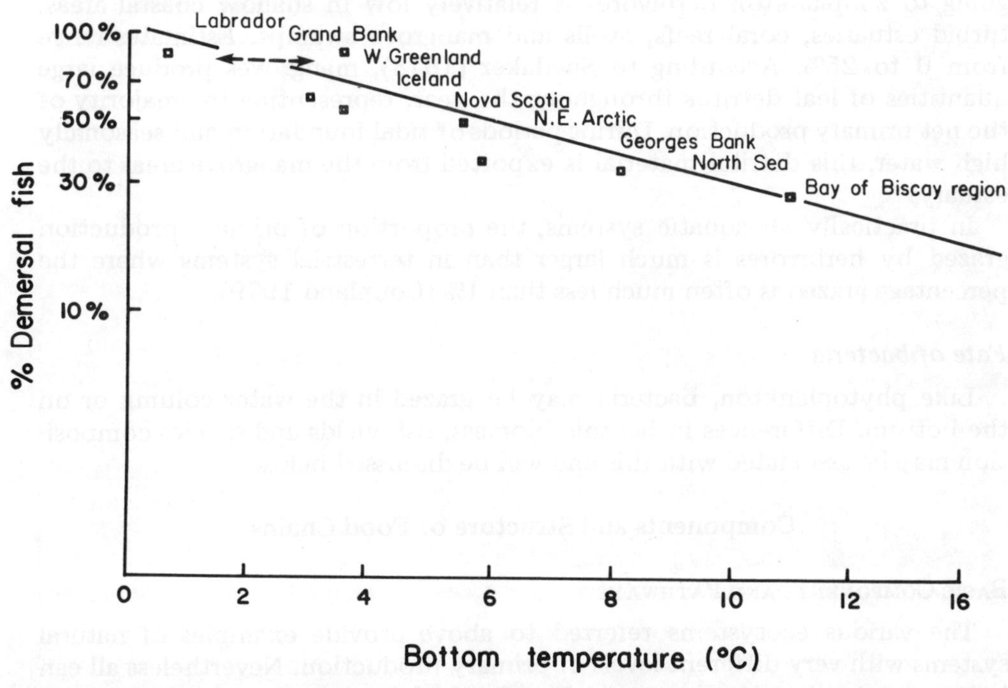

Fig. 3. Relationship between bottom temperature and percentage demersal fish in commercial landings—note logarithmic scale for percentage demersal fish.

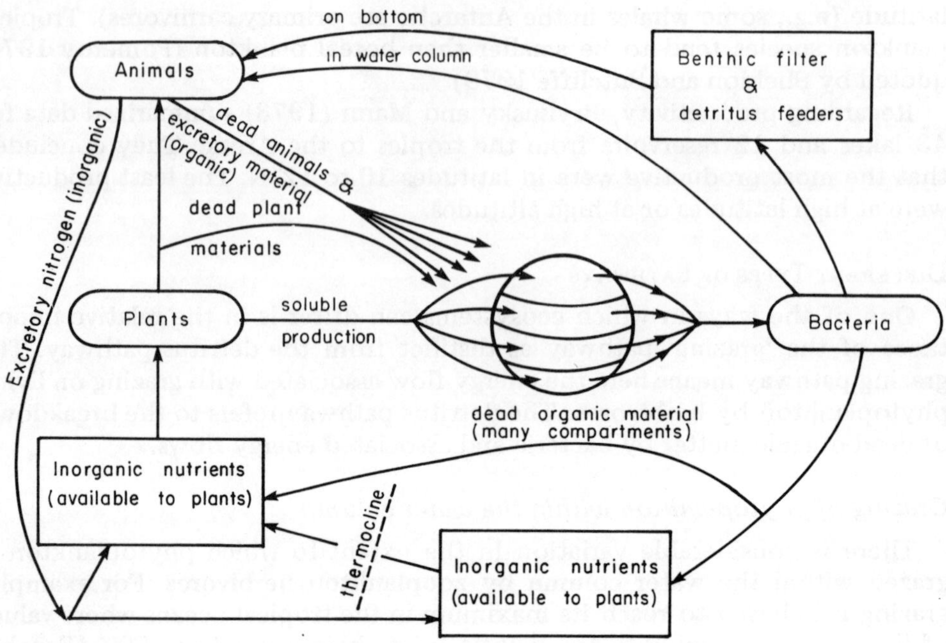

Fig. 4. Showing principle nitrogen compartments for a total ecosystem model.

According to Qasim (1973a, 1973b), the percentage of primary production going to zooplankton herbivores is relatively low in shallow coastal areas, turbid estuaries, coral reefs, atolls and mangrove swamps. Estimates range from 0 to 25%. According to Snedaker (1978), mangroves produce large quantities of leaf detritus throughout the year, representing the majority of the net primary production. During periods of tidal inundation and seasonally high water, this detrital material is exported from the mangrove areas to the estuary.

In practically all aquatic systems, the proportion of primary production grazed by herbivores is much larger than in terrestrial systems where the percentage grazed is often much less than 1% (Coupland 1979).

Fate of bacteria

Like phytoplankton, bacteria may be grazed in the water column or on the bottom. Differences in benthic biomass, fish yields and species composition may be associated with this and will be discussed below.

Components and Structure of Food Chains

Basic Components and Pathways

The various ecosystems referred to above provide examples of natural systems with very different levels of primary production. Nevertheless all can be represented by the diagram in Fig. 4, which represents a simple nitrogen model of a generalized ecosystem.

Principle components are plants, animals, dead organic material, bacteria and nutrients. Nutrients can be divided into nutrients in the euphotic zone (i.e., nutrients available to plants) and nutrients below the euphotic zone, which are not directly available to the plants.

Considerations of the circulation of energy show that there are two major pathways. One comprises available nutrients, plants and animals. The rate of turnover of nitrogen through this part of the system is relatively rapid, and nitrogen is returned to the nutrient pools via the excretion of ammonia or urea by the animals. The other major pathway involves the detritus food chain and is based on soluble and particulate dead organic material and its breakdown by bacteria.

Bacteria, either free living or associated with detritus, may be consumed in the water column or on the bottom. Bacteria in the water column are consumed by protozoans leading to a food chain that enters the animal pool at a relatively low trophic level (i.e., bacteria → protozoa → micro-zooplankton → carnivores). This energy source is unlikely to have much impact on the potential level of fish production. Bacteria associated with detritus that reach the bottom may be ingested by benthic detrital and filter feeders, many of which are large enough to be directly grazed by commercial fish, or by animals large enough to be eaten by fish. This food chain therefore enters the animal pool at a relatively high trophic level and has the potential to influence the overall level of fish production.

Nutrients are recycled by animals and by bacteria, either in the euphotic zone directly into the pool of available nutrient, or below the euphotic zone into the pool of non-available nutrient.

Factors affecting Primary Production

Extensive work, reported by many authors, suggests that the dynamics of the production process can be considered in three stages:

— The spring bloom. Available evidence suggests that the peak magnitude of plant biomass during the spring bloom is directly proportional to the nutrient concentration in the euphotic zone at the beginning of the annual cycle (Spencer 1954; Yentsch and Vaccaro 1958).
  The spring bloom typically ends when nutrient levels in the euphotic zone become extremely low. This is associated with a comparatively rapid fall-off in plant biomass and an increase in particulate matter in the water column. Calculations for the North Sea suggest that the increase in animal biomass is not likely to account for more than 40% of the nitrogen initially available in inorganic form. By the end of the spring bloom, therefore, as much as 60% of the nitrogen originally in inorganic form appears to have become transferred to the dead organic pool, either as soluble organic material released during the process of plant growth, or as particulate material emanating from plants and animals.

— Subsequent, or post-bloom production, is typically associated with low to very low levels of nutrient concentration in the euphotic zone. Available nutrient may be sufficient for only a few days production.

Post-bloom production can only continue, therefore, if the nutrient pool is continually supplemented by the regeneration of nutrient within the euphotic zone, or by the physical transfer of nutrient from the non-available pool to the available pool (Eppley and Peterson 1979). Simulation studies suggest that of the various factors that might affect the level of primary production, the most important appears to be the rate of transfer of nutrient from the non-available to the available nutrient pool. The relatively high productivity of very shallow lakes might thus be explained by the fact that mixing makes practically all nutrient available, without necessarily removing plants from the photic zone.

The high productivity of upwelling areas can be explained by the fact that there is a relatively high rate of transfer of nutrient to the available nutrient pool, without this necessarily being associated with the removal of plants from the photic zone.

In the freshwater literature, there are a number of references to the fact that the level of primary production tends to be inversely related to depth. Whilst this does not appear to be generally so for investigations over all depth ranges (Brylinsky and Mann 1973) it has been observed (Oglesby 1977) that the depth effect is only really noticeable for lakes less than 25 m deep.

In water deeper than about 25 m, the presence or absence of a thermocline, stratification and critical depth considerations appear to be the most important factors (Parsons and Le Brasseur 1968). In the North Sea for example, the relatively low primary production of the deep northern North Sea basin can be explained by the gradual development of a thermocline during the course of the summer. This presumably introduces a gradually increasing barrier to the transfer of nutrient from the non-available to the available pool. In New York Bight, autumn production is made possible as a result of the breakdown of the summer thermocline (Yentsch 1977).

The relatively low productivity of deep oceans appears to be largely due to the presence of a more or less persistent thermocline, minimizing the upward transfer of nutrients from the non-available nutrient pool in the deep water. Concentrations of nutrients in the euphotic zone are often below detection limits (Jackson 1980). According to Eppley and Peterson (1979), as much as 50% of the total primary production in the deep ocean is supported by nutrients coming from sources other than recycling by animals and bacteria.

Factors affecting Benthic Biomass

Variations in benthic biomass are presumably connected with the likelihood of bacteria and detritus reaching the bottom as distinct from being degraded in the water column. Depth and temperature are both factors that might be expected to influence this process.

For example, as water depth is increased, the chances of detritus being consumed by bacteria before reaching the bottom can be expected to increase. As far as temperature is concerned, bacterial action tends to be greater at higher temperature and this should make it more likely that detritus will be degraded in the water column before it has time to reach the bottom. In general, one might expect maximum water column degradation in deep warm water and minimum water column degradation in cold shallow water.

NOTES ON FOOD CHAINS

There is a limited amount of information in the literature from which food chains for various areas can be constructed.

North Sea

This is a well investigated area and several authors have drawn up food chains (Steele 1974; Jones 1978, Fig. 5). Harvesting occurs mainly at the primary and secondary carnivore levels.

Gulf of Thailand

Fig. 6 shows a food web for the Gulf of Thailand prepared by participants at the FAO/DANIDA seminar on the management of tropical demersal fisheries held in Bangkok, 22 October-2 November 1979. Some fish species are at least partial herbivores, but otherwise harvesting is at the primary and secondary carnivore level. Juveniles and even some larvae are taken commercially.

Yellow and East China Sea

Fig. 7 shows details based on Enomoto (1972). Chaetognaths and a small macruran shrimp, *Leptochela*, are important invertebrate carnivores intermediate between zooplankton and demersal fish. *Leptochela* is widespread in the southern areas of the Yellow Sea and East China Sea. It appears to migrate vertically from the surface to the bottom and is fed on by many fish species. Enomoto states that fish production amounts to about 0.00024% of the solar radiation energy.

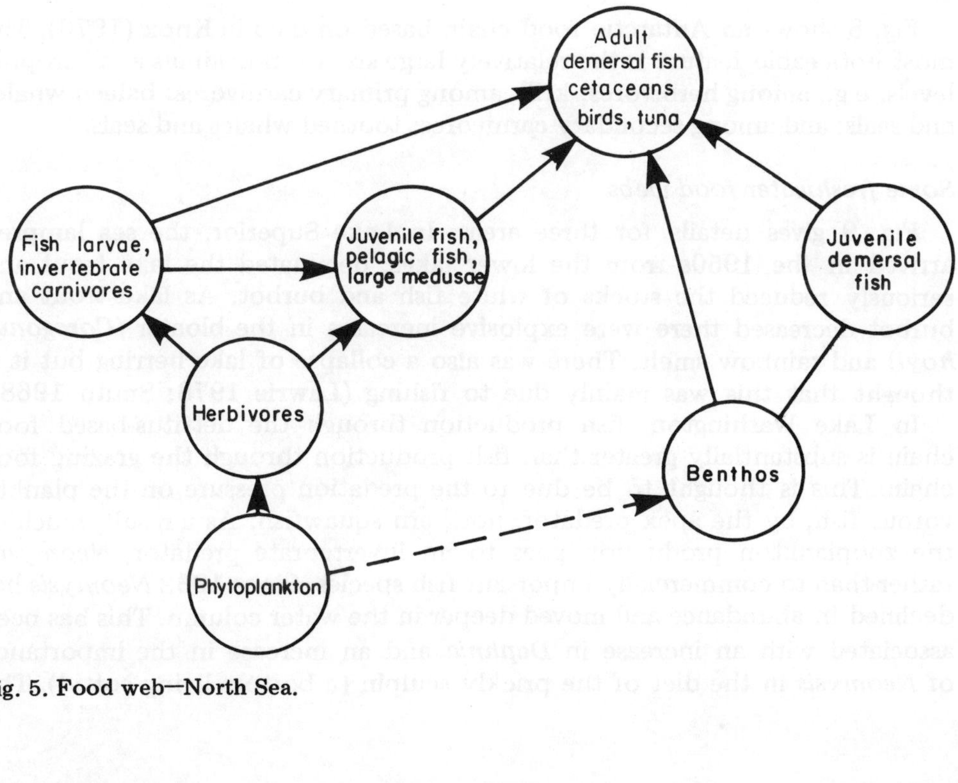

Fig. 5. Food web—North Sea.

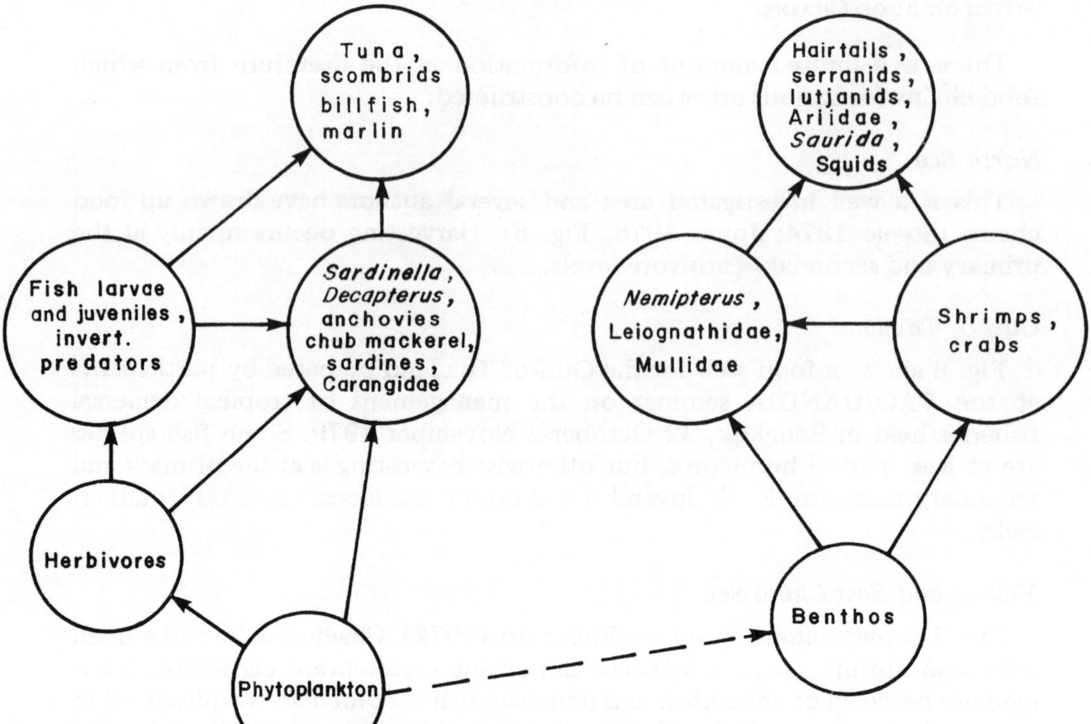

Antarctic (pack-ice zone)

Fig. 8 shows an Antarctic food chain based on data in Knox (1970). The most noticeable feature is the relatively large size of individuals at all trophic levels, e.g., among herbivores: krill; among primary carnivores: baleen whales and seals; and among secondary carnivores: toothed whales and seals.

Some freshwater food webs

Fig. 9 gives details for three areas. In Lake Superior, the sea lamprey arrived in the 1950s from the lower lakes, decimated the lake trout and seriously reduced the stocks of white fish and burbot. As lake trout and burbot decreased there were explosive increases in the bloater *(Coregonus hoyi)* and rainbow smelt. There was also a collapse of lake herring but it is thought that this was mainly due to fishing (Lawrie 1978; Smith 1968).

In Lake Washington, fish production through the detritus-based food chain is substantially greater than fish production through the grazing food chain. This is thought to be due to the predation pressure on the planktivorous fish, by the apex predator, northern squawfish. As a result, much of the zooplankton production goes to an invertebrate predator, *Neomysis*, rather than to commercially important fish species. Since 1962 *Neomysis* has declined in abundance and moved deeper in the water column. This has been associated with an increase in *Daphnia* and an increase in the importance of *Neomysis* in the diet of the prickly sculpin (a bottom-living cottid). The northern squawfish are important fish predators which feed heavily on juvenile sockeye salmon when these are abundant but switch to prickly sculpin, which is the most abundant and productive species in the lake, when the sockeye abundance is low (Eggers et al. 1978).

In Lake Windermere, the principal apex predator is the pike, of which the larger individuals (longer than 20 cm) feed almost entirely on fish—perch and trout predominantly, char considerably and stickleback and minnows to some extent (Frost 1954).

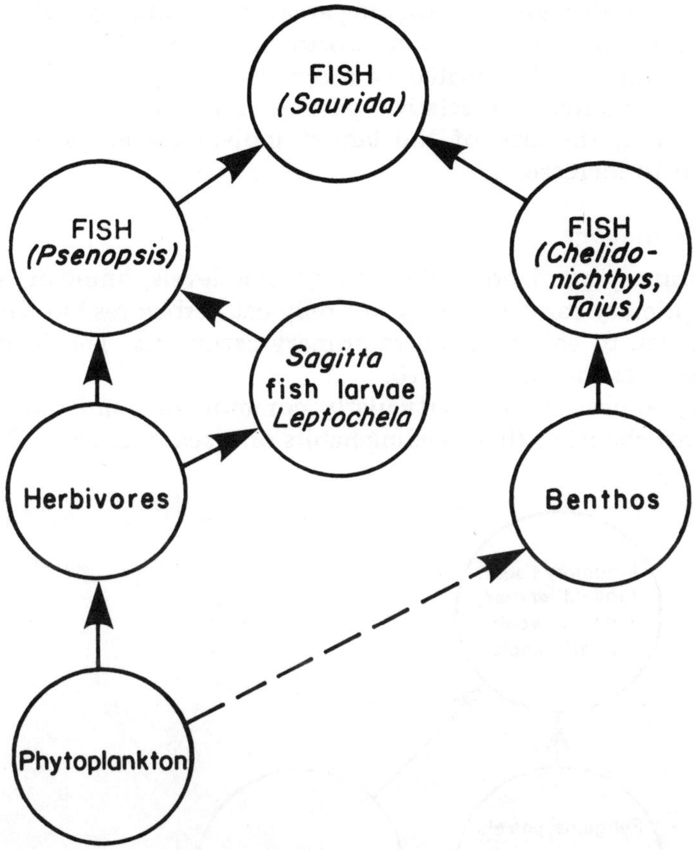

Fig. 7. Food web—Yellow and East China Sea (based on Enomoto 1972).

Theoretical Comparisons of Different Food Chains

All of the food webs considered above are of the form shown in Fig. 10. This comprises a pelagic food chain, consisting of plants, herbivores, primary and higher level carnivores. There is also a benthic food chain based on benthic detritus and filter feeders leading to apex predators, many of which are able to exploit both the pelagic and the benthic chains.

Comparison of the different systems suggests the following generalizations:

— With increasing primary production, there is not only an increase in the biomass of plants, but also an increase in individual cell size (i.e., when going from mid-ocean to coastal areas). At a high level of primary production, there may be a grouping of cells into colonies as may occur in upwelling areas and some productive lakes.

The tendency for the largest individuals at each trophic level to increase with increasing primary production appears to occur higher up the food chain also, so that commercial-sized fish appear at progressively lower points in the food chain. For example, in mid-ocean, starting from relatively small cellular plants, there appears to be four or even five steps from primary production to the fishery (Ryther 1969). In coastal areas, the largest primary carnivores include fish of harvestable size (e.g., adult herring and mackerel in the North Sea). At the highest levels of production, in upwelling areas and productive lakes, the largest herbivores are fish of a size worth harvesting. This suggests that a food chain may be treated as a series of simple parallel food chains classed according to relative particle size. With increasing primary production, the size of the largest individuals at each trophic level appears to increase.

The role of omnivory

At the primary carnivore and higher trophic levels, omnivory occurs, i.e., the largest primary carnivores may not only eat herbivores but they may also be large enough to eat the smallest primary carnivores. This leads to sets of food 'triangles' as shown in Fig. 10.

Omnivory appears to be particularly common in temperate waters, due to individuals changing their feeding habits on a seasonal basis. For example,

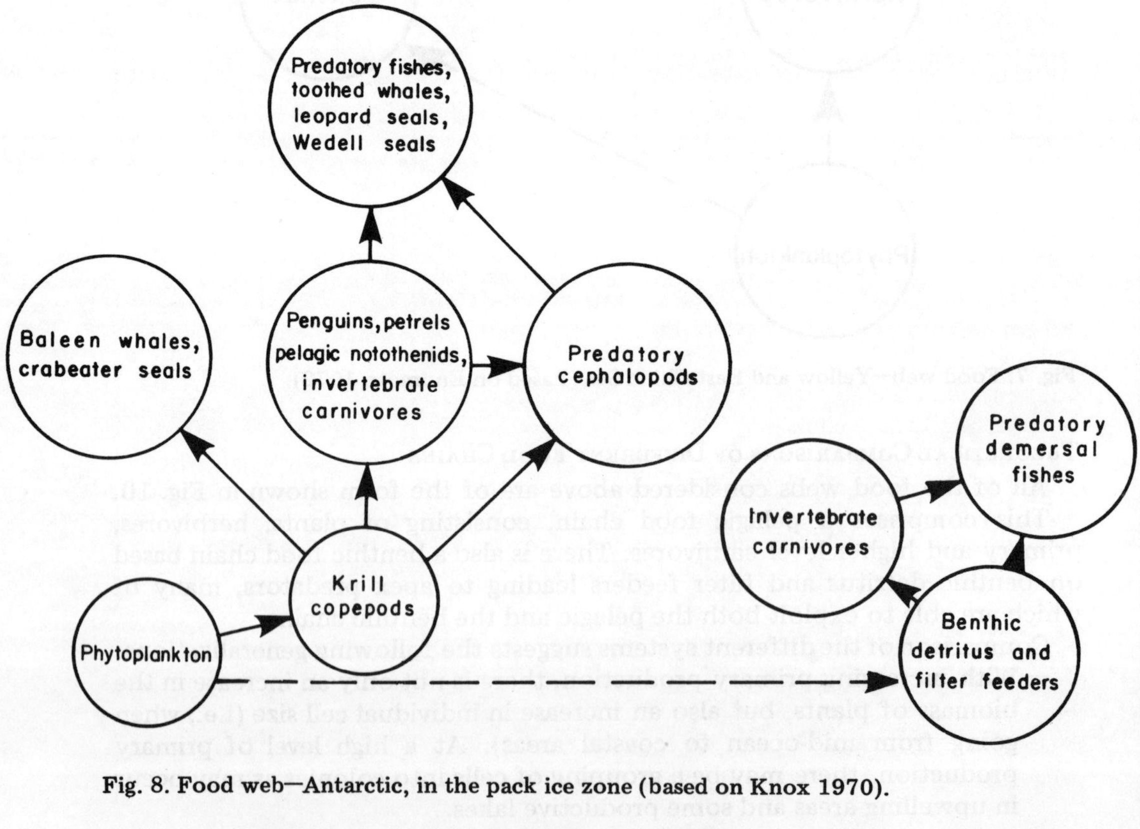

Fig. 8. Food web—Antarctic, in the pack ice zone (based on Knox 1970).

Sardina pilchardus off the coast of West Africa feeds primarily on chain-forming diatoms near the coast during the upwelling season and then moves offshore to eat zooplankton in the winter (Nieland 1980).

Recruitment

A particular feature of the fish recruitment process is that all marine teleosts appear to go through a stage in which the larvae and juveniles are pelagic and at the small-sized end of the primary carnivore level. This is true of demersal fish species and also appears to be true for herbivores such as anchovy, sardine and, in the case of freshwater species, of tilapias.

One effect of exploitation could be to make more food energy available for the small particle part of the food chain. This means that exploitation of large-sized adult species may in some instances be offset by an increase in energy flow to juvenile fish and hence by increases in recruitment. Some species may then respond to exploitation with an increased level of recruitment (e.g., pike in Lake Windermere (Le Cren et al. 1972)). In other cases, exploitation of one species may lead to an increase in recruitment of a different species.

Transfer and Ecological Efficiencies

Transfer Efficiency

In rapidly growing juvenile stages, assimilated energy can be converted into growth with an efficiency of as much as 30%. For older, and particularly for adult stages, the conversion efficiency will be much less and for the oldest animals may be zero. The net growth efficiency of a population depends on the relative proportions of juvenile and adult stages.

For the higher trophic levels, and particularly for homiotherms, adult stages tend to make up a relatively large part of the total biomass, and population growth efficiency is about 1.4 to 1.8% (McNeill and Laughton 1970).

For fishes the conventional value of 10% due to Slobodkin (1961) appears to be appropriate.

For zooplankton, estimates are variable and range up to about 20%. Simulation studies suggest that this may be too high for an average value but that values of this order may occur seasonally, whenever juveniles happen to make up a relatively large proportion of the total biomass.

For bacteria, values as high as 50% appear to be accepted by microbiologists. Values as high as this, however, presumably only apply when the bacteria are growing actively. For an annual average, some account must be taken of resting periods when bacterial growth is zero or negligible. The value of 50% is therefore likely to be an over-estimate for ecosystem calculations.

Fig. 11 shows the North Sea food web and computed transfer efficiencies.

Qasim (1972) quotes Phillipson (1970) to the effect that gross ecological efficiency is about 10%. He adds that Colman (1970) estimated growth efficiency for plaice as 15 to 26%.

Madhupratap et al. (1977) estimated the transfer coefficient from primary to secondary production for a tropical monsoonal estuary in a backwater

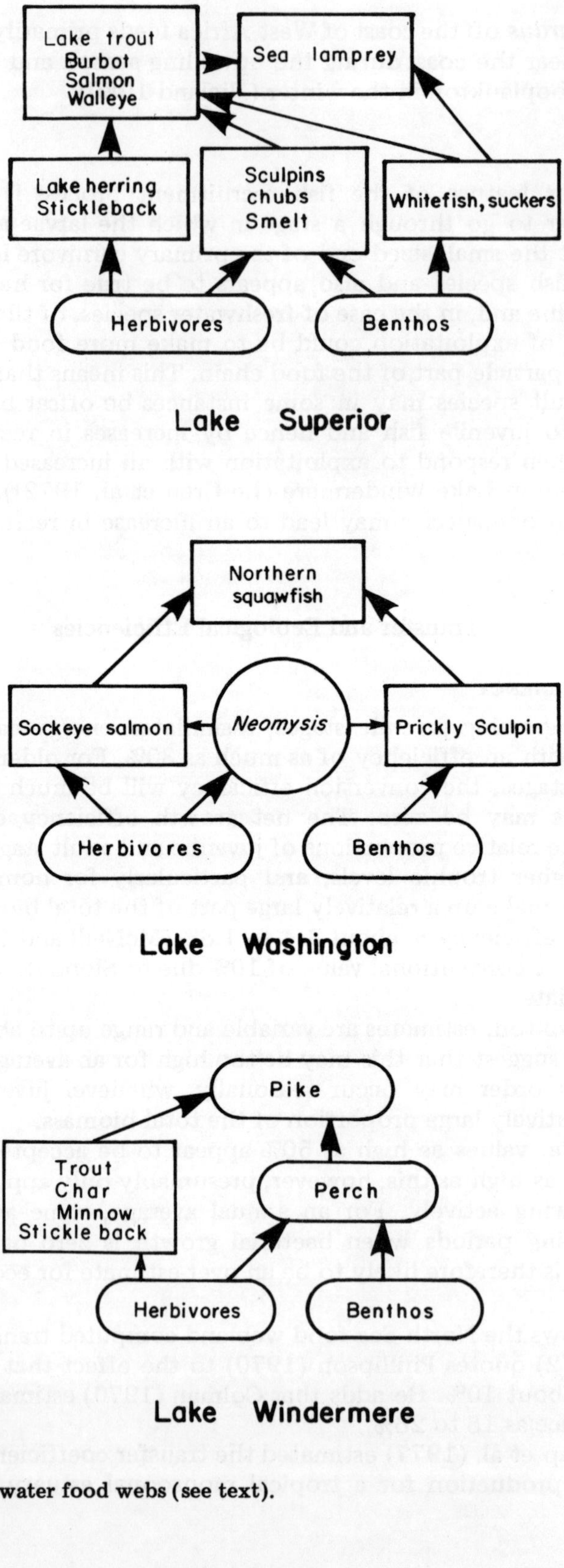

Fig. 9. Some freshwater food webs (see text).

near Cochin, India, and obtained a value of only 2.7%. They concluded that organic detritus plays an important role in the ecology of the system, as further evidenced from the large catches of detritus-feeding organisms in the estuary, such as prawns and clams.

Cushing (1973a) determined the ratio of zooplankton to phytoplankton production. Off Papua New Guinea, north of Marquesas, and in the eastern tropical Pacific, where primary production is only about 0.2 gC/m^2/day, the ratio was about 0.06%. Off Chile, Peru and California, where primary production is in the range 0.2 to 0.45 gC/m^2/day, he obtained a ratio of about 0.18%.

Pederson et al. (1976) determined transfer efficiencies between primary and secondary trophic levels for various lakes in the Lake Washington watershed. The values obtained for Findlay Lake, Chester Morse and Lake Sammamish were 13, 8 and 4%, respectively. They concluded that their results supported the hypothesis of Hillbricht-Ilkowska (1972) that energy transfer efficiency decreases with increasing trophic state.

Cushing (1973b) determined the transfer efficiency from primary production to secondary production in the Indian Ocean. Transfer coefficients declined with increasing production from about 20% to 5%; the mean value for the whole ocean for both monsoons is very close to 10%.

Wissmar and Wetzel (1978) examined five north American lake ecosystems and concluded that the efficiency of transfer of energy from primary to secondary consumers appears to be lower in the more productive systems. They suggested that as primary carbon supply increases, relatively more energy becomes diverted to the detrital or benthic microbial communities. For Lakes Findley and Marion, two of the less productive lakes, the efficiency of phytoplankton production transferred to zooplankton was as high as 20%.

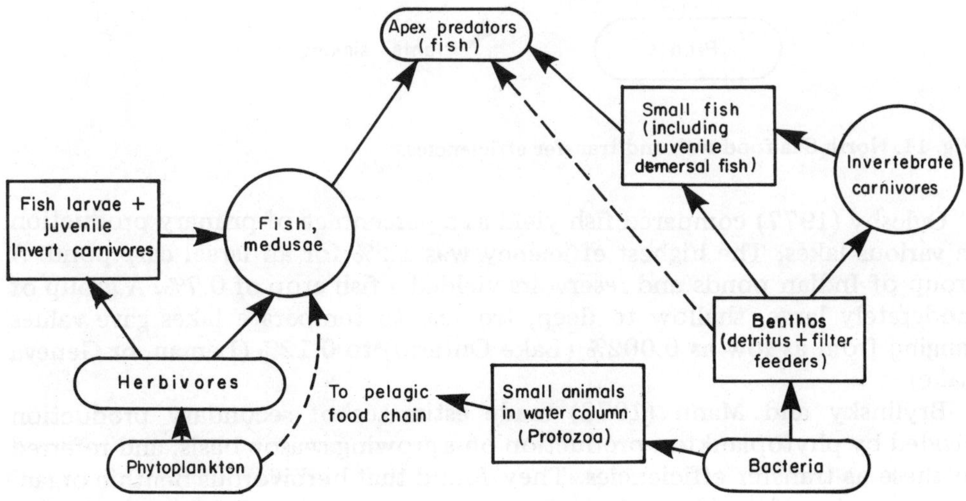

Fig. 10. Food web—generalized.

Slobodkin (1959) suggests that a growth efficiency of 20 to 28% may be considered universal. Reproductive efficiency should be slightly less than the square of the growth efficiency. He adds that at least 150 chemical steps are involved in the complete degradation of any food molecule and that there must be a sufficient number of exergonic steps in the sequence to guarantee unidirectionality of metabolism. The energy burned at each step is unavailable for growth. He calculates that to leave sufficient residual energy to permit an efficiency of much more than 25% requires an absurdly high mean efficiency per step.

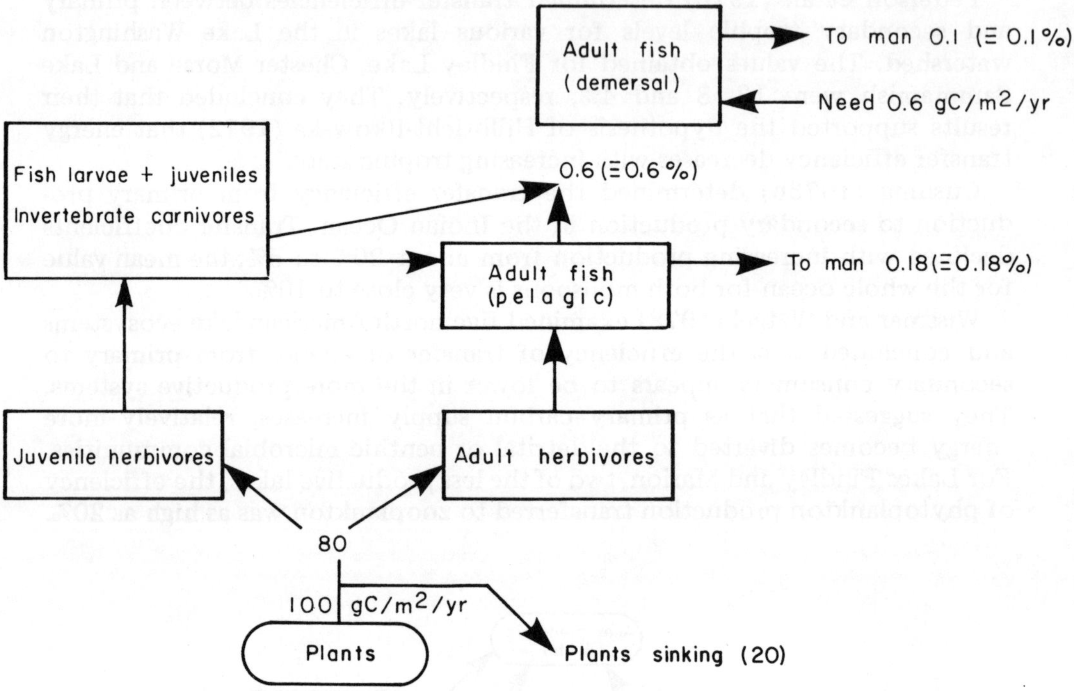

Fig. 11. North Sea food web and transfer efficiencies.

Oglesby (1977) compared fish yield as a percentage of primary production in various lakes. The highest efficiency was 1.5% for an Israel carp pond. A group of Indian ponds and reservoirs yielded a fish crop of 0.7%. A group of moderately large, shallow to deep, tropical to temperate lakes gave values ranging from as low as 0.002% (Lake Ontario) to 0.12% (Leman, or Geneva Lake).

Brylinsky and Mann (1973) made estimates of secondary production divided by phytoplankton production on a growing season basis, and referred to these as transfer efficiencies. They found that herbivorous benthic organisms have transfer efficiencies, on a percentage basis, of 0.5 to 9.5%. For herbivorous zooplankton, the range was 2.6 to 21% with a mean of 14%. For carnivorous zooplankton, the range was 0.2 to 2.5% with a mean of 1.4%. For carnivorous benthos, the range was 0.1 to 3.4% with a range of 0.9%.

Humphreys (1979) made an analysis of reported energy budgets to determine the relationship between annual production and respiration in natural populations of animals. Values of P/A were determined, where P = production and A = production + respiration. For fish, the mean value of P/A on a percentage basis was 9.8%. For non-insect invertebrate herbivores, carnivores and detritivores, the corresponding values were 21, 28 and 36%, respectively.

McNeill and Laughton (1970) also made estimates of P/A for natural populations. For homiotherms, the values were approximately constant for all levels of production varying only between 1.4 and 1.8% on a percentage basis. For poikilotherms in general, the situation is less clear. For comparatively short-lived poikilotherms, mean net population production efficiencies increased markedly at lower production levels. This may be applicable to species which over-winter as eggs, so that the respiration during the resting period is reduced to the minimum. Other species over-winter as juveniles and adults, and respiration continues at a comparatively high level while production is virtually zero. From the values given for roach, bleak, dace, perch and gudgeon, a mean production efficiency of 8% can be calculated.

According to Ricker (1969), growth efficiency in fish has a maximum of about 35% and declines continuously as fish grow older. He quotes Kevern (1966) as estimating growth efficiency for yearling carp in a lake of 8%; Lasker (1966) for estimating a value of 9% for euphausiids in the northeastern Pacific; Petipa (1967) for obtaining values of 15% and 25% for various copepod life stages; Pavlova (1967) for obtaining values of 12 to 16% for the adults of three cladocerans, and 40% for the young; and Richman (1958) for obtaining values of 4 to 17% for *Daphnia* fed on algae in the laboratory. Ricker concluded that the overall average gross efficiency of the plant eaters can hardly exceed 15%. For animals that eat animal food he considered that the average coefficient could be as high as 20%.

Ricker uses the term "ecotrophic efficiency" to describe the fraction of the annual production of the prey species that is consumed by predators. He uses a value of 70% for this. When this is taken into account, he arrives at net transfer efficiences of 15% for primary consumption and 20% for higher consumption levels.

Ricker uses the term "trophic levels" for successive steps on a food pyramid, and "ecological level" for groups of animals of generally similar size and habitat. He makes the point that cannibalism within each ecological level increases the average number of trophic steps from the primary plant food to any of the higher levels.

Ryther (1969) states that growth efficiency has been found by a large number of investigators and with a great variety of organisms to be about 30% in young, actively growing animals. The efficiency decreases to zero in thoroughly mature and old individuals.

Slobodkin (1961) considered that an ecological efficiency of about 10% was an appropriate average value, but Schaefer (1965) suggested that the figure may be as high as 20%.

For bacteria, both Fenchel and Jørgensen (1977) and Herbland (1978) suggest that as much as 50% of their detritus food could be converted into bacterial biomass.

Review of Anecdotal Information from the Literature

Life history strategy and natural selection

The objective of life history strategy is essentially, 'to gather food without being eaten in the process, in order to live long enough to reproduce successfully'. Therefore, the life of the individual maybe thought of as a continual compromise between food gathering on the one hand, and survival on the other. Further, whenever an animal takes in food, surplus to that needed for maintenance, it will be necessary to choose whether to increase somatic tissue, energy reserves or reproductive tissue. The life history therefore involves a continual succession of choices, and natural selection may be supposed to operate so as to favor that succession of choices which culminates in the optimum life history strategy, i.e., that strategy which maximizes the effective output of viable reproductive material.

Predation pressure

Because there are predators, there will be predation pressure, which means that an individual can be influenced not simply by the physical presence of a predator, but by the constant possibility that a predator may be present. Predation pressure can be expected to influence both distribution and behavior.

Level, sandy ocean bottoms are usually deserts (Stone 1971). Phillips (1977) observed blennies moving about in pools and noted that, in general, spaces adjacent to shelter were used more significantly than open space.

Vulnerability

A consequence of predation pressure is that even fit animals will experience different degrees of vulnerability according to the ways in which they behave, and the ways that they are distributed with reference to cover.

Ginetz and Larkin (1976) investigated the predation of rainbow trout on migrant sockeye salmon in experimental streams. They noted that fry that had survived exposure to predators for several days were less vulnerable to predation than naive fry. Successive exposure to predation further decreased vulnerability.

Compared to animals that are fit, unfit animals are likely to be more vulnerable. For example, Bethel and Holmes (1977) noted that gammarids infected with larval acanthocephalans were significantly more vulnerable to predation by mallards and muskrats than non-infected individuals.

Inexperienced animals may be more vulnerable than others. For example, Stirling and Archibald (1977) reported that in the Arctic, seal pups constituted at least half of the seals killed during the spring by polar bears. Symons (1974) noted that brook trout ate 26% of juvenile Atlantic salmon recently introduced into an artificial stream, but only 8% of salmon which had been present long enough to establish territories.

In midwater, shoaling can be an important factor affecting vulnerability. For example, Mylinski (1977) commented that 3-spine stickleback stragglers were attacked more often than sticklebacks in shoals. Neill and Cullen (1974) examined the hunting behavior of squid, cuttlefish and pike when offered shoals of live prey. For all species, increasing the shoal size decreased the success of the predators' attacks, per encounter with prey.

Vulnerability to predators may change with behavior during the day. For example, Major (1977) observed that predation on silversides was most intense during the twilight at times when shoals were either forming or breaking up.

Direct effects of predation

There are various experiments in which predators have been experimentally excluded from particular areas and in which there have been corresponding increases in prey abundance. Virnsterin (1977) investigated the effect of large predators on the abundance of benthic invertebrates in small areas. The blue crab and two species of bottom-feeding fishes were either excluded from, or confined to, small areas using wire-meshed cages. It was concluded that the population densities of the infauna under natural conditions were primarily limited by predation.

Paine (1971) described how a carnivorous starfish was removed manually, and kept removed, from a stretch of shore in New Zealand for a period of 9 months. This resulted in a species of mussel extending its vertical distribution by 40% of the available range.

Haven (1973) described how the removal of a gastropod from the intertidal zone in central California led to a significant increase in algal growth.

Reise (1977, 1978) described experiments in which epibenthic predators, shore crabs, shrimps and gobies, were experimentally excluded from areas of sea bottom by the use of cages. On a mud flat, the exclusion of these predators caused a 10-fold increase in the abundance of, and a four-times higher species density of, macrofauna. In a grass-bed on the other hand, there were only comparatively slight changes, presumably due to the protection naturally afforded by the nature of the substrate.

Effects of predation on zooplankton

Langeland (1977) described the relationships between char and zooplankton in 11 Norwegian lakes. In lakes with small populations of large-sized char, the most important prey species in the summer were *Daphnia, Dytotrephes, Holopedium* and *Bosmina.* In these lakes the spawning population could be caught mainly with floating nets. In lakes with high densities of small-sized char, the zooplankton were insignificant as food and the char were mainly caught on the bottom. He concluded that high densities of char can have a significant grazing effect on the large-sized zooplankton species.

Northcote and Hume (1977) described an experiment in which fish were introduced experimentally into three small fish-less lakes in coastal British Columbia between 1974 and 1975. One lake received about 16,000 cutthroat trout and a second received about 900 Dolly Varden *(Salvelinus malma).* The third lake served as a control. Zooplankton sampling indicated catastrophic declines in the abundance of mid-water populations of *Chaoborus* larvae in the early summer of 1975 in both fish lakes, but not in the control.

De Bernardi and Giussani (1978) described the effect of a mass fish mortality in a small Italian lake (Lago di Annone). The lake is formed by two basins connected by a narrow channel, and in August 1975 there was a natural mass mortality of the zooplankton-feeding fish *(Alburnus)* due to an infectious gill disease in the eastern basin. Comparison of the two basins showed that in the eastern basin where the fish mortality occurred, there was an increase in the maximum size of *Daphnia* and a rapid increase in its abundance during the spring of 1976. At the same time there was a decrease in the population of *Bosmina.*

Macan (1977) reported how the introduction of trout into an artificial pond was followed by a greater reduction in the numbers of tadpoles, certain beetles and *Notonecta,* all species normally found in open water. Of the species that sheltered in plant cover, the effect on some was a curtailment of the range. They concluded that only small invertebrates can survive predation in open water.

Eggers et al. (1978) described changes in Lake Washington associated with sewage treatment and a reduction in nutrient loadings. Since 1962, *Neomysis,* an important zooplankton predator, has decreased in abundance and has been found deeper in the water column (possibly associated with an increase in water transparency). It is now near enough to the bottom to constitute a significant proportion of the diet of the prickly sculpin (a bottom living cottid).

Fish removal

In some instances, the wholesale removal of a species can have considerable effects on other species. Johnson (1977) describes the effect of the removal of 85% of the estimated standing crop of adult white suckers in 1966. During the seven years after removal, catch indices for adult suckers remained far below pre-removal levels. However, juvenile sucker abundance increased about 17-fold. Yellow perch increased about 15-fold. Walleye standing crop increased by about one third, primarily due to an increase in the abundance of the newly recruited age groups.

Kipling and Frost (1972) describe the effects of the annual removal of pike from Lake Windermere by gill-netting. The result is that there is now about the same number of pike in the lake as there was in 1944, but they are very much younger fish. There has been some tendency for growth rate to increase, but the average size is still smaller. The main food of pike in summer is perch, and small pike eat young perch (Fig. 9). The increased number of small pike may account for the poor survival of young perch in most years. Large pike eat large perch and presumably the predation on the larger perch has declined as a result. There are two populations of char, one of which spawns in the shallow water in the autumn, and a totally separate population which spawns in the spring in deep water (Frost 1965). In November and December, the pike feed almost entirely on the autumn spawning char in the shallow water. There has been a definite trend upwards in the catch of breeding char in the year since pike netting began. Evidence is not good enough to state whether the trout population has increased or not.

Introductions of fish predators

Lawrie (1978) and Smith (1968) describe how in the 1950s, sea lampreys arrived in Lake Superior from the lower lakes and decimated the populations of lake trout and severely reduced the populations of white fish and burbot. At the same time there was a collapse in lake herring thought to be due mainly to overexploitation. These changes were associated with an explosive increase in bloater and rainbow trout (Fig. 9). The increase in smelt supported massive plantings of lake trout, migratory trout, Pacific salmon and other salmonids.

In Lake Washington, northern squawfish are important fish predators. It is concluded that they feed heavily on juvenile sockeye salmon when these are abundant but switch to prickly sculpin (the most abundant and productive species in the lake) when the sockeye abundance is low (Fig. 9). This provides a nice compensatory mechanism and is assumed to be the explanation of the fact that in years of high sockeye abundance, the sockeye experience much higher rates of mortality than in years of low sockeye abundance.

Juvenile sockeye salmon show a complex feeding behavior which can be explained as an adaptation to minimizing vulnerability to predation by northern squawfish, i.e., they limit the time spent foraging. They are generally found in regions of low light intensity, that is, away from regions of high zooplankton abundance and also engage in schooling behavior (which reduces their foraging efficiency). It is assumed that this is one of the reasons why the plankton eaters exploit only a part of the total zooplankton production in the lake.

Benthic fishes tend to be relatively large, and to consume relatively large prey items. However, their larvae and juveniles are extremely vulnerable to predators, and at this stage may have more potential predators than the plankton eaters.

In Lake Washington in general, the fish community is dominated by benthos-eating fish whereas the planktivores are predator-controlled and hence not able to fully exploit the smaller species of zooplankton (Fig. 9).

Eggers et al. (1978) point out that fish-eating predators can greatly affect the structure of fish communities. For example, the introduction of peacock bass *(Cicla)* in Lake Gatun, Panama, resulted in a virtual extinction of plankton-eating fish in a wide area of the lake (Zaret and Paine 1973). Similar stories have been reported from a number of other water bodies throughout the world.

Fish-eating fish have an important role in determining the community structure of African lakes. For example, in African lakes with large fish-eating predators, such as tiger fish *(Hydrocyon)* and Nile perch *(Lates)*, small fish are restricted to areas with submerged vegetation or very shallow areas (Jackson 1961). Jackson speculates that most of the reproductive behavior of fish in African lakes, such as mouthbrooding and the migration of adults to spawn in rivers, is adaptation against predation.

Lowe-McConnell (1975) considered that in tropical freshwater communities the evolution of the benthic-littoral and pelagic fish communities has led to two different ends which can be largely explained by predation. Benthic-littoral fish communities are characterized by a large number of cohabiting fish species with complex specializations. It is suggested that predation restricts fish to areas of cover, effectively preventing movement between habitats, and acting as an isolating mechanism. Competition within these micro-habitats further increases the species diversity and specialization. Pelagic environments, on the other hand, are characterized by a uniform prey resource of small individuals. There is no cover and the only effective refuge is behavioral, such as remaining in areas of low light and low density, forming schools, and feeding briefly (Eggers et al. 1978). Predation in the pelagic environment does not necessarily isolate populations, but forces more or less uniform behavioral responses.

Pelagic fish communities are, therefore, characterized by fewer species and a more uniform phenotype (small body to exploit zooplankton) and feeding behavior.

Emery (1973), from observations on 21 species of freshwater fish in Georgian Bay and Algonquin Park, Ontario, suggested that bold color patterns are disadvantageous when swimming at night, and that there is a strong evolutionary pressure for prey animals to disguise their body shape when resting at night.

Feeding behavior

Specializations by prey to avoid being eaten can be expected to have led to specialization by predators to locate and eat the prey. Fricke (1975) describes how the foraging behavior of *Balistes* is adapted to moving objects aside with the mouth to discover hidden prey.

Winslade (1974) investigated the behavior of the lesser sandeel, *Ammodytes marinus*. When food was present, there was a basic diurnal rhythm of activity, with the fish remaining buried in the sand during the dark but emerging to swim and eat during the light. In the absence of food the level of swimming activity during the light period was extremely low.

Karplus (1978) observed that groupers follow moray eels over coral reefs in the Red Sea. The eel flushes prey out from narrow fissures and burrows. It is suggested that the grouper follows the moray eel to catch small crustaceans and fish chased out of shelter by the moray eel, or to steal food from the moray eel.

Ogden (1977) observed shoals of juvenile grunts (< 12 cm) on a series of patch reefs surrounded by sea grass beds on the northeast coast of St. Croix, U.S. Virgin Islands. The grunts from large, inactive multispecies shoals in the reef areas by day. At dusk the shoals migrate some distance away to the sea grass-covered areas to feed on invertebrates. The routes that are followed into the sea grass beds are precise and persist over the years. At dawn they return to the reef along the same routes.

According to Hobson (1973), the pattern of migration of many tropical reef fishes is strongly influenced by the relative threat from predators at different periods of the diel cycle. For example, movements in the open regions adjacent to many reefs are not common in daylight, largely due to danger from predators. However, at night, reef fishes range freely into the open regions that are avoided in daylight, since most potential predators are visual feeders.

Hall et al. (1979) describe similar behavior for the golden shiner. During the day, golden shiners form schools in the littoral zone. Just after sunset, the schools break up and the fish migrate into open-water regions of the lake to feed on *Daphnia*. Most prey are obtained within an hour of sunset.

Ebeling and Bray (1976) found that kelp-bed fishes off Santa Barbara, California, fed and moved about in midwater, well above the reef during the day, but were found in holes and crevices in the reef at night.

Lack of overlap in diets and ecological separation

Kanamoto (1977) studied the ecology of many species of reef fish near Benten Island in Sendai Bay, Japan. The fish were divided into three major types: the "touching-type", the "floating-type" and the "swimming-type" fish. The touching-type fish spend long periods in a solitary state on the rocks or on the algae. The floating-type fish remain above the rocks or among the algae, some in the solitary state and some in a gregarious state. The swimming-type fish swim on the surface or in the water column, mostly in a gregarious state in daylight. There was little evidence of direct competition.

Caine (1977) describes how caprellid amphipods feed by browzing, filter feeding, predation, scavenging and scraping. Species with plumose setae obtain a significant amount of their diet from filtering and scraping, while species without setae usually rely on predation. Two primary predators, and eight primary filter-feeders and scrapers were investigated. Substrates inhabited by predators do not overlap. Some filter-feeders were found together. However, these either feed on different sized particles, or they filter at different heights from the substrate. Cohabiting species filtering the same-sized particles at the same height utilize alternate feeding mechanisms.

Keast (1973), investigating fish in some small Ontario waterways, concluded that cohabiting species show considerable levels of food specialization, and that much of the food overlap occurs when a particular resource becomes super-abundant. Sometimes fish species that are apparently taking the same kind of food are concentrating on larger or smaller-sized individuals. Alternatively, species may have different feeding periods, some being day feeders while others are night feeders. In general, cohabiting species tend to differ widely in body size and have a wide range of body and mouth morphologies. Keast concluded that they are structurally specialized for different feeding roles.

Tyler (1972) examined the division of food resources among 13 demersal fish species from Passamoquoddy Bay, New Brunswick, Canada. Over 100 prey species were found, but each predator took only three or four principal prey species, these species making up 70 to 99% of the food for each predator species. He concluded that there was relatively little overlap among the diets.

There are various degrees of ecological separation. Some species are territorial, such as coral reef fish on the southwest coast of Curaçao (Luckhurst and Luckhurst 1978). Sale (1976) investigated eight species of damsel fish that are territorial on dead coral rubble that is usually covered with a fine turf of filamentous algae, which is their principal diet. There was no place on the reef where a patch of rubble was not occupied by at least three species of the guild. The author suggests that this degree of apparent habitat overlap is possible because the reef environment is not stable but is subject to continuous growth and minor destruction.

Kellogg (1977) describes how seven species of hermit crabs collected in Beaufort Harbour are able to co-exist due to differences in shell-size and shell-species partitioning.

George and Hadley (1979) investigated juvenile rockbass and small mouth bass along the west shore of Grand Island in the upper Niagara river, New York. They concluded that ecological separation was by both food and habitat partitioning.

Wurtsbaugh and Brocksen (1975) examined the distribution of juvenile brook and rainbow trout in Castle Lake, California. The brook trout were located near the bottom and found primarily in shallow water on the eastern shore of the lake near springs. The rainbow trout were more pelagic and were found in the littoral areas along the entire shoreline. In study streams, brook trout and a cutthroat trout occupied different micro-habitats.

A Theoretical Treatment of Competition and Coexistence in relation to Predator-Prey Relationships (based on Jones 1979)

Competition and coexistence

A central problem in ecological theory is to reconcile the conflicting implications of competition and coexistence.

For example, if two predator species compete for a single prey species, relatively simple theoretical considerations suggest that eventually one predator species will eliminate the other (Fig. 12a). In practice, this does not appear to be the rule and the observation that many species are able to coexist is commonplace. One way of explaining this is to suppose that nature is never in a state of equilibrium and that high diversity is a consequence of perpetual change (Caswell 1978; Connell 1978). Another explanation is that predators partition their food resources in such a way that direct competition is avoided. Fig. 12b shows a situation in which there are two predators, each dependent on a different prey species. The difficulty with this model, however, is that food sharing is commonplace among marine fishes. Here, consideration is given to one way of combining the properties of the models in Fig. 12a and 12b so that both competition and coexistence can continue under theoretically stable conditions.

Vulnerability to mortality

One way of reconciling coexistence with the fact that food sharing appears to be commonplace is to adopt the Darwinian view that individuals are not all equal, but that some are more vulnerable than others. Differences in vulnerability might arise in several ways. For herbivores, some vegetation will be distributed near suitable refuges so that it can be grazed in relative safety. Other vegetation may be in exposed places so that it can only be grazed at a greater risk of being eaten by predators. Similarly, carnivores are likely to find that, to avoid the risk of being eaten by other predators, it is safer to concentrate on prey in certain places or at certain times. At it simplest, therefore, one may conceive of food that can be eaten with comparative safety, and food that can only be eaten at the risk of being eaten oneself (Fig. 13).

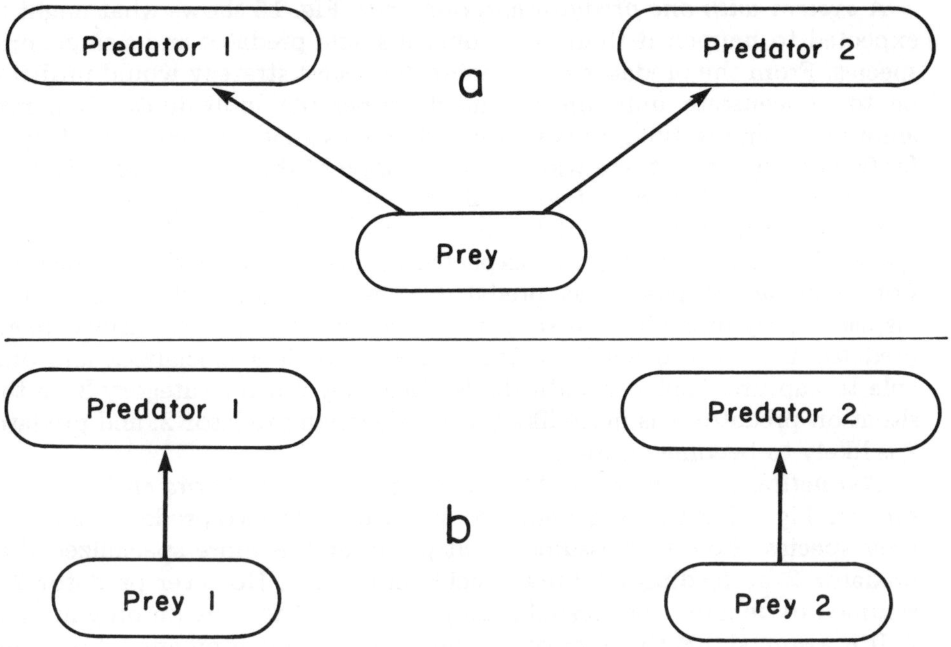

Fig. 12. a. Competition and food sharing. Showing two predators sharing a single prey food resource. b. Coexistence. Showing two predators each dependent on different prey food resources.

The concept of "displacement"

If populations did not grow, one might arrive at a situation in which prey species were only found in the 'safe food' region. In practice however, it is a property of living matter to grow (both by individual growth and by reproduction). A consequence of this is that interspecific competition for food in the safe resource will eventually force some individuals to try to take food in the region of the unsafe resource, i.e., population growth will lead to a displacement of individuals into less favorable regions. Fig. 14 shows a population consisting of non-displaced and displaced individuals. If there were no predators or disease or agents of mortality of any kind, individuals would eventually become displaced beyond the resource and would then die from starvation.

In practice, predators, disease and environmental effects all act as agents of mortality. In this case, an individual may be eaten long before it has had time to become displaced entirely beyond the food resource. The probability of observing a 'resource limited' individual, that is, an individual that is obviously suffering from lack of food, may then become negligible.

Resource limitation

Resource limitation in this model is thought of as limitation of just that part of the resource that can be exploited in comparative safety. The total resource may be very much greater and need not necessarily be limiting at all in a strictly "food available" sense.

A system with one predator and one prey. Fig. 15 shows what might be expected to happen if there were only a single predator and a single prey species. From the predator's viewpoint the easiest strategy would probably be to concentrate only on the most vulnerable individuals. This may amount to simply lying in wait for individuals that happened to stray so far from refuges that there was very little hope of them escaping predation.

Competition between two predators and one prey species. Fig. 16 shows what might happen if there were two predators and only one prey species. This situation might be expected to lead to predator specialization. For example, suppose that predator 1 becomes specialized to capture displaced prey individuals as soon as they reach category 1 of displacement. Predator 2 on the other hand is assumed to be less specialized and only able to capture displaced individuals once they reach a category 3. In this situation predator 1 is more likely to survive than predator 2, and predator 2 is likely to become extinct.

Competition combined with coexistence—two predators and two prey species. Fig. 17 shows a generalized situation with two predators and two prey species. Here it is assumed that predator 1 is more specialized than predator 2 at feeding on prey species number 1. However predator 2 is assumed to be more specialized than predator 1 at feeding on prey number 2. It is assumed that the price of specialization is that as an animal becomes more effective at feeding on one prey species, it becomes less effective at feeding on some other prey species. This arrangement combines the neces-

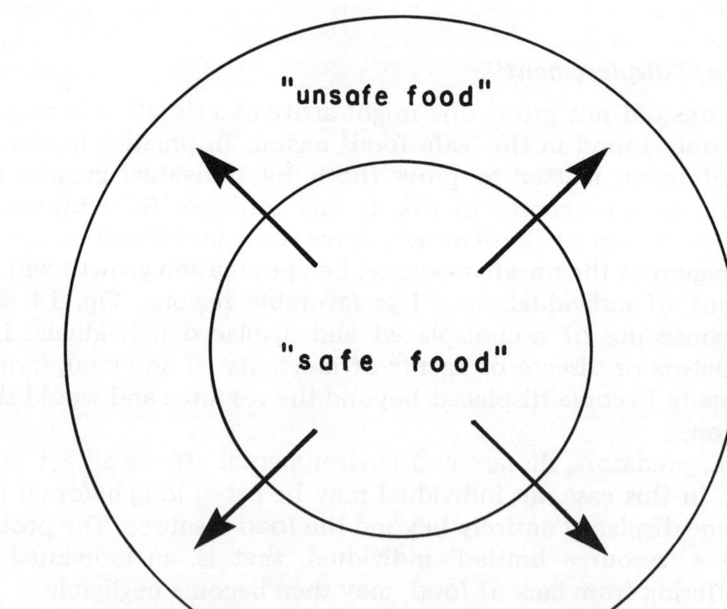

Fig. 13. Showing the subdivision of a resource into food that can be exploited with safety ("safe" food) and food that can only be exploited at the risk of being preyed upon ("unsafe" food). The arrows indicate a displacement pressure due to interspecific competition and population growth.

sary characteristics of Fig. 12a and 12b; coexistence without elimination is possible, since each predator is more specialized than the other at feeding on a particular prey resource. At the same time food sharing is possible since each predator is able to eat displaced individuals of both prey species.

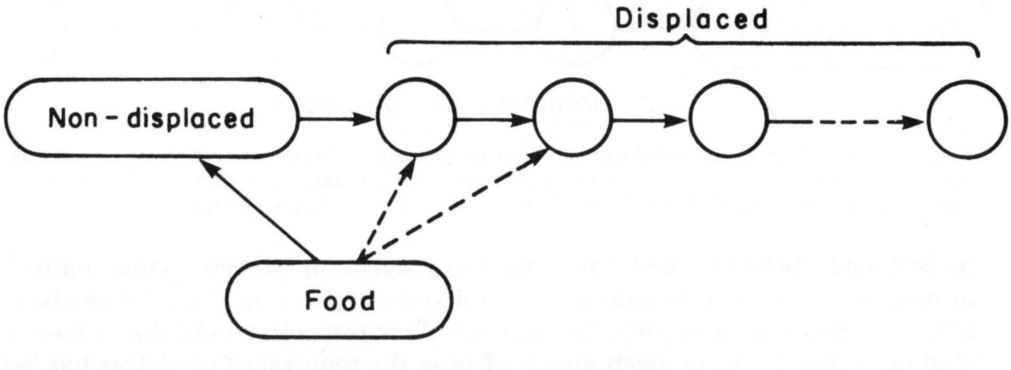

Fig. 14. Showing a population consisting of non-displaced individuals (dependent on safe food) and a succession of displaced individuals, i.e., dependent on food that can only be exploited with a progressively greater chance of being preyed upon.

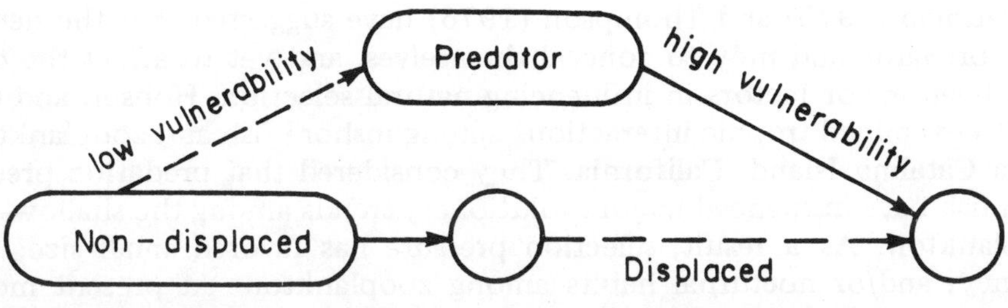

Fig. 15. A single predator preying upon a single prey species population. It is assumed that non-displaced individuals have the lowest vulnerability to predation whereas the most highly displaced individuals are most vulnerable to predation.

Predation Pressure and Species Replacement

Ideas about species replacement are based on the assumption that when one species is reduced in abundance, some other species will increase to take up the available food energy. Within a closed ecosystem this is clearly reasonable, the main problem being to decide which species are most likely to benefit in any particular situation. Some of the relevant literature was reviewed above, from which the following tentative conclusions have been drawn:

First, it is assumed that species extant today are the result of natural selection, and that this has operated so as to favor "successful" life history strategies. By a "successful" life history strategy is meant one that enables an individual to gather food, (in order to grow and reproduce successfully), without being eaten or succumbing to disease or environmental hazards in the process.

Predators are thought to have an influence on the outcome of natural selection not simply because they eat prey organisms, but because they exert predation pressure. The fact that predators are present, is in itself sufficient

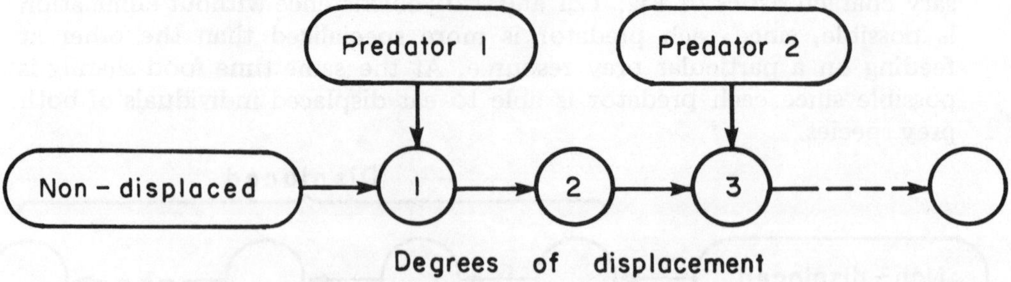

Fig. 16. Competition—two predators and one prey species. Competition between predators may be expected to lead to a situation in which one predator (predator 1) is able to exploit more recently displaced individuals than the other predator (predator 2).

to influence behavior and the long term selection process. Thus, natural mortality, even for a fit animal, will not only depend on the relative abundance of predator and prey, but also on their respective behavior patterns. Various examples were given above of how the removal of predators has led to increases in prey abundance, and also how the introduction of predators has caused reductions or changes in prey species composition or prey abundance.

Treisman (1975) and Thompson (1976) have suggested that the needs of both predator and prey to conceal themselves, and yet to affect the other, have been major factors in influencing natural selection. Hobson and Chess (1976) examined trophic interactions among inshore fish and zooplankton at Santa Catalina Island, California. They considered that predation pressures from fish have influenced major evolutionary trends among the shallow-water zooplankton. As a result, selection pressure has favored small size, transparency, and/or nocturnal habits among zooplankton. At present most of the zooplankton in the nearshore water column by day are very small (less than 2 mm in length), and any large zooplankton in the water column by day, for example chaetognaths, tend to be transparent. Zooplankton longer than 2 mm that are not transparent, tend to enter the water column only at night. As a result there have been adaptations among the fish that feed on the zooplankton. This includes specialized features, such as small, highly modified mouths for taking tiny organisms by day. Fish that prey on zooplankton at night tend to have larger mouths and also large eyes for detecting prey under poor light conditions.

There is a very considerable literature to suggest that coexistence of different species is associated with differences in food-partitioning and ways of exploiting different habitats.

Possible generalizations are:

— If a top predator is removed, it is unlikely that any other species at the same trophic level and of the same size will be able to completely replace it. It seems more probable that replacement will tend to be by individuals with a smaller body size.

— In tropical systems, because of the larger number of species, the removal of top predators may lead to increased competition among what were previously the prey species. This could be an important factor in influencing subsequent species interactions.

— If a fishery is directed at species other than top predators, it may be possible to increase yields by exploiting appropriate higher predators also. For example, Marten (1979) describes how yields of tilapias in Lake Victoria were higher in places where there was also fishing for large predators such as *Bagrus*.

For successive steps in a food chain, there appear to be two options:

— One is to adopt a transfer efficiency of about 10% for an overall average per step.

— The other is to use higher values, but to note that these must apply to the juvenile and therefore small particle-size part of the food web. This means that more steps may be needed to progress from a plant cell of about 100 μ to an animal the size of a harvestable fish.

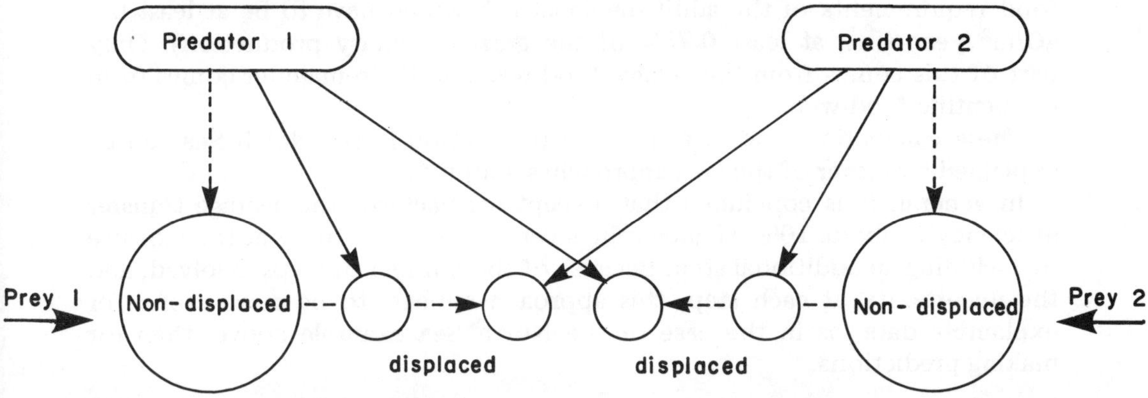

Fig. 17. Generalized situation with two predators and two populations of prey species. Predator 1 is assumed to be more specialized at feeding on prey species 1. Predator 2 is assumed to be more specialized at feeding on prey species 2.

For the North Sea, estimates are available of fish production and primary production, and it is useful to compare these (Fig. 11). Thus, the long term yield of pelagic fish prior to the 1960s was about one million tonnes equivalent to about 0.18 gC/m^2/year. If the average particulate production is taken as 100 gC/m^2/year and if about 80% of this is grazed (Steele 1974), the pelagic fish yield to man is equivalent to about 0.23% of the grazed particulate primary production, or 0.18% of the total particulate primary production.

This 80% grazing, plus two steps of 10% each, would suggest a production of primary carnivores equivalent to 0.8% of particulate primary production. If 0.18% goes to man, there would be a remainder of 0.6% to account for the production of larval and juvenile fish plus the production of invertebrate carnivores plus an allowance for the natural mortality of the harvestable stages of the pelagic fish. This appears not unreasonable.

Alternatively, one might consider two steps through juvenile forms (i.e., through juvenile, herbivores and then juvenile primary carnivores) plus one step through the adult stages of the pelagic fish. If one adopts efficiencies of 20%, 20% and 10% and allows for 20% not grazed, one arrives at a value of 0.32%. This would allow 0.18% for man and a balance of 0.14% for adult pelagic fish natural mortality.

Both approaches give results that appear reasonable and confirm that the two methods of calculation are not necessarily incompatible.

Similar calculations can be done for adult demersal fish in the North Sea. Sustainable yield prior to the 1960s was just over 0.5 million tonnes, equivalent to about 0.1 $gC/m^2/year$. Although adult demersal fish may be regarded as secondary carnivores, the juveniles obtain energy from lower trophic levels. Annual recruitment to the adult stock provides an energy input additional to that obtained solely from the food consumed by the adults themselves. At current levels of exploitation it is calculated that the ratio of adult fish production to food consumed by adult fish could be as high as 20% (as distinct from perhaps 5% in the virgin stock) (Jones 1982). If natural mortality of demersal fish is also taken into account, the food requirements of the adult demersal fish would have to be at least 0.6 $gC/m^2/year$ (i.e., at least 0.75% of the grazed primary production). Only part of this comes from the pelagic food web and the remainder comes from the benthic food web.

These calculations indicate that fish production in the North Sea can be explained by either of the two approaches adopted.

In general, it is concluded that, except for bacteria, the average transfer efficiency is about 10%. Higher values can be used, but only at the expense of including an additional step. Because of the number of steps involved, and the uncertainty at each step, this approach appears to be more useful for explaining data, as in the case of the North Sea example above, than for making predictions.

Relationship between Predation and Species Numbers in Tropical Waters

The dynamics of a temperate system are essentially the dynamics of a non-steady-state system, i.e., each year there is a renewed flow of energy to which all species are variously adapted. The relatively small number of species at the higher trophic levels may simply mean that there are relatively few strategies for dealing with this situation.

In tropical seas, one approximates more closely a steady state system and predation pressure may exist on a relatively more continuous basis. The existence of a relatively large number of species with none of them being particularly dominant, may be a consequence of this. It is possible that predation has the potential to hold a group down to a level at which there is more food than it can effectively utilize. This can be expected to lead to selection for prey species that can exploit potentially surplus food, without necessarily being eaten in the process. That could lead to a proliferation of feeding strategies and hence might explain the relatively large number of species. With an increasing number of species at the prey level, there would presumably be selection pressure for the evolution of predators able to exploit the increased range of prey species. The tendency for the largest individuals at each trophic level to increase in body size with increasing primary production may be a consequence of such a process. It seems reasonable to suppose that a process such as this would continue until all available food at each trophic level had been taken up. It is very difficult to conceive of surplus food occurring, other than in the short term.

Discussion (Management Implications)

Principal management implications from this study are:

— exploitation can be expected to reduce the biomasses of individuals at the higher trophic levels and at the large-particle parts of lower trophic levels;

— it seems unlikely that these would be replaced by individuals of a comparable size but more likely that there would be a redistribution of energy so as to benefit smaller-sized individuals;

— from the fishery viewpoint, the ideal situation is probably where the juveniles of the exploited species benefit from a redistribution of energy. In this situation, the direct effects of exploitation on a "per-recruit basis" could be partly offset by increased recruitment. The next best possibility is that the recruitment of other, potentially harvestable species is increased, leading to a change in species composition without necessarily a decline in overall catch. In the Gulf of Thailand, for example, it is possible that the increase in squid catch has been a direct consequence of exploitation of other species due to a reduction of predation on the eggs of squid (D. Pauly, pers. comm. and see Larkin and Gazey, this vol.). Of least benefit to a fishery is the situation where the redistribution of energy becomes dissipated at the small-particle end of the food web in ways that benefit only species of little harvestable value.

References

Alexander, V. and R.T. Cooney. 1978. Bering Sea ice edge ecosystem study: nutrient cycling and organic matter transfer. ERL Princ. Invent. Rep. Environ. Assess. Alaskan Cont. Shelf. p. 216-294.

Alton, M.S. 1974. Bering Sea benthos as a food resource for demersal fish populations, p. 257-277. *In* D.W. Hood and E.J. Kelly (eds.) Oceanography of the Bering Sea. NOAA, NMFS, North West Fish Center. Seattle, WA, U.S.A.

Beers, J.R. 1966. Studies on the chemical composition of the major zooplankton groups in the Sargasso Sea off Bermuda. Limnol. Oceanogr. 11(4): 520-528.

Beers, J.R., C.M.R. Stevenson, R.W. Eppley and E.R. Brooks. 1971. Plankton populations and upwelling of the coast of Peru, June 1969. Fish. Bull. 69(4): 859-876.

Belyayev, G.M. and N.G. Vinogradova. 1961. Quantitative distribution of benthic fauna in the north Indian Ocean. Dolk. Akad. Nauk. SSSR 138(5): 35-38.

Bethel, W.M. and J.C. Holmes. 1977. Increased vulnerability of amphipods to predation owing to altered behaviour produced by larval acanthocephalans. Can. J. Zool. 55(1): 110-115.

Brylinsky, M. and K.H. Mann. 1973. An analysis of factors governing productivity in lakes and reservoirs. Limnol. Oceanogr. 18(1): 1-14.

Burgis, M.J., J.P.E.C. Darlington, I.G. Dunn, G.G. Ganf, J.J. Gwahaba and L.M. McGowan. 1973. The biomass and distribution of organisms in Lake George, Uganda. Proc. R. Soc. London, ser. B. 184: 271-298.

Caine, E.A. 1977. Feeding mechanisms and possible resource partitioning of the Caprellidae (crustacea; amphiphoda) from Puget Sound, U.S.A. Mar. Biol. 42: 331-336.

Caswell, H. 1978. Predator mediated coexistence: a non-equilibrium model. Amer. Nat. 112(983): 127-154.

Colman, J.A. 1970. On the efficiency of food conversion in young plaice *(Pleuronectes platessa)*. J. Mar. biol. Ass. U.K. 50: 113-120.

Connell, J.H. 1978. Diversity in tropical rain forests and coral reefs. Nature (London) 199: 1302-1310.

Coupland, R.T. 1979. Grassland ecosystems of the world. I.B.P. Handbook 18. Cambridge Univ. Press, New York.

Crisp, D.J. 1975. Secondary productivity in the sea, p. 71-89. *In* Productivity of world ecosystems. National Academy of Sciences, Washington D.C.

Cushing, D.H. 1969. Upwelling and fish production. FAO Fish. Tech. Pap. No. 84. 40 p.

Cushing, D.H. 1973a. Introduction to pelagic food chains, p. 69-73. *In* J.H. Steele (ed.) Marine food chains. Oliver and Boyd, Edinburgh.

Cushing, D.H. 1973b. Production in the Indian Ocean and the transfer from the primary to the secondary level, p. 475-486. *In* B. Zeitzschel (ed.) The biology of the Indian Ocean. Chapman and Hall, London.

De Bernardi, R. and G. Giussani. 1978. The effect of mass fish mortality on zooplankton structure and dynamics in a small Italian Lake (Lago di Annone). Verh. Int. Ver. Theor. Angew. Limnol. 20: 1045-1048.

De Vooys, C.G.N. 1979. Primary production in aquatic environments, p. 259-292. *In* B. Bolin, E.T. Degen, S. Kempe and P. Ketner (eds.) The global carbon cycle. Wiley, Chichester, U.K.

Dugdale, R.C. and J.J. Goering. 1967. Uptake of new and regenerated forms of nitrogen in primary productivity. Limnol. Oceanogr. 12(2): 196-206.

Ebeling, A.W. and R.N. Bray. 1976. Day versus night activity of reef fishes in a kelp forest off Santa Barbara, California. Fish. Bull. 74(4): 703.

Eggers, D.M., N.W. Bartoo, N.A. Rickard, R.E. Nelson, R.C. Wissmar, R.L. Burgner and A.H. Devol. 1978. The Lake Washington ecosystem: the perspective from the fish community and forage base. J. Fish. Res. Board Can. 35: 1553-1571.

Emery, A.R. 1973. Preliminary comparisons of day and night habits of freshwater fish in Ontario Lakes. J. Fish. Res. Board Can. 30: 761-774.

Eppley, R.W., H. Renger, E.L. Venrick and M.M. Mullen. 1973. A study of phytoplankton dynamics and nutrient cycling in the central gyre of the north Pacific Ocean. Limnol. Oceancgr. 18(4): 534-551.

Eppley, R.W. and B.J. Peterson. 1979. Particulate organic matter flux and planktonic new production in the deep ocean. Nature (London) 282: 677-680.

Enomoto, V. 1972. On the food chain in the Yellow Sea and the East China Sea. Bull. Tokai Reg. Fish. Res. Lab. 69: 65-73.

Fenchel, T.M. and B.B. Jørgensen. 1977. Detritus food chains of aquatic ecosystems: the role of bacteria. *In* M. Alexander (ed.) Advances in microbial ecology Vol. 1. Plenum Press, New York and London.

Fricke, H.W. 1975. Solving of simple problems by a fish. Z. Tierpsychol. 38(1): 18-33.

Frost, W.E. 1954. The food and pike. *Esox lucius* L. in Windermere. J. Anim. Ecol. 23(2): 339-360.

Frost, W.E. 1965. Breeding habits of Windermere char (*Salvelinus willughbii* Gunther) and their bearing on speciation of these fish. Proc. R. Soc. London, Ser. B. 163: 232-284.

Fryer, G. and T.D. Iles. 1972. The cichlid fishes of the great lakes of Africa. Oliver and Boyd, Edinburgh.

Ganf, G.G. 1974. Rates of oxygen uptake by the planktonic community of a shallow equatorial lake (Lake George, Uganda). Oecologia (Berl.) 15: 17-22.

Ganf, G.G. 1975. Photosynthetic production and irradiance—photosynthetic from a shallow equatorial lake (Lake George, Uganda). Oecologia (Berl.) 18: 165-183.

George, E.L. and W.E. Hadley. 1979. Food and habitat partitioning between rock bass *(Ambloplites rupestris)* and small mouth bass *(Micropterus dolomieu)* young of the year. Trans. Amer. Fish. Soc. 108: 253-261.

Ginetz, R.M. and P.A. Larkin. 1976. Factors affecting rainbow trout *(Salmo gairdneri)* predation on migrant fry of sockeye salmon *(Oncorhynchus nerka)*. J. Fish. Res. Board Can. 33(1): 19-24.

Greenwood, P.H. 1976. Lake George, Uganda. Philos. Trans. R. Soc. London, Ser. B. 274: 375-391.

Grice, G.D. and H.D. Hart. 1962. The abundance, seasonal occurrence and distribution of the epizooplankton between New York and Bermuda. Ecol. Monogr. 32: 287.

Hall, D.J., E.W. Werner, J.F. Gilliam, G.G. Mittelbach, D. Howard, G.G. Dover, J.A. Dickerman and A.J. Stewart. 1979. Diel foraging behaviour and prey selection in the Golden Shiner *(Notemigonus crysoleucas).* J. Fish. Res. Board Can. 36: 1029-1039.

Hart, T.J. 1942. Phytoplankton periodicity in Antarctic surface waters. Discovery Rep. 21: 261-356.

Haven, S.B. 1973. Competition for food between the intertidal gastropods *Acnaea soabra* and *Acnaea digitalis.* Ecology 54(1): 143-151.

Heip, C., G. Hermann, J.C.R. Bisschop, H. Gobaere, D. Holvoet, D. VanDamme, C. Vandsmael, K.R. Willems and L.A. de Coninck. 1979. Benthic studies of the Southern Bight of the North Sea and its adjacent continental estuaries. ICES CM 1979/L;9 (mimeo) Copenhagen.

Herbland, A. 1978. Heterotrophic activity in the Maustarian upwelling in March 1973; assimilation and mineralisation of amino acids, p. 155-166. *In* R. Boje and M. Tomczak (eds.) Upwelling ecosystems. Springer Verlag, Berlin.

Hickman, M. 1979. Seasonal succession, standing crop and determinants of primary productivity of the phytoplankton of Ministik Lake, Alberta, Canada. Hydrobiologia 64(2): 105-121.

Hillbricht-Ilkowska, A. 1972. Interlevel energy transfer efficiency in planktonic food chains. International Biological Programme. Section PH, Dec. 13, 1972, Reading, England.

Hobson, E.S. 1973. Diel feeding migrations in tropical reef fishes. Helgol. Wiss. Meeresunters. 24(1-4): 361-370.

Hobson, B.S. and J.R. Chess. 1976. Trophic interactions among fishes and zooplankters nearshore at Santa Catalina Island, California. Fish. Bull. 74(3): 567-498.

Holme, N.A. 1953. The biomass of the bottom fauna in the English Channel off Plymouth. J. Mar. Biol. Assoc. U.K. 32: 1-49.

Houghton, R.W. and M.A. Mensah. 1978. Physical aspects and biological consequences of Ghanaian coastal upwelling, p. 167-180. *In* R. Boje and M. Tomczak (eds.) Upwelling ecosystems. Springer Verlag, Berlin.

Humphreys, W.F. 1979. Production and respiration in animal populations. J. Anim. Ecol. 48: 427-453.

Huntsman, S.A. and R.T. Barber. 1976. Primary production in the upwelling region off northwest Africa—a comparison with Peru. CUEA 5(1): 2-5.

Jackson, G.A. 1980. Phytoplankton growth and zooplankton grazing in oligotrophic oceans. Nature (London) 284: 439-441.

Jackson, P.N.B. 1961. The impact of predation, especially by the tigerfish *(Hydrocyon vittatus)* on African freshwater fishes. Proc. Zool. Soc. London 136: 603-622.

Johnson, F.H. 1977. Responses of walleye *(Stizostedion vitreum vitreum)* and yellow perch *(Perca flavescens)* populations to removal of white sucker *(Catostomus).* J. Fish. Res. Board Can. 34(10): 1633-1642.

Jones, R. 1978. Further observations on the energy flow to the major fish species in the North Sea. Intern. Counc. Explor. Sea. CM 1978/Gen. 6 (Symp.) (mimeo) 18 p. Copenhagen.

Jones, R. 1979. Predator-prey relationship with particular reference to vertebrates. Biol. Rev. 54: 73-97.

Jones, R. 1982. Species interactions in the North Sea, p. 48-63. *In* M.C. Mercer (ed.) Multispecies approaches to management advice. Can. Spec. Publ. Fish. Aquat. Sci. 59. 169 p.

Kanamoto, Z. 1977. On ecology of hexagrammid fish. 3. Niches of *Agrammus agrammus* (Temminck et Schlegel) and *Hexagrammos otakii* Jordan Starks and the mode of life of some reef fish. Jap. J. Ecol. 21(3): 215-226.

Karplus, I. 1978. A feeding association between the grouper *Epinephelus fasciatus* and the moray eel *Gymnothorax griseus.* Copeia 1: 164.

Keast, A. 1973. Food specialisations and bio-energetics interrelations in the fish faunas of some small Ontario waterways, p. 377-414. *In* J.H. Steele (ed.) Marine food chains. Oliver and Boyd, Edinburgh.

Kellogg, C.W. 1977. Co-existence in a hermit crab species ensemble. Biol. Bull. 153: 133-144.

Kevern, N.R. 1966. Feeding rate of carp estimated by a radioisotope method. Trans. Amer. Fish. Soc. 95(4): 363-371.

Kipling, C. and W.E. Frost. 1972. Certain effects of an annual removal of pike *(Esox lucius)* from Windermere. Verh. Int. Ver. Theor. Angew. 18: 1122-1127.

Knox, G.A. 1970. Antarctic marine ecosystem, p. 69-96. *In* H.W. Holdgate (ed.) Antarctic ecology, Vol. 1. Academic Press, London and New York.

Langeland, A. 1977. The effect of fish *(Salvelinus alpinus*, char) predation on the zooplankton in 11 Norwegian lakes. In SIL 20 Congress Copenhagen published by DIS Congress Services, 3 Knabrostraede, DK 1210 Copenhagen, Denmark, 306 p.

Lasker, R. 1966. Feeding, growth and carbon utilisation by a euphausiid crustacean. J. Fish. Res. Board Can. 23(9): 1291-1317.

Lawrie, A.H. 1978. The fish community of Lake Superior. J. Great Lakes Res. Int. Assoc. Great Lakes Res. 4(3-4): 513-549.

Le Cren, E.D., C. Kipling and J. McCormack. 1972. Windermere: effects of exploitation and eutrophication on the salmonid community. J. Fish. Res. Board Can. 29: 819-832.

Longhurst, A. 1957. Density of marine benthic communities off West Africa. Nature (London) 179: 542-543.

Longhurst, A. and R. Williams. 1979. Materials for plankton modelling: vertical distribution of Atlantic zooplankton in summer. J. Plankton Res. 1(1): 1-37.

Lowe-McConnell, R.H. 1975. Fish communities in tropical freshwaters. Longman, New York, N.Y.

Luckhurst, B.E. and K. Luckhurst. 1978. Diurnal space utilisation in coral reef communities. Mar. Biol. 49: 325-332.

Macan, T.T. 1977. The influence of predation on the composition of freshwater animal communities. Biol. Rev. 52(1): 45-70.

Madhupratap, M., T.S.S. Rao and P. Haridas. 1977. Secondary production in the Cochin backwaters, a tropical monsoonal estuary, p. 515-519. *In* Proceedings of the symposium on warm water zooplankton, 14-19 October 1976. Goa, India. Oceanography, Goa, India.

Major, P.F. 1977. Predator-prey interactions in schooling fishes during periods of twilight. A study of the silverside *Pranesus insularum*, Hawaii. Fish. Bull. 75(2): 415-426.

Mare, M.F. 1942. A study of a marine benthic community with special reference to the micro-organisms. J. Mar. Biol. Assoc. U.K. 25: 517-554.

Marten, G.G. 1979. Predator removal: effect on fisheries yields in Lake Victoria (East Africa). Science 203: 646-647.

McIntyre, A.D. 1958. The ecology of Scottish inshore fishing grounds. I. The bottom fauna of east coast grounds. Mar. Res. No. 1. 24 p.

McIntyre, A.D. 1961. Quantitative differences in the fauna of boreal mud associations. J. Mar. Biol. Assoc. U.K. 41: 599-616.

McNeill, S. and J.H. Laughton. 1970. Annual production and respiration in animal populations. Nature (London) 225: 472-474.

Mills, E.L. 1975. Benthic organisms and the structure of marine ecosystems. J. Fish. Res. Board Can. 32: 1657-1663.

Mokyevsky, O.B. 1960. Geographical zonation of marine littoral types. Limnol. Oceanogr. 5(4): 389-396.

Moriarty, D.J.W., J.E.P.C. Darlington, I.G. Dunn, G.G. Ganf, J.J. Gwahaba, L.M. McGowan and M.P. Tevlin. 1973. Feeding and grazing in Lake George, Uganda. Proc. R. Soc. London, Ser. B. 184: 299-319.

Mylinski, M. 1977. Do all members of a swarm suffer the same predation? Z. Tierpsychol. 45(4): 373-388.

Neill, St. J.S.R. and J.M. Cullen. 1974. Experiments on whether schooling by their prey affects the hunting behaviour of cephalopods and fish predators. J. Zool. 172(4): 549-567.

Nieland, H. 1980. Die Nahrung von Sardinen, Sardinellen und Maifischen vor der Westküste Afrikas. Ber. aus dem Inst. Meereskunde am der Christian-Albrechts Universität Kiel. 137 p.

Northcote, T.G. and G.M.B. Hume. 1977. Initial impact of experience fish introductions on macro-zooplankton communities of small oligotrophic lakes. Presented at 20 Limnological Congress, Copenhagen, 7 September 1977. *In* SIL 20 Congress Copenhagen. DIS Congress Service, Denmark.

Ogden, J.C. 1977. The behaviour of heterotypic tending schools of juvenile grunts (Pomadasyidae). Mar. Biol. 42: 273-280.

Oglesby, R.T. 1977. Relationship of fish yield to lake phytoplankton standing crop, production, and morphoedaphic factors. J. Fish. Res. Board Can. 34: 2271-2279.

Ortner, P.B., P.H. Wiebe and J.L. Cox. 1980. Relationships between oceanic epizooplankton distributions and the seasonal deep chlorophyll maximum in the Northwestern Atlantic Ocean. J Mar. Res. 38(3): 507-531.

Paine, R.T. 1971. A short-term experimental investigation of the resource partitioning in a New Zealand rocky intertidal habitat. Ecology 52: 1096-1106.

Parsons, T.R. and R.J. LeBrasseur. 1968. A discussion of some critical indices of primary and secondary production for large-scale ocean surveys. Calif. Mar. Res. Comm. CALCOFI Rep. 12: 54-63.

Pavlova, E.V. 1967. Food requirements and transformation of energy by *Cladocera* populations in the Black Sea, p. 66-85. *In* Struktara i Dinamiko Vodnykh Soobshchestv i Populyatsii. Institut Biologii Yuzhnykn Morei Akademiya Nank Ykrainskoi SSR (In Russian)

Pederson, G.L., E.B. Welch and A.H. Litt. 1976. Plankton secondary productivity and biomass: a relation to lake trophic state. Hydrobiologia 50(2): 129-144.

Petersen, G.H. and M.A. Curtis. 1980. Differences in energy flow through major components of subarctic, temperate and tropical marine shelf systems. Dana 1: 53-64.

Petipa, T.S. 1967. On the efficiency of utilisation of energy in the pelagic ecosystem of the Black Sea, p. 44-65. *In* Struktara i Dinamiko Vodnykh Soobshchestv i Populyatsii. Institut Biologii Yuzhnykn Morei Akademiya Nank Ukrainskoi SSR (In Russian)

Phillips, R.R. 1977. The relationship between social behaviour and the use of space in the benthic fish *Chasmodes bosquianus* Lacèpède (Teleostei, Blenniidae). 4. Effects of topography on habitat selection and shelter choice. Behavior 60(1-2): 1-27.

Phillipson, J. 1970. Ecological energetics. The English Language Book Society and Edward Arnold (Publishers) Ltd.

Phleger, F.B. 1970. Some general features of coastal lagoons. Contributions from the Scripps Institution of Oceanography, La Jolla, California 40(2865): 1601-1621.

Platt, T. and D.V. Subba Rao. 1975. Primary production of marine microphytes, p. 249-280. *In* J.P. Cooper (ed.) Photosynthesis and productivity in different environments. International Biological Programme Ser. no. 3. Cambridge Univ. Press., N.Y.

Pomeroy, L.R. 1974. The ocean's food web, a changing paradigm. BioScience 24: 499-504.

Qasim, S.Z. 1972. The dynamics of food and feeding habits of some marine fishes. Indian J. Fish. 19(1): 11-28.

Qasim, S.Z. 1973a. Some problems related to the food chain in a tropical estuary, p. 46-51. *In* J.H. Steele (ed.) Marine food chain. Oliver and Boyd, Edinburgh.

Qasim, S.Z. 1973b. Productivity of backwaters and estuaries, p. 143-154. *In* B. Zeitzchel (ed.) The biology of the Indian Ocean. Chapman and Hall, London.

Reise, K. 1977. Predation pressure and community structure of an intertidal soft bottom fauna, p. 513-519. *In* B.F. Keegan, P.O. Ceidigh and P.J.S. Bowden (eds.) Biology of benthic organisms. Pergamon Press, Oxford.

Reise, K. 1978. Experiments on epibenthic predation in the Wadden Sea. Helgol. Wiss. Meeresunters. 31(1-2): 55-101.

Richman, S. 1958. The transformation of energy by *Daphnia pulex*. Ecol. Monogr. 28: 272-291.

Ricker, W.E. 1969. Food from the sea, p. 87-108. *In* Resources and man: a study and recommendations by the Committee on Resources and Man. W.H. Freeman, San Francisco. U.S.A.

Riley, G.A. 1956. Oceanography of Long Island Sound 1952-54. IX. Production and utilisation of organic matter. Bull. Bingham Oceanogr. Coll. 15: 324-343.

Rowe, G.T., P.T. Polloni and S.G. Horner. 1974. Benthic biomass estimates from the northwestern Atlantic Ocean and the northern Gulf of Mexico. Deep Sea Res. 21: 641-650.

Ryther, J.H. 1969. Photosynthesis and fish production in the sea. Science 166: 72-76.

Sale, P.F. 1976. Reef fish lottery. Nat. Hist. (N.Y.) 85(2): 61-64.

Sale, P.F. 1980. The ecology of fishes on coral reefs. Oceanogr. Mar. Biol. 18: 367-421.

Savich, M.S. 1972. Quantitative distribution and food values of benthos from the west

Pakistan Shelf. Okeanologiya 12(11): 113-119.
Schaefer, M.B. 1965. The potential harvest of the sea. Trans. Amer. Fish. Soc. 94: 123-128.
Sheldon, R.W. and W.H. Sutcliffe, Jr. 1978. Generation time of 3h for Sargasso Sea microplankton determined by ATP analysis. Limnol. Oceanogr. 23(5): 1051-1055.
Slobodkin, L.B. 1959. Energetics in *Daphnia pulex* populations. Ecology 40: 232-243.
Slobodkin, L.B. 1961. Growth and regulation of animal populations. Holt, Rinehart and Winston, New York.
Smayda, T.J. 1973. A survey to photoplankton dynamics in the coastal waters from Cape Hatteras to Nantuetat. Occas. Publ. Grad. School Oceanogr. 5(3): 1-100.
Smith, S.H. 1968. Species succession and fishery exploitation in the Great Lakes. J. Fish. Res. Board Can. 25: 667-693.
Snedaker, S.C. 1978. Mangroves: their value and perpetuation. Nat. Resour. 14(3): 6-13.
Spencer, C.P. 1954. Studies on the culture of a marine distom. J. Mar. Biol. Assoc. U.K. 33: 265-290.
Steele, J.H. 1958. Plant production in the northern North Sea. Mar. Res. 7. 36 p.
Steele, J.H. 1974. The structure of a marine ecosystem. Cambridge, Mass: Harvard University Press. 128 p.
Steele, J.H. and D.W. Menzel. 1962. Conditions for maximum primary production in the mixed layer. Deep Sea Res. 9: 39-49.
Stirling, J. and W.R. Archibald. 1977. Aspects of predation of seals by polar bears. J. Fish. Res. Board Can. 34: 1126-1129.
Stone, R.B. 1971. Recent developments in artificial reef construction. *In* Notes on 7th Annual Conf. and Exposition of the Marine Technology Society, Washington D.C. Aug. 16-18, 1971.
Symons, P.E.K. 1974. Territorial behaviour of juvenile Atlantic salmon reduced predation of Brook trout. Can. J. Zool. 52(6): 677-679.
Talling, J.F., R.B. Wood, M.V. Prosser and R.M. Blaxter. 1977. The upper limit of photosynthetic productivity by phytoplankton: evidence from Ethiopian soda lakes. Freshwater Biol. 3: 53-76.
Thompson, J.M. 1976. Prey strategies in evolution and ecology: or how to stay alive long enough to fertilise some eggs. Environ. Biol. Fishes 1(1): 93-100.
Thorson, G. 1957. Bottom communities (sublittoral or shallow shelf), p. 461-534. *In* J.W. Hedgpeth (ed.) Treatise on marine ecology and paleoecology. Mem. Geol. Soc. Amer. 67. Geological Society of America, New York.
Treisman, M. 1975. Predation and evolution of gregariousness. 1. Models for concealment and evasion. Animal Behaviour 23: 779-800.
Tyler, A.V. 1972. Food resource tradition among northern marine, demersal fishes. J. Fish. Res. Board Can. 29(7): 997-1003.
Venrick, E.L., J.A. McGowan and A.W. Hantyla. 1973. Deep maxima of photosynthetic chlorophyll in the Pacific Ocean. Fish Bull. 71(1): 41-52.
Virnsterin, R.W. 1977. The importance of predation by crabs and fishes on benthic infauna in Chesapeake Bay. Ecology 58(6): 1199-1217.
Walsh, J.J. 1971. Relative importance of habitat variables in predicting the distribution of phytoplankton of the ecotone of the Antarctic Upwelling System. Ecol. Monogr. 41: 291-309.
Warwick, R.M. and R. Price. 1975. Macrofauna production in an estuarine mid-flat. J. Mar. Biol. Assoc. U.K. 55: 1-18.
Watson, S.W. 1978. Role of bacteria in an upwelling ecosystem, p. 139-154. *In* R. Boje and M. Tomczak (eds.) Upwelling ecosystems. Springer Verlag, Berlin.
Whitledge, T.E. 1978. Regeneration of nitrogen by zooplankton and fish in the northwest Africa and Peru upwelling ecosystems, p. 90-100. *In* R. Boje and M. Tomczak (eds.) Upwelling ecosystems. Springer Verlag, Berlin.
Winslade, P. 1974. Behavioural studies on the lesser sandeel *Ammodytes marinus* (Raitt). I. The effect of food availability on activity and the role of olfaction in food detection. J. Fish. Biol. 6(5): 565-576.
Wissmar, R.C. and R.G. Wetzel. 1978. Analysis of five North American lake ecosystems. VI. Consumer community structure and production. Verh. Int. Ver. Theor. Angew. Limnol. 20: 587-597.
Wium-Anderson, S. 1979. Plankton primary production in a tropical mangrove bay at

the south west coast of Thailand. Ophelia 18(1): 53-60.

Wurtsbaugh, W.A. and R.W. Brocksen. 1975. Food and distribution of underyearling brook and rainbow trout in Castle Lake, California. Trans. Amer. Fish. Soc. 104(1): 88.

Yentsch, C.S. 1977. Plankton production. MESA New York Bight Atlas. Monograph No. 12 (77-007).

Yentsch, C.S. and R.F. Vaccaro. 1958. Phytoplankton nitrogen in the oceans. Limnol. Oceanogr. 3: 443-448.

Zaret, T.M. and R.T. Paine. 1973. Species introduction in a tropical lake. Science 182: 449-455.

Zenkevitch, L. 1963. The biology of the seas of the USSR. Translated by S. Botcherskeya. George Allen and Unwin, London.

Discussion of Mr. Jones' Paper

The discussion dealt primarily with the following points:

— Data presented showed a logarithmic decline in the percentage of demersal fish in commercial landings from the north Atlantic, when plotted against increasing temperature. The discussion made it clear, however, that this should be interpreted with care. On the one hand, one would expect less organic material to reach the bottom in warm water. This should be associated with a poorer benthic production and hence could be used to argue that there should be a poorer food supply for demersally-feeding fish in warm water. On the other hand, however, the majority of demersal fish in northern waters are cod, and although these pass through a demersally-feeding stage, much of their food as adults consists of fish, such as capelin, derived from the pelagic food chain. The picture is further confused by the fact that in tropical waters, such as the Gulf of Thailand, it is difficult to distinguish between fish that feed demersally, and those that are classed as "demersal", in the official statistics.

— Regarding transfer efficiencies, Mr. Jones explained that there would be a relatively high ratio of adult fish production/food consumed by adults in an exploited stock. This is because adult production depends on recruitment, as well as on the conversion of food into growth. In stocks exploited at North Sea levels of intensity, ratios of adult fish production/ food consumed by adults might attain as much as 25%.

— On the question of the food consumed by larval and juvenile stages, it was pointed out that although the biomasses of these stages were not necessarily large, they were stages that accounted for relatively large energy flows. This is because of the relatively high growth rates of larval and juvenile stages (both of fish and zooplankton) compared with the growth rates of adults.

Questioned on models, and their possible usefulness for prediction, Mr. Jones expressed the view that for the time being, models should be used as an aid to understanding the workings of the ecosystem. It could be a big step from this to the stage when models could be used to make predictions of, for example, the effect of exploitation on species composition.

— With regard to Mr. Jones' theoretical consideration on predation, it was pointed out that larval biomass was relatively small and hence that larval feeding was not likely to have a very large impact on food supply. In reply, it was pointed out that there is an important difference between biomass and flow. Since larvae have the potential to grow relatively rapidly (up to 10% per day in body weight) it is possible for a relatively small biomass of larvae to have a significant impact on potential food.

Commenting on predator-control versus food limitation, it was pointed out that predation might reduce a population below a point at which they were able to have a significant impact on their food. There might even be the possibility of a surplus of food. It was agreed that this might be so with only one predator. With many predator species however, it was suggested that some species should always be able to adapt to take up any potential food surplus. It was difficult to conceive of any surplus food in the long term, although, of course, this could occur locally or for short periods of time.

One participant pointed out that the real world does not necessarily have to be in a situation of stable equilibrium but that a situation of apparent equilibrium could persist for a very long time. Important factors then would not necessarily have to be biotic. Food, predators and environmental factors may all take turns at influencing population size at different times and different places and in ways quite unrelated to life stages. The need to necessarily believe in stable equilibrium and the competitive exclusion principle was therefore questioned, especially as the only evidence for it appears to have come "from aquarium experiments conducted in the 1920s in German laboratories."

In reply, it was agreed that stable equilibrium, mediated through interspecific competition, did not necessarily have to be the whole of the explanation all of the time. It is just one way of reconciling competition, coexistence and food sharing in a situation where a number of predators and prey all live together under conditions of what might appear to be stable equilibrium.

Another participant pointed out that if individuals were to be regarded as unequal, the really important thing was their vulnerability, regardless of whether this was due to food or shelter, and that the displacement theory was equivalent to a shelter-limited system rather than necessarily to a food-limited system. Mr. Jones did not disagree with this but added that displacement was a consequence of population growth, as well as of a limited resource (i.e., the nature of the resource is not as important as the fact that it is not infinite and that the population is growing). Although most of the examples given were terrestrial ones, it seemed reasonable to assume that the sea is a non-homogenous environment and that vulnerability need not necessarily be constant.

The question of the relationship between natural mortality and body weight was introduced and discussed with reference to the question of adaptation in an evolutionary sense. As a comment on this, it was pointed out that time scales of 10 to 15 years are important and that we are not in a position to make long-term predictions.

Mr. Jones stated that in his theory of predation, natural mortality is assumed to be dependent on two separate functions. One is a predator-prey interaction term, and the other a function of the prey/food abundance relationship. By adding this second term, a much greater flexibility is given to the equations and this makes it possible to satisfy a wider range of conditions than is possible with simple Lotka-Volterra equations, which only contain the predator-prey interaction terms.

The Structure and Dynamics of Coral Reef Fish Communities

PETER F. SALE

School of Biological Sciences
The University of Sydney
Sydney, N.S.W., Australia 2006

SALE, P.F. 1982. The structure and dynamics of coral reef fish communities, p. 241-253. *In* Pauly, D. and G.I. Murphy (eds.) Theory and management of tropical fisheries. ICLARM Conference Proceedings 9, 360 p. International Center for Living Aquatic Resources Management, Manila, Philippines and Division of Fisheries Research, Commonwealth Scientific and Industrial Research Organisation, Cronulla, Australia.

Abstract

This paper briefly summarizes available data on the ecology of fish communities on coral reefs, and attempts to relate them in a context useful for fisheries management. The primary conclusion is that reef fish ecologists have paid little attention to those attributes of fish–growth rate, fecundity, mortality and recruitment–likely to be most important to fisheries biologists. Data on production rates are virtually non-existent.

The available evidence indicates that reef fish assemblages are loosely structured species groupings, and lack the high degree of stability anticipated by ecologists on theoretical grounds. Management schemes which assume that these are strongly interactive systems, or which assume that they possess mechanisms enabling them to recover from, or to resist external perturbations, such as enhanced fishing effort, seem unlikely to be successful.

Introduction

The aim of this paper is to provide information about the ecology of reef fishes in a context that will be useful for fisheries management. The field has been reviewed recently by Ehrlich (1975), Goldman and Talbot (1976) and Sale (1980a). The latter should be read for amplification of points raised here.

Study of the ecology of coral reef fish is a relatively recent activity which has developed contemporaneously with the availability of scuba equipment. The major methodologies of the field have more in common with those of ecologists interested in bird communities than with those of ecologists interested in fish. This is due to two features of reef fish communities. Fish on reefs can be observed like birds in a forest, which is possible only with far more difficulty and discomfort in most other fish habitats, and is impossible in some. Many of the standard techniques for sampling fish populations can only be used with great difficulty in reef environments. This is not to imply that traditional methods have not been profitably used in the study of reef fish ecology. Highly creditable examples of such 'traditional' approaches include those of Moe (1969), Munro (1976), Munro et al. (1973) and Thompson and Munro (1978). However, such approaches remain atypical.

Considering the affinities with terrestrial (particularly bird) ecology, it is not surprising that reef fish ecology has been concerned primarily with different questions from those central to fish ecology in general. In particular there has been a lack of attention to questions of relevance to management, in part because of a lack of western-style commercial exploitation of reef species. Thus, while considerable information is available on the intimate details of reproductive behavior and feeding ecology of numerous species, rather little is known about growth rate, survivorship, fecundity or recruitment, the raw data of population dynamics.

Another feature of reef fish ecology, which may be of major significance in any attempt to incorporate its data into fisheries management programs, is the small spatial scale at which most studies have been done. Reef fish ecologists provide substantial evidence of the factors organizing assemblages of fish on small patch reefs or on demarcated study sites a few hundred square meters or less in area. However, we have little detailed information from sampling on a broader scale. Some of us believe extrapolation to whole reefs or to whole reef systems is possible. There is always danger in such a procedure although information on this larger scale is needed if reef fisheries are to be managed.

In the following, I will summarize what is known about the dynamics of coral reef species, and discuss the nature of fish assemblages on coral reefs. Their organization, stability and predictability of structure, both on the small scale for which some data exist, and on the broader scale where information is needed, will be considered.

Dynamics of Reef Fish Populations

The typical reef fish is a strongly site-attached, sedentary animal which lives to breed a number of times and exports all its offspring as pelagic larvae. Very few species are atypical in these characteristics (Ehrlich 1975; Johannes 1978; Sale 1980a).

Among small species, 10 to 15 cm or less in length, the area used by an individual is seldom as much as 10 m^2 and in many species may not extend beyond the confines of a single coral colony or other microhabitat (Sale 1977, 1978). Larger, more commercially attractive species such as scarids, siganids and coral trout move about more, but even here movements are seldom extensive (Ogden and Buckman 1973; Robertson and Choat 1974; Goeden 1978; Sale 1978). Some of these larger species do make longer migrations (often in large aggregations) when spawning (Smith 1972; Warner et al. 1975; Johannes 1978), or among herbivores particularly, in order to feed at high tide in shallow regions inaccessible at low tide (Ogden and Buckman 1973; Winn and Bardach 1960).

Apart from instances of such migrations, reef fish are members of a mixed-species assemblage which they joined on recruiting from the plankton, and they will breed in their turn with neighboring individuals of their own species. The opportunity exists, and seems frequently to be used, for the development of complex, long-standing social organizations among these sedentary individuals. Such social structures can include interspecific as well as intraspecific relationships (Itzkowitz 1977; Sale 1978).

Despite early expectations by some workers, and with the exception of a small number of species, breeding appears to be a seasonal phenomenon. The seasons are lengthy, however, (commonly 3 to 4 months), and in many species *some* breeding can take place most of the year (Munro et al. 1973; Sale 1977, 1978; Thresher, in press). Within the breeding season, production of multiple broods is usual, with daily, weekly, biweekly and monthly spawning cycles all being common (examples are Randall 1961; Keenleyside 1972; Warner et al. 1975; Doherty 1980; reviews in Sale 1977; Johannes 1978; Thresher, in press). A variety of breeding habits exist (reviewed in Breder and Rosen 1966; Thresher, in press). Most species spawn in pairs and shed their pelagic eggs in midwater. A lesser number, particularly of the families Apogonidae, Pomacentridae, Pseudochromidae, Gobiidae, Blenniidae and Syngnathidae show parental care of demersal eggs, although the eggs hatch within 1 to 7 days into larvae which are themselves pelagic. A very small number of aberrant forms show more intensive care. Pairs of the pomacentrid *Acanthochromis polyacanthus* care for the hatched young for up to two months by which time the juveniles are able to take up independent life on the reef (Robertson 1973; Thresher, pers. comm.). No other species is known not to possess a pelagic larval phase although some Brotulidae are live-bearers, producing well-advanced fry which are probably not pelagic in habit (Wourms and Bayne 1973).

Johannes (1978, 1980) has classified reef fish into four groups on the basis of reproductive behavior. He has argued cogently for the importance of planktivorous predators on the reef as an agent selecting for the particular reproductive strategies found. These all achieve a rapid export of the eggs or larvae from the immediate vicinity of the reef. He argues that seasonality of reproduction has been determined by this need to export eggs, but have them return later. Fish which spawn at the time of year when currents are favorable for this will be selected. His hypothesis remains untested because of the paucity of data on water movements in reef areas, but it is more parsimonious than hypotheses based on temperature regimes or phytoplankton availability to account for seasonality. Seasons are generally more or less coincident among the species at a reef but can differ markedly in timing in sites only 1,000 km apart (e.g., a late winter season in Jamaica) (Munro et al. 1973) but spring and autumn seasons in Curaçao (Luckhurst and Luckhurst 1977).

Only limited data on the fecundity of reef fish are available. The tendency to spawn many times in a season obviously hampers collection of such data. Estimates of numbers of eggs in a single spawning range from a few hundred in many pomacentrids to a million in some siganids (Sale 1980a). Even in those species such as pomacentrids and blennies which produce small clutches, however, the total number of eggs produced in a season can be several tens of thousands (Doherty 1980; Dybdahl 1978).

The pelagic larval period is the one for which least information is available (but see Leis and Miller 1976; Leis 1977). Recent developments in the interpretation of daily and even subdaily growth increments in otoliths (Brothers et al. 1976) can be expected to add substantially to our knowledge (see Munro, this vol.), at least in determining growth rate during, and duration of the larval phase. Results to date (Brothers and McFarland, in press; Brothers, Sale and Williams, unpub. data) appear generally to confirm previous, less direct estimates of the duration of larval life (7 to 20 days in species like apogonids and pomacentrids; 30 to 60 days or more in many labrids, scarids and acanthurids). The main significance of this lengthy larval period is that the spatial scale on which many adult fish operate must be very much smaller than that over which their offspring are dispersed. The means by which larvae return to the reef are completely unknown.

Survivorship during the pre-recruitment larval phase appears to be very low, as is usual for most fish. The numbers of recruits are very much fewer than the number of eggs produced each season. Several recent studies have measured recruitment to monitored sites (Luckhurst and Luckhurst 1977; Russell et al. 1977; Sale 1979; Doherty 1980; Williams 1980; Williams and Sale 1981). For the majority of species, recruits appear singly on the reef but appear to survive well from then on. For some, the density of recruits arriving in a season may approach the adult density, but high juvenile mortality cuts this number back over the next 6 months (e.g., *Acanthochromis polyacanthus* and *Chromis nitida*, Williams 1980). Still other species, particularly among the Apogonidae, Gobiidae and Scaridae at One Tree Reef, recruit in densities three or four times greater than that of adults, but suffer high mortality during the first post-recruitment season (Williams and Sale, pers. obs.).

More important than the mean density of recruitment is the variability in time and space of recruitment. Williams and Sale (1981) monitored recruitment to identical sets of colonies of three species of coral at four similar sandy sites within the One Tree lagoon. The four sites were no more than 2 km apart. Corals were kept clear of fish with removals on a daily or near-daily basis during the height of the recruitment season. Over two years a total of 7,000 new recruits were collected. Order of magnitude differences existed for many common species in the number of recruits settling at each of the four sites or at particular sites in the two seasons. Relative abundance across sites also changed substantially between years in many species. Doherty (1980) monitored recruitment of *Pomacentrus wardi* to a total of 2,635 m^2 of habitat distributed through the One Tree lagoon over three years. Total numbers of recruits were 105, 141 and only 14 for 1977, 78 and 79 respectively. Earlier, Russell et al. (1977) and Talbot et al. (1978) had monitored recruitment to artificial reefs at One Tree Reef over a 3½ year period. They demonstrated for many species 3 to 4 fold differences between years in given months, and up to 10 fold differences between whole years in the numbers of recruits arriving. Kami and Ikehara (1976) used 13 years' creel census data to document the recruitment of *Siganus spinus* and *S. argenteus* to shallow reef flats at Guam. Yearly catch (= recruitment) varied from 0.1 to 16 tonnes, and there was no significant auto-correlation between years.

Such variability in recruitment will not be surprising to fisheries biologists. However, its existence further emphasizes the lack of coupling of the local adult populations, the subject of any commercial fishery, and the recruitment to them. Sedentary adult populations are replenished by a patchy and temporally variable recruitment derived predominantly from breeding carried on by fish living elsewhere.

Little is known of the survivorship of reef fish following recruitment. This is primarily because of the difficulty of aging fish living in the relatively aseasonal tropics. Usual methods for aging fish using annual marks in scales or bones, or the tracking of annual size classes, are difficult or impossible to use when growth is not markedly seasonal and recruitment is not confined to a short season each year (Moe 1969; Stark and Schroeder 1970; Thompson and Munro 1978).

Data derived from small numbers of tagged fish, anecdotes concerning recognized individuals which occupy particular sites, and aquarium records indicate many reef species can live 5 to 10 years. Mean survivorship has rarely been determined with any accuracy, however. Thompson and Munro (1978) suggested *Mycteroperca venosa* in Jamaica requires 10 years to reach its modal adult length of 70 cm. This is in reasonable agreement with other age estimates for serranids of similar size (e.g., Moe 1969; Goeden 1978).

Among smaller species, mean survivorship of 3 to 5 years after recruitment is probably common (Randall 1961; Stark and Schroeder 1970; Ross 1978, Sale 1979; Doherty 1980) although even annual species exist.

Biomass and Production

A number of estimates have been made of the fish biomass supported by coral reefs (Brock 1954; Bardach 1959; Clark et al. 1968; Goldman and Talbot 1976; Odum and Odum 1955; Randall 1963). These have mostly been based on collections using rotenone or explosives, although Brock (1954) derived his estimates from visual census data. The estimates vary by a factor of 10, and there is as much variation among habitats at a single location as among locations. Goldman and Talbot (1976) reported biomasses ranging from 175 kg/ha on the upper reef slope to 1,950 kg/ha at the reef base at One Tree Island. This range encompasses the estimates of other workers at other sites. Insufficient data are available to permit examination of the effect of fishing pressure on biomass present, or the temporal constancy of biomass present at unfished sites.

Only one estimate has been made of the rate of fish production on reefs and more work on this question is clearly needed. Bardach (1959) estimated this rate as 2.2×10^5 kcal/ha/year for a patch reef in Bermuda. Much of this would be in species that are not commercially attractive. Trophic structures have been examined by a number of workers (Bakus 1966; Gladfelter and Gladfelter 1978; Goldman and Talbot 1976; Talbot 1965). The main point to emerge is that herbivores are substantially more important on reefs, both in numbers and number of species, than in temperate communities. On average they comprise about 25% both by weight and number of fish species on coral reefs (Bakus 1966).

Structure of Reef Fish Assemblages

A primary feature of reef fish assemblages is their high diversity of species. This high diversity has made them attractive subjects of study by ecologists interested in such systems and in the coexistence of similar, potentially competing species. A major thrust in reef fish ecology since the early 1960s has been in elucidating the structure of such assemblages and in identifying the factors responsible for that structure.

From what has been said already, it is clear that on the small spatial scale at which ecologists have worked, reef fish assemblages are composed entirely of individuals derived from a largely externally-generated larval pool. In turn, members of the assemblages export their reproductive products to that same pool. In the establishment of such a recruitment-derived set of individuals, there are numerous ways whereby biotic interactions might be important factors in shaping the composition, the diversity, and the numerical abundance of the fish which come to live together.

It was perhaps not surprising, given the ecological climate of the 1960s and 1970s that most of us assumed the structure of reef fish assemblages would be rather constant, probably also stable, and certainly predictable. These features were seen primarily as consequences of the action of interspecific competition for limiting resources. Particular emphasis was put on competition for space on the reef. Space was seen as the resource most likely in short supply, and assemblages were expected to be usually limited by their space resources (Sale 1974, 1977; Smith and Tyler 1972, 1975). In addition, predictable predatory, commensal and other affiliative interactions were expected among the component species, all acting in communities at equilibrium in a benign and stable environment (Sale 1980a).

In fact, this now appears not to be the case, at least at the scale at which ecologists have worked. Studies of the recovery of assemblages following the removal of all, or nearly all fish, have given mixed results except for one feature. Where the species composition of the assemblage has been accurately measured, the re-formed assemblage is considerably different to that which had been removed (Sale 1980b; Sale and Dybdahl 1975, 1978). In cases where the recovered fauna is reported to be very similar to that which it replaced, emphasis appears to have been placed on species presence only (Gundermann and Popper 1975; Smith and Tyler 1975). When examined with respect to the relative abundances of the species present in the two assemblages, similarity is found to be low (Sale 1980a).

Additional support for this result comes from studies of small artificial reefs. Recruitment to replicate artificial reefs leads to assemblages which differ substantially from one another (Bohnsack and Talbot 1980; Molles 1978; Russell et al. 1974; Talbot et al. 1978). Talbot et al. (1978) showed similarities in faunas on replicate concrete block reefs built at the same time of year to be only 32 to 41% when relative abundances of species were taken into account.

Furthermore, monitoring studies of natural and artificial reefs have shown substantial variation in species' relative abundance through time, in the absence of any experimental interference (Sale 1979, 1980a; Talbot et al. 1978; Williams 1980). These results indicate that on this small spatial scale, undirected change in assemblage structure will naturally occur through time.

Monitoring studies and studies of the colonization of artificial reefs or denuded natural reefs have disclosed substantial seasonal and non-seasonal variations in both a) numbers of fish colonizing sites (Doherty 1980; Luckhurst and Luckhurst 1977; Russell et al. 1977; Williams 1980; Williams and Sale 1981) and b) the standing crop of fish resident on the sites (Bohnsack and Talbot 1980; Doherty 1980; Russell et al. 1974; Sale 1980b; Talbot et al. 1978; Williams 1980). Numbers of fish resident on a patch reef can vary seasonally by a factor of two, although a coefficient of variation of about 0.4 is more usual (Sale and Douglas, in preparation). The extent of variation in standing crop and recruitment varies substantially among species.

Factors Responsible

On this local scale reef fish assemblages are diverse, but vary substantially through time in their composition and in the numerical abundance of the fish which comprise them. Ecologists working on reef fishes expected to find structures that were temporally more stable. They also expected these structures to be determined largely through interactions among the component species, particularly competitive interactions that would lead to niche diversification (Smith and Tyler 1972). This has been the general view concerning diverse tropical communities (Diamond 1975; Sanders 1969).

The few attempts to detect experimentally interspecific, competitive interactions responsible for maintaining this structure, have been largely unsuccessful. Sale and Dybdahl (1975, 1978) were not able to demonstrate for any of 15 common species, a single pair in which one avoided, or excluded the other from similar coral heads, and Talbot et al. (1978) could find only one pair of species which colonized the same type of artificial reef but avoided or excluded each other. Futhermore, Sale (1977, 1979) has argued that even if competitive interactions occurred between reef fish they would not necessarily lead to niche partitioning. The lack of association inherent in the reef fish life cycle between adult competitive success and subsequent recruitment to that site would lead to a competitive lottery rather than exclusion.

While loose commensal relationships among reef fish are known (Ehrlich and Ehrlich 1973; Itzkowitz 1974, 1977), and at least cleaning symbiosis has been claimed to have a structuring role (Slobodkin and Fishelson 1974) none of them appear sufficiently strong to determine local species composition (Gorlick et al. 1978; Talbot et al. 1978). Again, on theoretical grounds, the structuring of diverse tropical assemblages could have been expected to be based to an extent on such interactions (Diamond 1975).

The remaining type of interspecific interaction, predation, has yet to be examined in this context. My intuition is that this factor, too, will be found not to be important in determining species composition of assemblages.

At present, and this is a judgment made on the basis of very little, and largely negative evidence, I expect that the structure of these small reef fish assemblages is a consequence of a series of causal factors which independently determine the likelihood of each species colonizing any given site. These include the habitat requirements of the colonizing larva and of the resident adult, and their mortality rate and chances of colonizing the site. The latter is itself a function of that species' fecundity, larval mortality, larval dispersal, and availability as breeding groups in nearby sites. Importantly, there is no indication of the assemblage of fish being any more than the sum of its component species.

STRUCTURE ON A LARGER SPATIAL SCALE

If the interpretation offered above is a correct assessment of the structure of local fish assemblages, and if larger regions can be viewed as simply the sum of a number of adjacent, small patches of habitat, certain predictions can be made about the structure of reef fish communities on this larger scale.

—To the extent that there is variation from year to year in rate of colonization of different species there will be changes in relative abundance as certain species increase while others decrease in numbers. However, since rate of colonization varies in substantially different ways at similar small sites as little as 1 km apart (Williams 1980; Williams and Sale 1981), much of this variation will not be apparent on the larger spatial scale. On the other hand seasonal variation in colonization, and thus in numerical standing crop is sufficiently pronounced that this seasonal variation in total density of fish can be expected on the larger scale as well. Since the seasonal pulse in numbers is primarily due to the influx of numerous juveniles, it is unlikely to be mirrored in any substantial seasonal variation in biomass of fish present.

—Variation in species composition can be expected to be much less when sampling is done on a larger spatial scale. Each of the species present in the larval pool available to colonize a particular type of habitat has some determinate chance of colonizing a particular small site in that habitat at any given time. When many such sites are included in the area sampled, all species present in the pool can be expected to be present in one or other site. Monitoring studies will indicate a greater degree of constancy of composition than has been found in the smaller scale studies.

—There is no *a priori* reason to expect that this greater constancy in composition, or, barring the seasonal fluctuations, in numbers, is an indication of great stability, or of a structure determined by strong interspecific interactions of the type sought but not found on the smaller scale. Non-experimental studies on this larger spatial scale will not be capable of assessing the degree of stability of structure present. The constancy they seem likely to detect may be deceiving.

Implications for Fisheries Management

I stress that the following list is based on a limited amount of experimental data collected in studies carried out on a very local scale. Extrapolation and speculation are all that can be managed at present, and much more research effort is clearly needed.

—If the reef fish community really is relatively constant but not particularly stable in structure, fishing activity is a perturbation for which compensatory responses will not be present. Removal of a proportion of the standing crop of one or a few species seems unlikely to result in their enhanced subsequent recruitment. A permanent change in the structure of the community is likely because of the effective increase in the mortality of the fished species. We lack evidence adequate to assess the impact of such changes.

—The reef fish assemblage is sufficiently weakly structured, and the diversity of species provides so many ecological analogues, that attempts to model a multispecies reef fishery using extended sets of Lotka-Volterra competition and predation equations, as has been done by May et al. (1979) for the Antarctic community, seem unlikely to be fruitful. Such an approach assumes a far more tightly interactive system than the experimental evidence to date indicates for reef fish.

—Given that for many of the larger, more commercially attractive species, mean survivorship may be around 10 years with relatively few recruits each year, sustainable rates of fishing may be substantially lower than an initial inspection of standing crops suggests. The data needed to calculate sustainable yields for most reef species are simply not available.

—Intense fishing in one site may have marked effects on the recruitment to other sites many kilometers downstream. It is important that fishing not be centered on spawning aggregations where they occur (Johannes 1978), and that the effects on other sites of any heavy fishing be monitored. Notwithstanding this, the patchy nature of most coral reef regions does lend itself to a program of management by alternately heavily fishing and then closing a succession of nearby sites. Before such a program could be established as policy, substantial additional information would be needed on water circulation, larval dispersal and recruitment. The durations of closed seasons in such a management program might be in the order of 10 years or more. Recruitment overfishing may be more easily prevented using this procedure than with a more uniform fishing effort.

—Particular species with complex social structures, with obligate size-related sex-reversal, or with patterns of growth which strongly skew the sex-ratio at larger sizes, may all be particularly susceptible to fishing pressure. All occur among reef species.

Acknowledgments

This paper is a contribution from The University of Sydney's One Tree Island Field Station, at which much of the experimental work referred to here was carried out. My research cited here has been done with support from the Australian Research Grants Committee, the Great Barrier Reef Marine Park Authority and The University of Sydney.

References

Bakus, G.J. 1966. Some relationships of fishes to benthic organisms on coral reefs. Nature (London) 210: 280-284.

Bardach, J.E. 1959. The summer standing crop of fish on a shallow Bermuda reef. Limnol. Oceanogr. 4: 77-85.

Bohnsack, J.A. and F.H. Talbot. 1980. Species-packing by reef fishes on Australian and Caribbean reefs: an experimental approach. Bull. Mar. Sci. 30: 710-723.

Breder, C.M., Jr. and D.E. Rosen. 1966. Modes of reproduction in fishes. Natural History Press, New York.

Brock, V.E. 1954. A preliminary report on a method of estimating reef fish populations. J. Wildl. Manage. 18: 297-308.

Brothers, E.B., C.P. Mathews and R. Lasker. 1976. Daily growth increments in otoliths from larval and adult fishes. Fish. Bull. 74(1): 1-8.

Brothers, E.B. and W.N. MacFarland. Correlations between otolith microstructure, growth, and life-history transitions in newly-recruited French grunts (*Haemulon flavolineatum* (Desmarest), Haemulidae). Contrib. to I.C.E.S. Symp. on Early Life History of Fish.

Mar. Biol. Lab., Woods Hole, Mass., April 1979. (In press)

Clark, E., A. Ben-Tuvia and H. Steinitz. 1968. Observations on a coastal fish community, Dahlak Archipelago, Red Sea. Sea Fish. Res. Stn. Haifa Bull. 49: 15-31.

Diamond, J.M. 1975. Assembly of species communities, p. 342-444. *In* M.L. Cody and J.M. Diamond (eds.) Ecology and evolution of communities. Belknap Press. Harvard University Press, Cambridge, Massachusetts.

Doherty, P.J. 1980. Biological and physical constraints on the populations of two sympatric territorial damselfishes on the southern Great Barrier Reef. University of Sydney, N.S.W. 183 p. Ph.D. dissertation.

Dybdahl, R.E. 1978. The ecology and behaviour of a mimetic pair of coral reef fishes, *Meiacanthus lineatus* and *Petroscirtes fallax* (Pisces: Bleniidae). University of Sydney, N.S.W. 121 p. M.S. thesis.

Ehrlich, P.R. 1975. The population biology of coral reef fishes. Annu. Rev. Ecol. Syst. 6: 211-247.

Ehrlich, P.R. and A.H. Ehrlich. 1973. Coevolution: heterotypic schooling in Caribbean reef fishes. Amer. Nat. 107: 157-160.

Gladfelter, W.B. and E.H. Gladfelter. 1978. Fish community structure as a function of habitat structure on West Indian patch reefs. Rev. Trop. Biol. 26 (Suppl. 1): 65-84.

Goeden, G.B. 1978. A monograph of the coral trout, *Plectropomus leopardus* (Lacepède). Queensl. Fish. Serv. Res. Bull. 1: 1-42.

Goldman, B. and F.H. Talbot. 1976. Aspects of the ecology of coral reef fishes, p. 125-154. *In* O.A. Jones and R. Endean (eds.) Biology and geology of coral reefs, Vol. III. Biology 2. Academic Press, New York.

Gorlick, D.L., P.D. Atkins and G.S. Losey, Jr. 1978. Cleaning stations as water holes, garbage dumps, and sites for the evolution of reciprocal altruism? Amer. Nat. 112: 341-353.

Gundermann, N. and D. Popper. 1975. Some aspects of recolonization of coral rocks in Eilat (Gulf of Aqaba) by fish populations after poisoning. Mar. Biol. 33: 109-117.

Itzkowitz, M. 1974. A behavioural reconnaissance of some Jamaican reef fishes. Zool. J. Linn. Soc. 55: 87-118.

Itzkowitz, M. 1977. Social dynamics of mixed-species groups of Jamaican reef fishes. Behav. Ecol. Sociobiol. 2: 361-384.

Johannes, R.E. 1978. Reproductive strategies of coastal marine fishes in the tropics. Environ. Biol. Fishes 3: 65-84.

Johannes, R.E. 1980. Using knowledge of the reproductive behavior of reef and lagoon fishes to improve fishing yields, p. 247-270. *In* J.E. Bardach, J.J. Magnuson, R.C. May and J.M. Reinhart (eds.) Fish behavior and its use in the capture and culture of fishes. ICLARM Conference Proceedings 5. International Center for Living Aquatic Resources Management, Manila.

Kami, H.I. and I.I. Ikehara. 1976. Notes on the annual juvenile siganid harvest in Guam. Micronesica (J. Coll. Guam) 12: 323-325.

Keenleyside, M.H.A. 1972. The behaviour of *Abudefduf zonatus* (Pisces, Pomacentridae) at Heron Island, Great Barrier Reef. Anim. Behav. 20: 763-774.

Leis, J.M. 1977. Systematics and zoogeography of the porcupinefishes (*Diodon*, Diodontidae, Tetraodontiformes), with comments on egg and larval development. Fish. Bull. 76: 535-567.

Leis, J.M. and J.M. Miller. 1976. Offshore distributional patterns of Hawaiian fish larvae. Mar. Biol. 36: 359-367.

Luckhurst, B.E. and K. Luckhurst. 1977. Recruitment patterns of coral reef fishes on the fringing reef of Curacao, Netherlands Antilles. Can. J. Zool. 55: 681-689.

May, R.M., J.R. Beddington, C.W. Clark, S.J. Holt and R.M. Laws. 1979. Management of multispecies fisheries. Science 205: 267-277.

Moe, M.A., Jr. 1969. Biology of the red grouper *Epinephelus morio* (Valenciennes) from the eastern Gulf of Mexico. Prof. Pap. Ser. Mar. Lab. Fla. 10: 1-95.

Molles, M.C., Jr. 1978. Fish species diversity on model and natural reef patches: experimental insular biogeography. Ecol. Monogr. 48: 289-305.

Munro, J.L. 1976. Aspects of the biology and ecology of Caribbean reef fishes: Mullidae (goat fishes). J. Fish Biol. 9: 79-97.

Munro, J.L., V.C. Gaut, R. Thompson and P.H. Reeson, 1973. The spawning seasons of

Caribbean reef fishes. J. Fish Biol. 5: 69-84.

Odum, H.T. and E.P. Odum. 1955. Trophic structure and productivity of a windward coral reef community on Eniwetok Atoll. Ecol. Monogr. 25: 291-320.

Ogden, J.C. and N.S. Buckman. 1973. Movements, foraging groups, and diurnal migrations of the striped parrotfish *Scarus croicensis* Bloch (Scaridae). Ecology 54: 589-596.

Randall, J.E. 1961. Overgrazing of algae by herbivorous marine fishes. Ecology 42: 812.

Randall, J.E. 1963. An analysis of the fish populations of artificial and natural reefs in the Virgin Islands. Caribb. J. Sci. 3: 31-47.

Robertson, D.R. 1973. Field observations on the reproductive behaviour of a pomacentrid fish, *Acanthochromis polyacanthus*. Z. Tierpsychol. 32: 319-324.

Robertson, D.R. and J.H. Choat. 1974. Protogynous hermaphroditism and social systems in labrid fish, p. 217-225. *In* Proc. 2nd Int. Coral Reef Symp., Vol. 1. GBRC, Brisbane.

Ross, R.M. 1978. Reproductive behavior of the anemone fish *Amphiprion melanopus* on Guam. Copeia 1: 103-107.

Russell, B.C., F.H. Talbot, and S. Domm. 1974. Patterns of colonisation of artificial reefs by coral reef fishes, p. 207-215. *In* Proc. 2nd Int. Coral Reef Symp., Vol. 1. GBRC, Brisbane.

Russell, B.C., G.R.V. Anderson and F.H. Talbot. 1977. Seasonality and recruitment of coral reef fishes. Aust. J. Mar. Freshwater Res. 28: 521-528.

Sale, P.F. 1974. Mechanisms of co-existence in a guild of territorial fishes at Heron Island, p. 194-206. *In* Proc. 2nd Int. Coral Reef Symp., Vol. 1. GBRC, Brisbane.

Sale, P.F. 1977. Maintenance of high diversity in coral reef fish communities. Amer. Nat. 3: 337-359.

Sale, P.F. 1978. Reef fishes and other vertebrates: a comparison of social structures, p. 313-346. *In* E.S. Reese and E.J. Lighter (eds.) Contrast in behavior. Adaptations in the aquatic and terrestrial environments. John Wiley and Sons, New York.

Sale, P.F. 1979. Recruitment, loss, and coexistence in a guild of territorial coral reef fishes. Oecologia (Berlin) 42: 159-177.

Sale, P.F. 1980a. The ecology of fishes on coral reefs. Oceanogr. Mar. Biol. Annu. Rev. 18: 367-421.

Sale, P.F. 1980b. Assemblages of fish on patch reefs—predictable or unpredictable? Environ. Biol. Fishes 5: 243-249.

Sale, P.F. and R. Dybdahl. 1975. Determinants of community structure for coral reef fishes in an experimental habitat. Ecology 56: 1343-1355.

Sale, P.F. and R. Dybdahl. 1978. Determinants of community structure for coral reef fishes in isolated coral heads at lagoonal and reef slope sites. Oecologia (Berlin) 34: 57-74.

Sanders, H.L. 1969. Benthic marine diversity and the stability-time hypothesis. Brookhaven Symp. Biol. 22: 71-81.

Slobodkin, L.B. and L. Fishelson. 1974. The effect of the cleaner-fish *Labroides dimidiatus* on the point diversity of fishes on the reef front at Eilat. Amer. Nat. 108: 369-376.

Smith, C.L. 1972. A spawning aggregation of Nassau Grouper, *Epinephelus striatus* (Bloch). Trans. Amer. Fish. Soc. 101: 257-261.

Smith, C.L. and J.C. Tyler. 1972. Space resource sharing in a coral reef fish community. Nat. Hist. Mus. Los Angeles Cty. Sci. Bull. 14: 125-170.

Smith, C.L. and J.C. Tyler. 1975. Succession and stability in fish communities of dome-shaped patch reefs in the West Indies. Amer. Mus. Novit. 2572: 1-18.

Stark, W.A., II. and R.E. Schroeder. 1970. Investigations on the grey snapper, *Lutjanus griseus*. Stud. Trop. Oceanogr. 10: 1-224.

Talbot, F.H. 1965. A description of the coral structure of Tutia Reef (Tanganyika Territory, East Africa) and its fish fauna. Proc. Zool. Soc. London 145: 431-470.

Talbot, F.H., B.C. Russell and G.R.V. Anderson. 1978. Coral reef fish communities: unstable high-diversity systems? Ecol. Monogr. 49: 425-440.

Thompson, R. and J.L. Munro. 1978. Aspects of the biology and ecology of Caribbean reef fishes: Serranidae (hinds and groupers). J. Fish Biol. 12: 115-146.

Thresher, R.E. Reproduction in reef fishes. T.F.H. Publications, Neptune City, New Jersey. (In press)

Warner, R.R., D.R. Robertson and E.G. Leigh, Jr. 1975. Sex change and sexual selection. Science 190: 633-638.

Williams, D. McB. 1980. Dynamics of the pomacentrid community on small patch reefs in One Tree Lagoon (Great Barrier Reef). Bull. Mar. Sci. 30: 159-170.

Williams, D. McB. and P.F. Sale. 1981. Spatial and temporal patterns of recruitment of juvenile coral reef fishes to coral habitats within "One Tree Lagoon," Great Barrier Reef. Mar. Biol. 65: 245-253.

Winn, H.E. and J.E. Bardach. 1960. Some aspects of the comparative biology of parrot fishes at Bermuda. Zoologica 45: 29-34.

Wourms, J.P. and O. Bayne. 1973. Development of the viviparous brotulid fish, *Dinematichthys ilucoeteoides.* Copeia 1973: 32-40.

Discussion on Dr. Sale's Paper

Dr. Murphy opened the discussion with the comment that he too had felt for some time that a systematic routine of successive "clearcutting" might be a viable management option for coral reefs, and possibly for other fisheries such as that on Australia's northwest shelf. Dr. Sale emphasized that prior to embarking on such a scheme of management for reefs there was a need to know substantially more about the extent of dispersal of larval fish, and about the age structure of reef communities. Only with such information could appropriate spatial and temporal scales be chosen for the operation. Dr. Gulland noted that such an approach would permit mistakes to be made on a small scale instead of a large one.

Mr. Jones questioned whether recruitment to the adult population as opposed to colonization of the reef, was independent of the size of the adult population. Data on this point were limited; only three experiments had been done. Dr. Sale had denuded small patches of habitat of adult fish and had some evidence of enhanced colonization of them, and of improved survivorship of juveniles. His patches were surrounded by a less satisfactory habitat, but some colonization was by young fish already settled from the plankton. By contrast, D.M. Williams and P.J. Doherty had used patches of habitat sufficiently isolated by open sand from other reef that all colonization was direct from the plankton. They also had more replicates of their experiment. In both, colonization was not enhanced by removal of adults. Doherty's data also showed, convincingly, that survivorship through the first year of post-larval life was not different in the presence or absence of adults—indicating that recruitment to the adult stock was also not influenced in his system by the size of that stock.

Some of Doherty's experiments used juveniles experimentally added to sites from elsewhere. While it is true that survival of these added fish was poor during the first 24 hours if adults were present, survivorship over the next year was independent of the adult stock. The poor early survivorship is interpreted as an artifact due to the different behavior of added juveniles and naturally colonizing larvae. The latter have less pigmentation, attract less attention from adults, and retreat towards the substratum when threatened. Juveniles added from elsewhere tend to flee when attacked.

Dr. Pauly argued that the data presented were on such a small spatial scale, and involved such small numbers that the variability present was entirely consistent with the operation of a stable system. Mr. Jones suggested that the use of the term, 'persistence', rather than 'stability', might be more appropriate in this context. Dr. Marten noted the work of Sewall Wright on genetic drift as an analogous study in that drift occurred only in small populations.* He wondered whether results of small-scale studies had any direct relevance to the operation of larger-scale systems.

Dr. Sale agreed with the usefulness of the term 'persistence' which he defined as meaning that a system remained relatively constant. To him 'stability' implied additionally that the system was organized in such a way that it remained constant when not disturbed, resisted disturbance, and recovered after disturbance—in other words was equilibrial.

He stressed that nothing in the results of the small-scale studies he had discussed contradicted an impression of persistence at the larger scale. Indeed, the results predicted such persistence. But they did not predict stability, and since there is no evidence available of the kinds of interactions necessary to create stability, as opposed to persistence, then it is only blind faith that leads one to assume tropical communities are stable. Future experimental research may change this situation, but at present, persistence rather than stability should be assumed. The kinds of models developed will be quite different under the two assumptions, so the distinction between them is important.

Dr. James suggested that such techniques as time series analysis would be appropriate to analyze the small numbers inherent in the kinds of experiment Dr. Sale had discussed.

Dr. Larkin asked why the experiments discussed had been done on such a small scale. Dr. Sale replied with a plea for manipulative investigations as the most powerful tools for testing opposing hypotheses. In the Australian University context, resources were insufficient for experimental studies on a larger scale, and, in fact, coral reefs lent themselves to small-scale studies because most adult fish were relatively sedentary, and because a reef consists of a heterogeneous set of different types of small patches of habitat which support different assemblages of fish.

Mr. Simpson sought clarification concerning movement of adult fish, and their role in colonizing denuded or artificial reefs. Dr. Sale indicated that when habitat patches, such as artificial reefs or small patch reefs were separated from other rocky habitat by 4 to 5 m sand, virtually all colonization was by newly settling post-larvae. While there is some evidence that individual fish remain sedentary on continuous coral tracts, denuding of small patches in such areas leads to substantial infiltration by juveniles and young adults living nearby. Larger species of scarid, lutjanid, serranid, and labrid tend to range further, and will visit even quite isolated patch reefs. In addition to these movement patterns associated with feeding, some species move conspicuously when spawning. *Thalassoma bifasciatum*, a labrid 15 to 20 cm long, moves as much as 0.5 km each day in order to spawn at the down-stream end of large patch reefs at San Blas, Panama (data from Warner et al. 1975). Only anecdotal information seems to be available concerning movements of sharks, jacks and other reef-associated forms. At least some reef sharks are localized in their movements, having particular caves in which they shelter each day.

---

*Editors' note: the paper referred to is Wright, S. 1931. Evolution in Mendelian populations. Genetics 16: 97-157.

A Comparative Study of Fish Yields from Various Tropical Ecosystems

GERALD G. MARTEN

Environment and Policy Institute
East-West Center
Honolulu, Hawaii 96848

AND

JEFFREY J. POLOVINA

Southwest Fisheries Center Honolulu Laboratory
National Marine Fisheries Service, NOAA
Honolulu, Hawaii 96812

MARTEN, G.G. AND J.J. POLOVINA. 1982. A comparative study of fish yields from various tropical ecosystems, p. 255-289. *In* Pauly, D. and G.I. Murphy (eds.) Theory and management of tropical fisheries. ICLARM Conference Proceedings 9, 360 p. International Center for Living Aquatic Resources Management, Manila, Philippines and Division of Fisheries Research, Commonwealth Scientific and Industrial Research Organisation, Cronulla, Australia.

Abstract

There is an immense variation in actual and potential fish yields, both within and between different kinds of aquatic tropical ecosystems. The finfish catches from lakes, reservoirs, rivers, continental shelves and coral reefs all fall in approximately the same range of 0.1 to 30 tonnes/km^2/year, with most catches falling in the range of 1 to 10 tonnes/km^2/year and modal values falling in a range of 3 to 6 tonnes/km^2/year. The yields from estuaries and ponds can be considerably higher (up to 120 tonnes/km^2/year without supplementary feeding), and the yields from open ocean are considerably lower (.002 to .05 tonnes/km^2/year). Within the tropics, the best physical indicator of a fishery's potential yield is water depth, and the best biological indicator is primary productivity. Some ecosystems such as rivers, estuaries and ponds have relatively high fish yields for their primary productivity, presumably because of organic matter inputs. Other ecosystems such as coral reefs and open oceans have relatively low fish yields, presumably because of long food chains. However, no single indicator, including primary productivity, can predict fish yields very precisely. More refined yield assessments will have to be empirical in nature, recognizing the importance of fishing practices in assessing potential yields and giving particular attention to how the composition of fish communities in different habitats is altered by fishing and other human activities.

Introduction

There is an immense variation in fish yields in different parts of the tropics. The annual yield from a square kilometer of intensively managed fishponds can be as much as a million times the average annual fish catch from a square kilometer in the middle of the ocean. Even fisheries which are not intensely managed can vary by a factor of more than ten thousand in their yields, depending upon where they are located and the ecological conditions that prevail.

Sufficient information has accumulated on these fisheries to attempt an assessment of why they vary so much. This review examines the extent to which different fish yields are associated with different aquatic ecosystems such as lakes, reservoirs, rivers, estuaries, continental shelves, coral reefs and open ocean. Also, this review examines the extent to which fish yields can be explained by biological or physical characteristics of the different ecosystems and how these characteristics differ from one ecosystem to another.

One of the motivations for such an assessment is to assist inventories of the potential of tropical fisheries. Because it has not been feasible to conduct stock assessments and establish commercial catch record systems of the sort desired for management in all of the numerous lakes, rivers, estuaries, etc., that are situated in the tropics, there is a need for methods with limited data requirements to assess fishery potentials and the yield implications of present fisheries practices.

Methods

The basic information for this review comes from records of finfish yields (wet weight) in various ecosystems throughout the tropics. The analysis is not precise because the information is not precise. Large errors can be expected in fish yield statistics for a variety of reasons, many deriving from the artisanal character of multispecies tropical fisheries. One of the weakest points is the estimation of fishing effort, which is important to yield estimates because such estimates are customarily made by multiplying catch per unit effort (from a sampling of fish landings) by the total effort in the fishery. It can be extremely difficult, however, to know the effort of a fishery that consists of thousands of individual fishermen operating out of canoes along a complex shoreline where transportation and communications may be poor or non-existent. This may be further complicated if a fishery has restrictive regulations which encourage the fishermen to avoid reporting their catch faithfully.

It is important to appreciate the spatial scale of the yield estimates, which most often are highly aggregated. A single number may be used to represent the yield from a lake whose fishery production is ten times as great in its shallow inshore areas as its offshore areas. The estimate for a river may embrace both the highly productive flood plains and relatively sterile headwaters. The estimate for a coral reef fishery may cover an area that is a patchwork of highly productive coral and less productive sandy areas. Estimates from the open ocean come from areas that are thousands of square kilometers in extent and may be very heterogeneous.

The lower end of the scale of fish yields from each ecosystem is not well defined. The less productive fisheries, which by necessity have fewer fishermen, tend not to have catch records. In one sense the lower end of the scale is zero because there are places where each of the ecosystems is not fished at all. Even with regard to potential yield, the lower end of the scale is very low if the spatial scale is fine enough.

The upper end of the scale is reflected in maximum sustainable yields (MSYs). There are limitations in estimating the upper end of the scale because the MSY of a particular fishery can be inferred reliably only from actual yield experiences. However, in many instances none of the yields have been large enough to give suitable information concerning the maximum. This may be because the fishery has never been intensive enough to approach the maximum. The small-scale fisheries that are found on many tropical lakes and seashores are restricted to the inshore areas, so that offshore areas remain unexploited.

It may also be that the stocks have not been fished in the intricate fashion that would evoke the highest yield. This may involve the kinds of fishing gear that are used and the species of fish that are being harvested. Any history of yields, including the maximum from that history, is specific to a particular technology and may be considerably below the maximum possible yield.

Whenever possible, MSYs were estimated by tabulating the total annual finfish catches (summed over all species) for different years or different locations and plotting the catches against the fishing effort in those years or locations. In the absence of such information, the yields of fisheries with intensive fishing were considered to be representative of their MSY. It is possible that the yields of some of the intensely fished fisheries were depressed by overfishing, but we do not consider this to be a serious error because heavy fishing of a multispecies fishery does not usually result in significant reduction of the total catch, provided fishing activities do not lead to habitat destruction or other degradation of the resource base. Finally, in the absence of intensive fishing, MSY was estimated to be one-half the virgin biomass multiplied by natural mortality (when available).

Among indicators of potential yield, particular attention is given in this report to primary productivity because of its biological relationship to fish yield, and where possible we have tabulated fish yield and primary productivity figures from the same fisheries. The estimates are highly imprecise because of the spatial and seasonal variation in primary productivity, but the relationship seems to be strong enough to show through. For some ecosystems we did not find primary productivity and fish yield figures for the same fisheries. In those cases we assembled information on the range of primary productivities in the ecosystem in order to compare it with the range of fish yields.

Results

Lakes, Reservoirs and Ponds

Table 1 lists a number of tropical lakes (primarily African) in order of their catch per unit area. There is a range of 0.1 to 23 t/km^2/year.

Toews and Griffith (1979) reported a significant negative correlation between fish catch per unit area and the size of African lakes. This relation is probably due in part to the fact that smaller lakes tend to have a higher percentage of shallow water, and shallow water generally has higher fish production than deep water. This is reflected in the observations of Kud-

Table 1. Fish yields from tropical lakes.

Lake	Location	Area (km²)	Catch (t/km²/yr)	Source
Upemba	Zaire	530	23	Henderson and Welcomme (1974)
Yercaud*	India	.08	22	Oglesby (1977)
Kyoga	Uganda	2,700	18	Henderson and Welcomme (1974)
George*	Uganda	250	15	Ganf (1975)
Mwadingusha	Zaire	393	13	Henderson and Welcomme (1974)
Malombe	Malawi	390	13	Henderson and Welcomme (1974)
Guiers	Benin	170	13	Henderson and Welcomme (1974)
Ooty*	India	0.34	10	Oglesby (1977)
Tanganyika* (north)	Tanzania	3,575	8.9	Mann and Ngomirakiza (1973)
Mweru	Zaire	4,580	6.8	Henderson and Welcomme (1974)
Edward	Uganda	2,300	6.8	Henderson and Welcomme (1974)
Victoria* (inshore)	Kenya	1,300	6.2	Melack (1976)
Chilwa	Mozambique	1,750	5.6	Henderson and Welcomme (1974)
Kodaikanal*	India	0.26	5.3	Oglesby (1977)
Rukwa	Tanzania	2,000	4.9	Henderson and Welcomme (1974)
Lanao*	Philippines	—	4.8	Frey (1969)
Chiuta	Mozambique	113	4.4	Henderson and Welcomme (1974)
Albert*	Uganda	5,600	4.2	Cadwalladr and Stoneman (1966)
Mweru-Wa-Ntipa	Zaire	1,520	3.8	Henderson and Welcomme (1974)
Kitangiri	Tanzania	1,200	3.4	Henderson and Welcomme (1974)
Baringo*	Kenya	160	3.3	Melack (1976)
Malawi (Southern portion)	Malawi	6,000	2.5	Turner (1977a, b)
Victoria* (offshore)	Kenya, Uganda, Tanzania	41,200	2.1	Kudhongania and Cordone (1974)
Chad*	Chad	16,000	1.4	Welcomme (1972a)
Bangweulu*	Zambia	2,733	1.3	Toews and Griffith (1979)
Tumba	Zaire	1,767	0.6	Henderson and Welcomme (1974)
Maji Ndombe	Kenya	1,300	0.5	Henderson and Welcomme (1974)
Rudolf	Kenya	7,200	0.3	Henderson and Welcomme (1974)
Kiru	Tanzania	2,699	0.1	Henderson and Welcomme (1974)

*Catch and primary productivity data also used in Fig. 3.

hongania and Cordone (1974) on Lake Victoria and Turner (1977a, b) on Lake Malawi that fish stocks are significantly larger in the shallow peripheral portions of the lake than they are in the deep, central area.

The main reason that shallow water has higher fish production is that primary productivity is higher in shallow water due to the availability and recycling of nutrients for photosynthesis. It is quite likely, however, that an even higher percentage of the primary production passes to fish in shallow water because of the more significant role of the benthic food chain. This is because plankton and detritus have a better chance of sinking to the bottom without being captured by the pelagic food chain if the water is shallow (see Jones, this vol.). Benthic food chains tend to be shorter than pelagic food chains because many benthic invertebrates which eat sediment are large enough to be fish food, whereas pelagic food chains start with microscopic organisms and pass through a series of progressively larger organisms before reaching a size sufficient for fish food. More of the primary production is translated to fish production when the food chain is short.

Small lakes also tend to be more productive because the nutrients and detritus which pass into them from outside can make a significant contribution to the lake's productive capacity in comparison with the biological production that originates in the lake itself. Finally, small lakes tend to be more intensively exploited, whereas the offshore areas of a large lake are not easily accessible to the small-scale fisheries which exploit them.

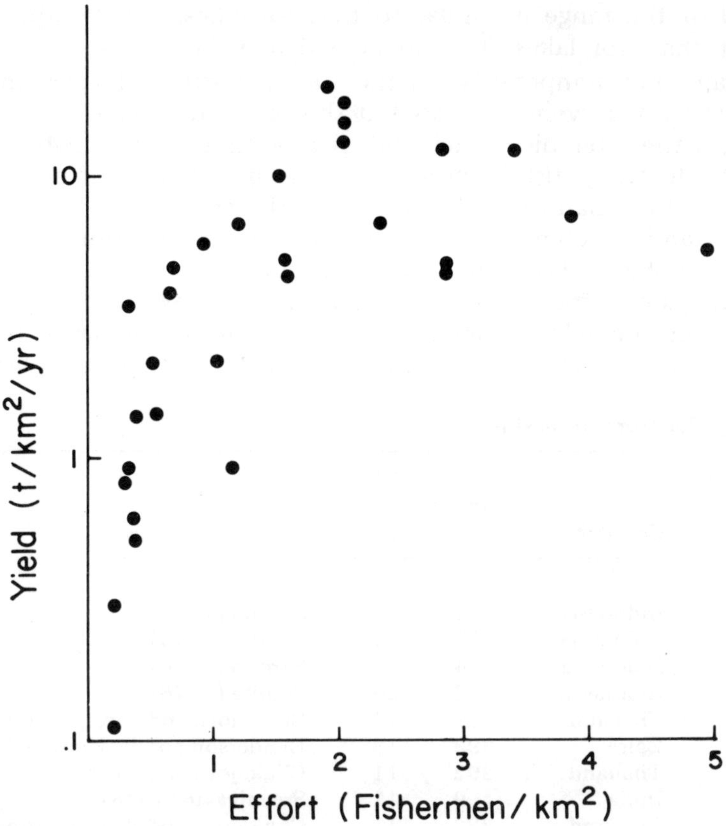

Fig. 1. Fish yields and fishing effort on African lakes (from Henderson and Welcomme 1974).

Although the yields in Table 1 vary by a factor of more than 100, the lower part of the range appears to correspond to lakes where the fishing intensity is not sufficient to yield a harvest near the maximum. Fig. 1 shows the relationship between fishing intensity (as measured by the number of fishermen per unit area) and catch per unit area for some African lakes. Since different lakes have different MSYs, it is not possible to infer the MSY simply as the maximum of the cluster of points in Fig. 1. However, Fig. 1 gives the impression that the range of MSYs may be approximately 1 to 23 t/km^2/year.

The main difference between reservoirs and lakes is that a reservoir was once a river or dry land before the river was dammed. As a consequence, a reservoir does not possess a full complement of lake fauna, even though the physical conditions of the reservoir are the same as those of a lake. This means that unless new fish are introduced, there may be parts of the reservoir that are virtually uninhabited by fish and their food, and the efficiency of translation of the reservoir's biological production to fish yields may be correspondingly reduced.

Fernando (1976) and Welcomme (1979) have reviewed the fish yields of reservoirs, and Table 2 lists the fish yields of some tropical reservoirs. The bottom end of the range is similar to that for lakes, but the upper end is 50% greater than for lakes. The upper end may be higher because many reservoirs can draw temporarily on the standing stock of trees and other plant materials which were inundated at the time the reservoir was formed. In addition, if the reservoir has an ecological vacuum, it is possible to introduce high-productivity fish such as the African cichlids and attain higher levels of production than would be possible with native fauna.

Henderson and Welcomme (1974) demonstrated a relationship between the morphoedaphic index (total dissolved solids/average depth) and fish yields in a series of African lakes. Although Fig. 2 shows a definite relationship between morphoedaphic index and yield, it is also apparent that most of the variation in yield is not explained by the morphoedaphic index.

Table 2. Fish yields from tropical reservoirs.

Reservoir	Country	Area km^2	Catch (t/ km^2/yr)	Source
Pacal	Indonesia	4	35	Sarnita (1976)
Pening	Indonesia	22	32	Sarnita (1976)
Dorma	Indonesia	4	28	Sarnita (1976)
Jombor	Indonesia	2	20	Sarnita (1976)
Lam praloung	Thailand	19	14	Chukajorn and Pawapootonan (1976)
Mwadingusha	Zaire	393	13	Henderson and Welcomme (1974)
Sirin thorn	Thailand	292	11	Chukajorn and Pawapootonan (1976)
Amaravathy*	India	9	11	Sreenivasan (1978)
Lam Poa	Thailand	230	11	Chukajorn and Pawapootonan (1976)
Sathanur*	India	13	10	Sreenivasan (1978)
Nzilo	Zaire	280	10	Henderson and Welcomme (1974)
Ayame	Ivory Coast	135	7.4	Henderson and Welcomme (1974)
Nam pung	Thailand	21	6.2	Chukajorn and Pawapootonan (1976)
Ubolratana	Thailand	410	6.0	Bhukuswan and Pholprasith (1976)
Prijetan	Indonesia	2	4.8	Sarnita (1976)
Volta*	Ghana	8,482	4.7	Henderson and Welcomme (1974)
Sentir	Indonesia	1	4.3	Sarnita (1976)
Kainji	Nigeria	1,270	4.2	Henderson and Welcomme (1974)
Kalen	Indonesia	1	4.0	Sarnita (1976)
Chulaporn	Thailand	12	3.3	Chukajorn and Pawapootonan (1976)
Tirumoorthy	India	5	2.8	Sreenivasan (1978)
Sandy Nulla	India	3	2.6	Sreenivasan (1978)
Stanley*	India	147	2.3	Sreenivasan (1978)
Lam takong	Thailand	44	2.2	Chukajorn and Pawapootonan (1976)
Jatiluhur	Indonesia	83	2.2	Sarnita (1976)
Nasser	Egypt	3,330	2.1	Henderson and Welcomme (1974)
Bhavanigasar*	India	79	2.0	Sreenivasan (1978)
Nam oon	Thailand	86	1.7	Chukajorn and Pawapootonan (1976)
Krishnagiri*	India	13	1.0	Sreenivasan (1978)
Kariba	Zambia	5,364	0.8	Henderson and Welcomme (1974)
Ghandi Sagar	India	660	0.7	Dubey and Chatterjee (1976)
Rihand	India	302	0.7	Natarajan (1976)
Nagaryanasagar	India	184	0.5	Natarajan (1976)
Konar	India	15	0.2	Natarajan (1976)
Tana	Kenya	3,500	0.1	Henderson and Welcomme (1974)

*Catch and primary productivity data also used in Fig. 3.

The morphoedaphic index has been reviewed by Ryder et al. (1974). The value of the morphoedaphic index for predicting fish yields presumably derives from its relationship with primary productivity. Total dissolved solids are related to the supply of nutrients required for primary production, and the depth of the water reflects the extent to which the nutrients are available at the lake surface where most of the primary production occurs. Of the two components of morphoedaphic index, depth is a better indicator of fish production than total dissolved solids (Matuszek 1978).

Melack (1976) and Oglesby (1977) have demonstrated a relationship between fish yields and primary productivity. We ran a multiple regression analysis of yield versus mean depth and primary productivity for the lakes and reservoirs in Table 3. The results indicated primary productivity to be a better predictor than depth. Comparing Fig. 3 with Fig. 2 suggests that primary productivity is also a better predictor of fish yields than the morphoedaphic index.

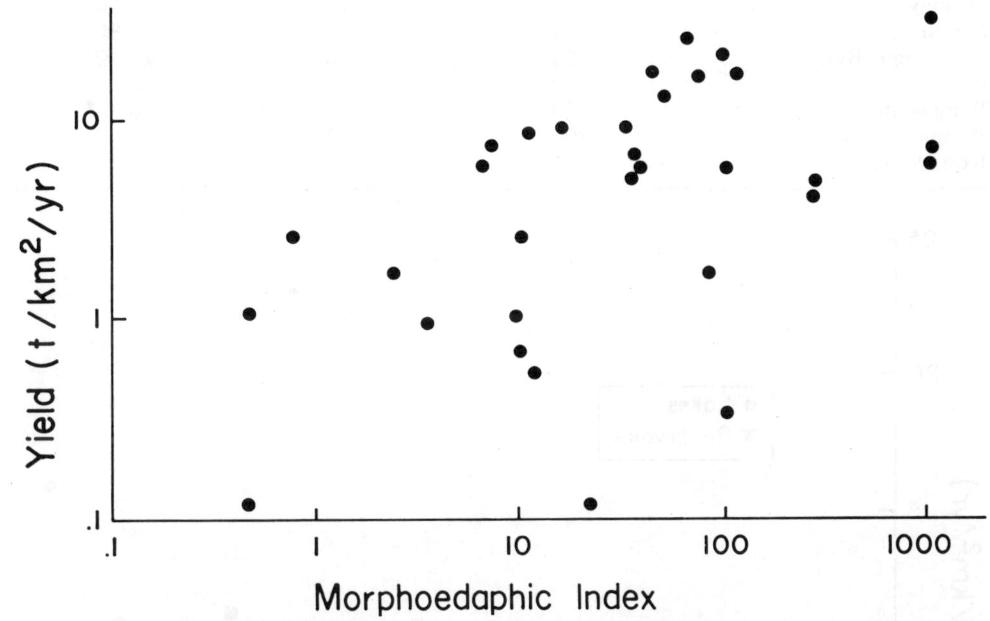

Fig. 2. Fish yields and morphoedaphic index (from Henderson and Welcomme 1974).

If a curve is put through the points in Fig. 3, it appears not to pass through the origin, suggesting that fish yields greater than zero are only possible once primary production exceeds a threshold. This would be expected from food-chain theory (Haussman 1971). There is also a suggestion of an upward curvature in Fig. 3, which implies that the efficiency of translation of primary production to fish production may be greater at higher levels of primary productivity. This may be because phytoplankton tend to be larger under highly productive conditions, requiring a shorter food chain (as particle size increases up the chain) in order to reach fish food size.

There is considerable variation in the fish yield (about fivefold) for any given level of primary productivity in Fig. 3. Part of the variation is undoubtedly due to errors in estimating yield and primary productivity, but it

Table 3. Physical and biological characteristics of some tropical lakes and reservoirs.

Lake	Primary productivity (gC/m²/yr)	Depth (m)	Morphoedaphic index
George	1,500	2	83
Ooty	1,380	3	48-158
Amaravathy	1,175	12	4
Yercaud	1,060	2	55-100
Volta Res.	1,040	10	3
Tanganyika	1,040	330	2
Victoria (offshore)	1,000	63	2
Kainji	890	10	7
Albert	970	25	29
Volta	930	18	2
Victoria (inshore)	850	6	2
Bhavanigasar Res.	850	12	4-17
Stanley Res.	849	17	7-15
Sathanur	844	10	32-80
Krishnagiri Res.	565	5	47-108
Chad	370	4	103
Bangweulu	270	4	6
Baringo	200	7	83
Kodaikanal	160	2	14

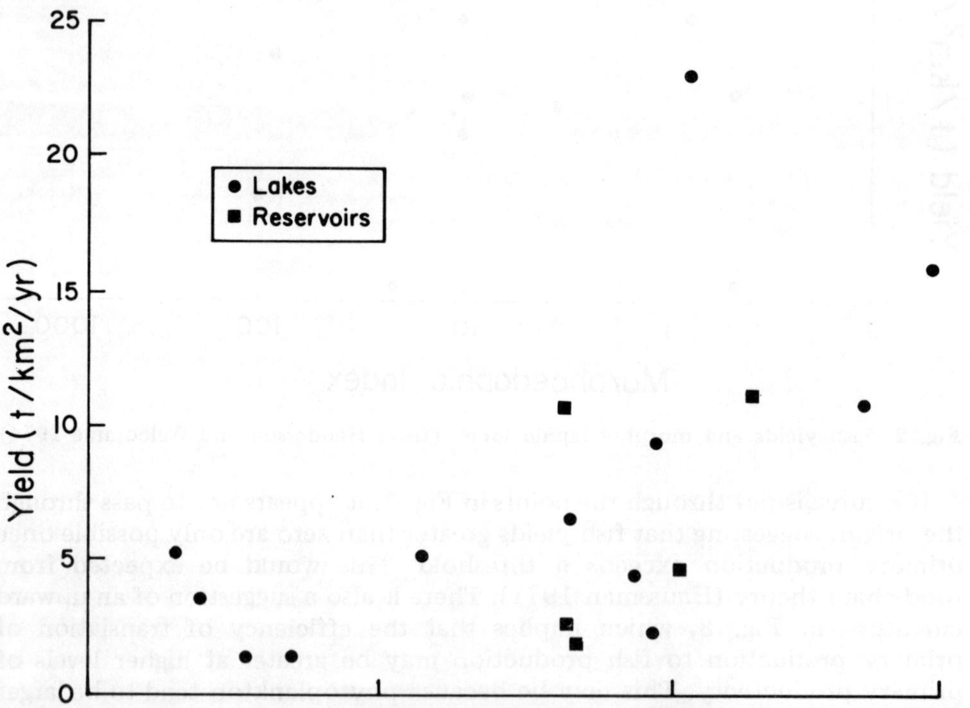

Fig. 3. Fish yields and primary productivity in tropical lakes and reservoirs, based upon lakes and reservoirs in Tables 1, 2 and 3 (r = .61).

appears the prediction would not be highly precise even in the absence of such errors. The loose relationship between primary productivity and fish yields is due in part to variation in the efficiency with which primary production is translated to fish yields. This may be due to the efficiency of the fishery (the fact that yield may be below MSY) or the efficiency with which primary production is converted to fish production through the food chain. Another limitation of primary productivity as a predictor is that the productivity of small lakes can derive in large measure from detritus and other organic materials that come from outside.

Oglesby (1977) explored other measures of the food available to fish as predictors of fish yield and found a closer correlation between yield and the standing crop of benthic fauna in temperate lakes. A similar relationship could have been sought for zooplankton and pelagic fish. The value of fish food abundance as a yield predictor may be limited, however, because the same standing stock of fish food could have a low or high production depending upon whether or not it is intensely harvested by the fish (Hayne and Ball 1956).

Fishponds, intensively managed with fertilization but without food supplementation, can show fish yields as high as 120 t/km^2/year (Bardach et al. 1972). This is partly due to the fact that the primary productivity of intensively managed ponds can be as much as three times the maximum primary productivity of lakes. It is also because fishponds can be stocked with fishes, such as carp and tilapias, which are highly efficient at translating primary production to yields. The yield from ponds can also be very high because they lack the predators which compete with fishermen for the fish harvest in natural water bodies such as lakes and rivers. The relationship between yield and primary productivity can be very close in fishponds (Fig. 4).

Rivers

Welcomme (1979) has reviewed the fisheries ecology of rivers. Table 4 lists the yields of some tropical rivers. Fig. 5 shows the relation between fish yields and the number of fishermen on some tropical rivers.

Rivers show a broader range of yields than lakes. The low end of the range in yields for rivers is lower than for lakes, and is associated with headwaters in areas of highly weathered soils which have a very low primary productivity and a correspondingly low fish production. The high end of the range for rivers is slightly higher than for lakes, even though the primary productivity of rivers is generally lower than that of lakes. The explanation may lie in the fact that rivers receive from the large watershed area that surrounds them a significant quantity of nutrients and organic material which contributes to the biological productivity of the fishery. Moreover, many of the rivers with high yields are flood-plain rivers that can draw upon the terrestrial productivity of the areas they flood, and the highest yields in Table 4 are probably due in large measure to the quantity of municipal sewage received by those rivers. Welcomme (1979) found the best predictor of fish yields from a river to be the total area of its drainage basin.

Rivers may also have relatively high yields because even large rivers are easier for fishermen to exploit fully than are large lakes. Furthermore, many of the fish that are caught in rivers have moved into them from lakes or the sea, so their growth has occurred primarily in ecosystems outside the river.

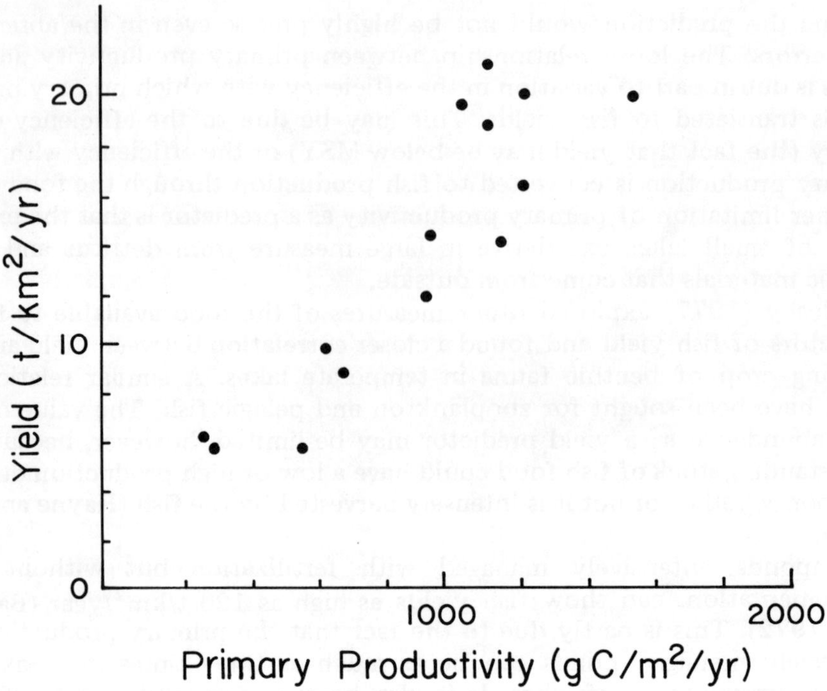

Fig. 4. Fish yields and primary productivity in tilapia ponds, based upon Almazan and Boyd (1978) (r = .89).

Table 4. Fish yields from tropical rivers.

River	Location	Catch[a]	Source
Ganges	Bangladesh	78	FAO (1976)
Niger	Benin[b]	44	Welcomme (1972b)
Lower Mekong	Viet Nam	41	R. Welcomme, pers. comm.
Mahaweli	Sri Lanka	34	Indrasena (1970)
Lubuk Lampan	Indonesia	24	Arfin and Arfin (1976)
Shire	Mozambique	13	R. Welcomme, pers. comm.
Kamulondo	Zaire	11	Poll and Renson (1948)
Oueme	Benin[b]	10	CTFT (1957)
Oueme	Benin[b]	6.5	Welcomme (1972b)
Niger	Niger	5.2	Dobrovici (1971)
Central Delta	Niger	4.5	Konare (1977)
Senegal	Senegal	4.3	Reizer (1974)
Pendjari	Benin[b]	3.5	Welcomme (1972b)
Magdalena	Colombia	3.3	Bazigos et al. (1977)
Benue	Nigeria	3.1	Mothwani (1970)
Niger	Nigeria	3.0	Mothwani (1970)
Kafue	Zambia	2.0	Zambia (1965)
Rufiji	Tanzania	1.9	R. Welcomme, pers. comm.
Kafue	Zambia	1.6	Zambia (1971)
Cross	Nigeria	1.0	R. Welcomme, pers. comm.
Barotse	Zambia	0.7	Zambia (1974)
Okawango	Namibia[c]	0.5	R. Welcomme, pers. comm.
Upper Amazon	Brazil	.02	FAO (1979)

[a]In t/km of river reach.
[b]Formerly Dahomey.
[c]Formerly S.W. Africa.

Estuaries and Lagoons

Estuaries and coastal lagoons have a mixture of freshwater and saltwater that may vary considerably in salinity at different times of the year. An estuary is subject to tidal mixing whereas a lagoon is not. Estuaries can have considerable impact on the adjacent marine fisheries because they serve as nursery areas for many fish that move out to sea at a later stage in life. Estuaries and lagoons are usually shallow, only a few meters in depth, and secondary production is often concentrated in the sediment (with its short benthic food chains) rather than in the water column. The sediment has an important role, even though it often does not itself contribute the major portion of the primary production (particularly when the water is turbid).

Saila (1975) has reviewed the ecology of estuarine fish production and Hickling (1970) has reviewed estuarine fish culture. Table 5 lists the fish yields of some tropical coastal lagoons and estuaries. The yields from both estuaries and lagoons are generally higher than the yields from lakes and rivers. This is due partly to the shallowness of estuaries and lagoons and partly to the nutrients they receive from rivers. It may also be due to the large quantity of plant materials they receive from their shoreline. This is particularly so for estuaries that are bordered by mangroves. The highest fish yields in estuaries and lagoons occur where intensive aquaculture is practiced.

Continental Shelf

Table 6 shows the demersal and pelagic MSYs of the fisheries of continental shelves on a broad geographic scale. In general, the pelagic and demersal components make similar contributions. The highest MSYs occur with anchovies and herring, which are species with short food chains, in upwelling areas which show the highest levels of primary production recorded in oceanic waters.

Table 7 shows the estimated MSYs of some demersal multispecies fisheries in tropical coastal areas, and Fig. 6 illustrates the information behind some of the catch-effort MSY estimates in Table 7. A regression analysis of MSY against primary productivity and depth showed a significant relation only with depth (Fig. 7), the higher MSYs appearing at depths less than 50 m. There was a weak positive correlation between primary productivity and fish yields ($r = 0.24$), but primary productivity explained none of the variation in yields beyond what was explained by depth. It is worth noting that Qasim (1979) found the average primary production of Indian coastal waters shallower than 50 m to be about six times that in waters deeper than 50 m. The MSYs above and below 50 m in Table 7 differ in about the same proportion.

Table 8 presents estimated MSYs for some tropical continental shelf pelagic fisheries. These MSYs have a significant relationship with primary productivity (Fig. 8), but not with mean depth. It is interesting to note that this is the opposite of demersal fisheries, where mean depth rather than primary production best predicted yields. An explanation for the stronger correlation between pelagic yields and primary productivity may be the direct connection of pelagic fish to the planktonic food chain (Petersen and Curtis 1980), whereas demersal fish may have much less direct connection to

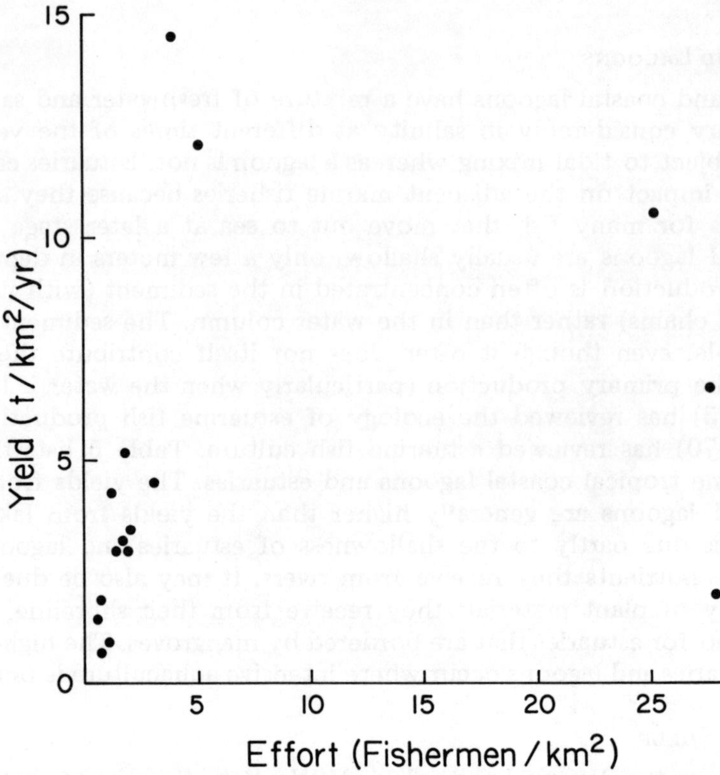

Fig. 5. Fish yields and fishing effort on tropical rivers (from Welcomme 1976).

Table 5. Fish yields from tropical estuaries and coastal lagoons.

Name of water body	Country	Area (km^2)	Catch (t/ km^2/yr)	Source
—	Taiwan (China)*		94-250	Lin (1968)
Aheme	Benin*, **	85	86	Welcomme (1972b)
—	India*		86-124	Pakrasi et al. (1964)
Nakove/Pio Novo	Benin*, **	157	56	CTFT (1969)
—	Philippines*		50-100	Frey (1947)
—	Philippines*		50	Tang (1967)
—	Philippines*		47	Rabanal (1961)
Ovidah, Grand Popo, Awo channel	Benin*, **	14	28	Welcomme (1972b)
—	Singapore*		25	Le Mare (1949)
—	Hawaii*		20	Cobb (1901)
Ebrié	Ivory Coast	556	16	Durand et al. (1978)
Sakumo	Ghana	1	15	Pauly (1976)
—	Java*		14-63	Schuster (1952)
Unare	Venezuela	54	14	Okuda (1965)
—	Morocco		13	Belloc (1938)
Cienaga Grande	Colombia	450	12	INDERENA (1974)
—	India		11-17	Pillay (1954)
Tacarigua	Venezuela	63	11	Gamboa et al. (1971)
Piritu	Venezuela	22	5.8	Carvajal (1972)
Mandapam	India	4	5.6	Tampi (1959)
Tamiahua	Mexico	659	4.7	Garcia (1975)
Chillca	India	1,036	3.7	Jhingran and Natarajan (1969)
Pangalanes	Madagascar	98	3.7	Lasserre (1979)
Anony	Madagascar	23	2.8	Moulherat and Vincke (1968)
Pulicat	India	392	2.6	Jhingran and Gopalakrishnan (1973)
Maracaibo	Venezuela	14,344	1.9	Nemoto (1971)
Jiquilisco	El Salvador	121	1.7	Hernandez and Calderon (1974)

*Includes aquaculture (without fertilization or feeding).
**Formerly Dahomey.

the primary production in the water column above. The lack of correlation between pelagic yields and depth may also be because coastal pelagic fishes probably range over a larger geographic area than demersal fish, so the depth in the area they are fished may not accurately represent the mean depth of their entire habitat.

Yesaki (unpub. data) observed a positive association between total multispecies fish yields and primary productivity when comparing a number of tropical and temperate continental shelf fisheries. He also observed a negative association between total yields and the number of species in the fishery.

Coral Reefs

The coral reef is an ecosystem in which the reef surface provides a substrate for growth of algae, both free-living and symbiotic with coral polyps. This fosters the maximum biological production possible within the limits of nutrients available in the surrounding water.

Table 6. Estimated maximum sustainable yields (t/km^2/year) for tropical continental shelf areas (from Gulland 1971 except where noted otherwise).

		Minimum			Maximum	
Region	Pelagic	Demersal	Total	Pelagic	Demersal	Total
NE Atlantic	0.8	0.6	1.4	7.3	8.0	15.3
NW Atlantic	1.8	1.8	3.6	7.0	7.0	14.0
NW Pacific	0.6	1.7	2.3	8.5	4.0	12.5
Indian Ocean[a]	0.7	1.4	2.1	4.3	5.5	9.8
E Central Atlantic	4.0	1.2	5.2	5.0	2.5	7.5
South China Sea[b]	0.2	0.8	1.0	2.4	4.3	6.7
W Central Atlantic	0.7	0.2	0.9	3.2	2.5	5.7
			Upwelling			
SW Atlantic	1.5	6.0	7.5	17.5	10.0	27.5
Peru[c]	—	—	—	1.0	21.9	22.9

[a]Yesaki (unpub. data).
[b]Menasveta et al. (1973).
[c]Murphy (1972).

Table 9 lists the fish yields of some coral reefs. The range is similar to that for other continental shelf fisheries, despite the higher primary productivity of coral reefs. Marshall (1980) reviewed potential fish yields from coral reefs and found a range of 0.8 to 5 t/km^2/year (see Fig. 9). He noted the observation of 14-20 t/km^2/year* for an intensively exploited reef in the Philippines (Alcala 1981) but questioned the generality of that observation. However, recent work by Wass (in press) on an intensively exploited reef in Samoa has estimated the finfish yield to be 18 t/km^2/year, suggesting that some of the lower estimates from previous studies may not reflect the yields of fully exploited reef systems. Fig. 10 shows the relation between the number of fishermen and the yield which has been realized on some coral reef fisheries.

*Editorial note: Marshall (1980) cites Alcala as having suggested a figure of 15 tonnes/km^2/year; it is the actual figures given by Alcala (1981) which are given here.

Table 7. Estimated maximum sustainable yield of tropical demersal marine fisheries.

Location	MSY ($t/km^2/yr$)	Depth (m)	Area (km^2)	Primary productivity ($gC/m^2/yr$)[a]	Estimation method	Source
North Coast Gulf of Mexico	6.7	0-110	111,210	135	Research Survey (0.5 B_o)	Klima (1977)
U.S. Atlantic Coast (N. Carolina—Florida)	5.5	0-110	125,000	135	Research Survey (0.5 B_o)	Klima (1977)
Gulf of Thailand	3.9	0-50	179,000	365	Catch vs. Effort (Fig. 6a)	SCS (1978)
Philippines	2.8	0-200	152,700	135	Research Survey (0.5 B_o)	Aoyama (1973)
North Coast of Java	2.6	0-50	26,160	180	Catch vs. Effort (Fig. 6b)	SCS (1979)
Sunda Shelf—South	2.3	0-50	267,900	75	Research Survey (0.482 B_o)	SCS (1978)
South China Sea	2.0	0-200	500,000	45	Research Survey (0.5 B_o)	Aoyama (1973)
West Coast of Florida	1.8	0-110	179,280	135	Research Survey (0.5 B_o)	Klima (1977)
Sunda Shelf—NW Borneo	1.7	0-50	62,900	75	Research Survey (0.482 B_o)	SCS (1978)
South Coast of Kalimantan (Borneo)	1.5	0-50	113,590	135	Research Survey (0.5 B_o)	SCS (1979)
Sunda Shelf—NW Borneo	1.1	50-200	69,700	75	Research Survey (0.482 B_o)	SCS (1978)
Gulf of Thailand	1.1	50-200	126,700	365	Research Survey (0.482 B_o)	SCS (1978)
Sunda Shelf	0.8	50-200	398,500	135	Research Survey (0.482 B_o)	SCS (1978)
U.S. Atlantic Coast (N. Carolina—Florida)	0.5	110-548	20,480	45	Research Survey (0.5 B_o)	Klima (1977)
North Gulf of Mexico	0.5	110-548	44,070	45	Research Survey (0.5 B_o)	Klima (1977)
West Coast of Florida	0.4	110-548	72,120	45	Research Survey (0.5 B_o)	Klima (1977)

[a]Primary productivity estimated from a world map in FAO (1972).

As a reef fishery is generally a patchwork of coral reef (which is highly productive) and sandy bottom (which is not so productive); the yield per unit area that is calculated for a reef can depend very much upon the size of the area and the percentage of that area which is actually covered by coral or other hard substrate. Some fisheries records cover a large area of many square kilometers, only part of which is actually covered by coral, whereas other records of fish yields apply to very small areas that are entirely coral reef. The productivity of a reef may also vary with the complexity of its vertical structure.

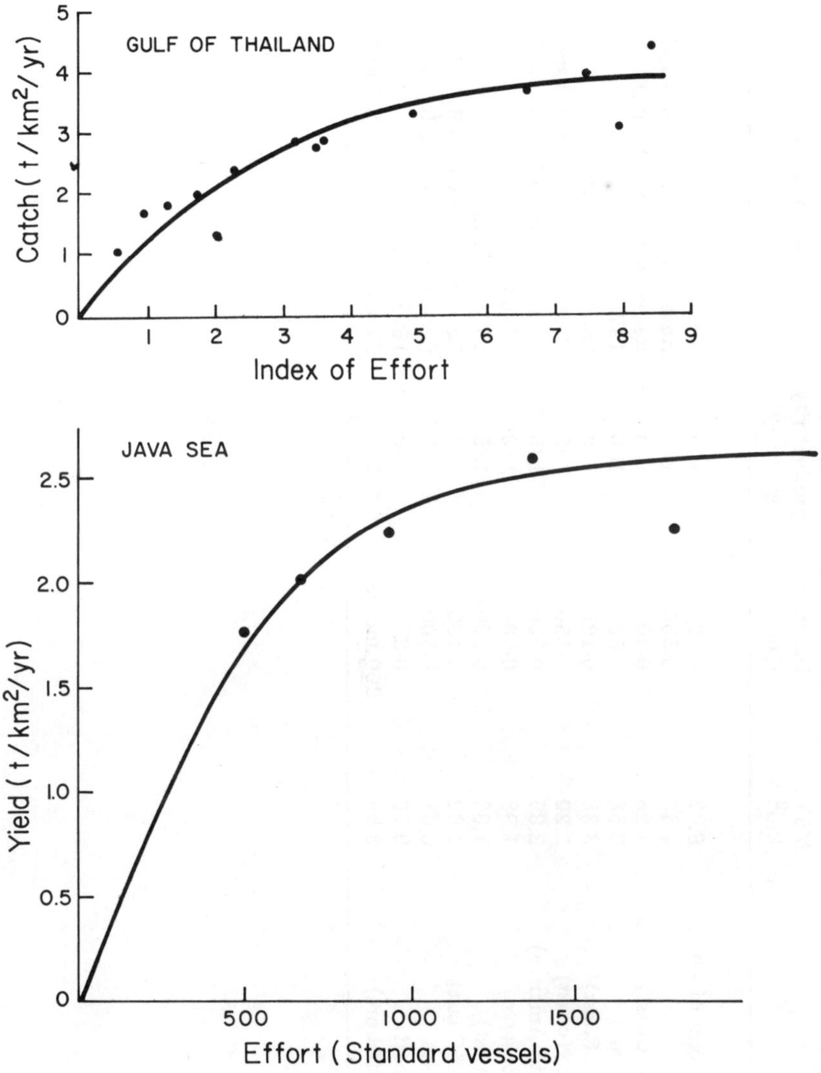

Fig. 6. Estimation of maximum sustainable yield for continental shelf demersal fisheries (from SCS 1978 and SCS 1979). The points in the graph represent different years in the history of the fishery.

Table 8. Estimated maximum sustainable yields of tropical pelagic fisheries.

Location	MSY ($mt/km^2/yr$)	Depth (m)	Primary productivity ($gC/m^2/yr$)	Estimation method	Source
Java Sea (N. Coast of Java)	6.02	0-50	180	0.5 B_o	SCS (1979)
South Atlantic	4.43	0-550	135	Research Survey (0.5 B_o) and Landings	Klima (1977)
Sumatra (West Coast)	4.28	0-40	130	Research Survey (0.5 B_o)	SCS (1979)
India (West Coast)	3.58	0-50	180	1978 Landing	Anon. (1979)
Malaysia (West Coast)	3.38	0-100	130	Landing 1973-1974	Yesaki (unpub. data)
Gulf of Mexico (coast)	3.20	0-550	90	Research Survey (0.5 B_o) and Landings	Klima (1977)
Atlantic (South America)	2.35	0-550	135	Research Survey (0.5 B_o) and Landings	Klima (1977)
India (East Coast)	1.98	0-50	180	1978 Landing	Anon. (1979)
India (West Coast)	1.03	0-200	135	1978 Landing	Anon. (1979)
Thailand (West Coast)	1.02	0-100	55	Research Survey (0.5 B_o)	SCS (1976b)
South China Sea	0.81	0-500	45	Research Survey (0.5 B_o) and Landings	SCS (1973)
India (East Coast)	0.70	0-200	90	1978 Landing	Anon. (1979)
Philippines (offshore)	0.55	200 and more	110	Research Survey (0.5 B_o) and Landings	Menasveta et al. (1973)

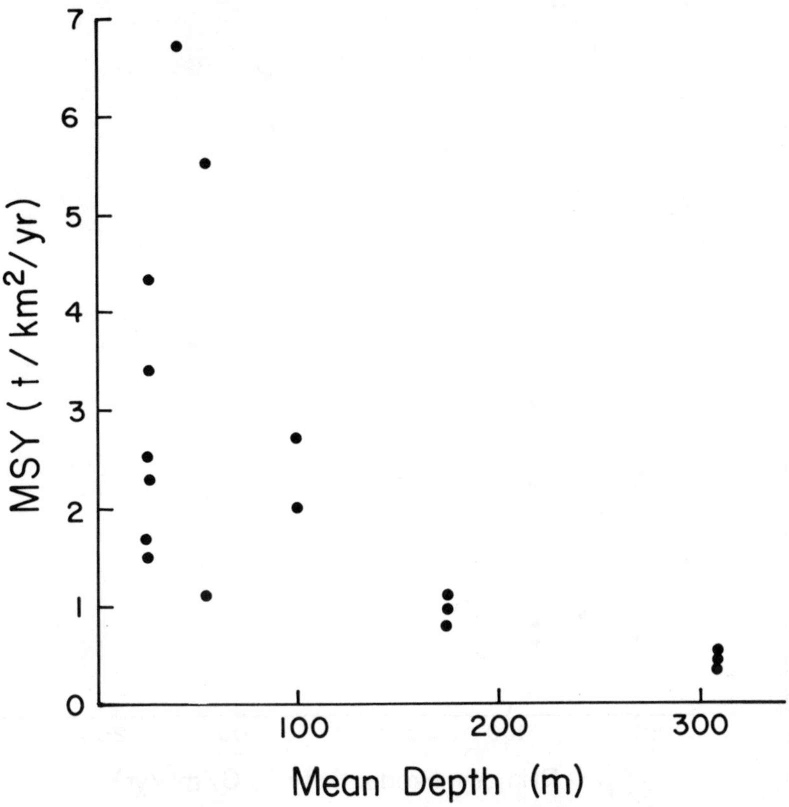

Fig. 7. Maximum sustainable yields and depth of continental shelf demersal fisheries, based on Table 7 (r = .86).

Open Ocean

Annual catches of tunas and billfishes by Japanese vessels in the western tropical Pacific, aggregated by 10-degree squares (data on file at the National Marine Fisheries Service, Southwest Center, Honolulu Laboratory) show a range of yields from 0.0025 to 0.04 t/km^2/year, with an average yield of 0.016 t/km^2/year. (This is an underestimate of the total catch because Korean vessels also fish this region but are not included in the statistics.) Catches of tunas and billfishes in the eastern tropical Pacific (Calkins 1975) suggest a range of yields from 0.002 to 0.04 t/km^2/year, with an average of 0.024 t/km^2/year. Finally, catches of tunas and billfishes in the 10-degree squares in the tropical Atlantic off the coast of Africa (ICCAT 1980) are as high as 0.05 t/km^2/year. The range of existing yields in the open ocean does not extend much below the range of MSYs for that ecosystem because much of the open ocean, like much of the continental shelves, is fished intensively by sophisticated fishing fleets.

Discussion

Relationship between Yields and Ecosystem Types

Fig. 11 summarizes the ranges of primary productivities and fish yields encountered in various tropical ecosystems. The range of fish yields is

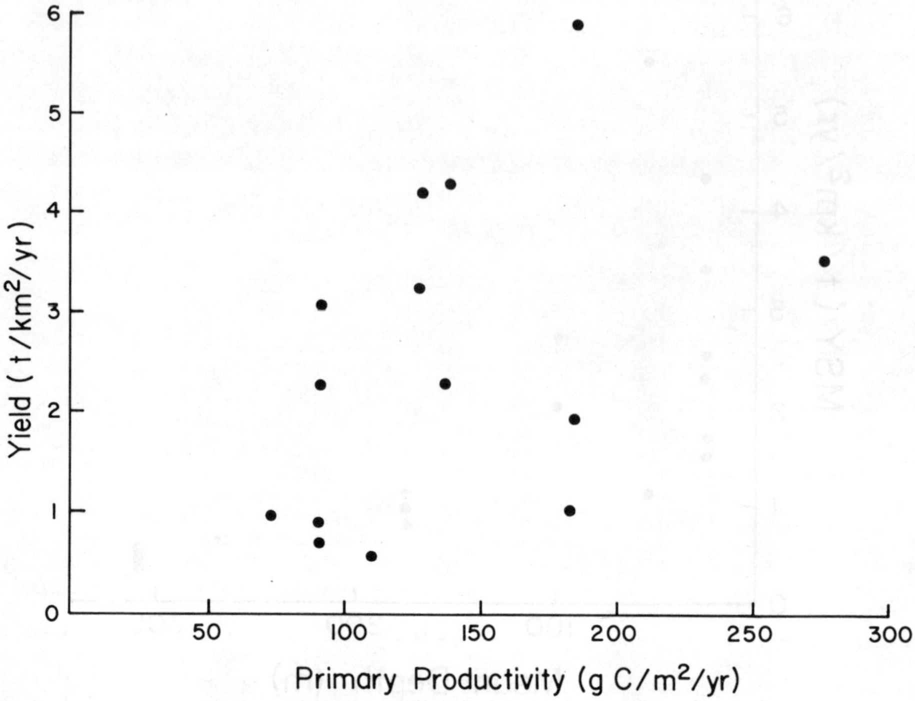

Fig. 8. Maximum sustainable yield and primary productivity of continental shelf pelagic fisheries, based on Table 8 (r = .61).

Table 9. Yields from coral reef fisheries.

Location	Area (km^2)	Catch (t/km^2)	Source
Samoa*	3	18[b]	Wass (in press)
Philippines	1	18[b]	Alcala (1981)
Samoa		8[b]	Hill (1978)
Ifaluk* (Pacific)	6	5.1	Stevenson and Marshall (1974)
East Africa		5[c]	Gulland (1979)
Mauritius*	350	4.7[a]	Wheeler and Ommanney (1953)
Fiji		4.4[a]	Bayliss-Smith (pers. comm.)
Jamaica*	2,860	4.1[c]	Munro (1978)
Bahamas		2.4[a]	Gulland (1971)
Puerto Rico*	2,300	0.8[b]	Juhl and Suarez-Caabro (1972)
Kapingamaringi* (Pacific)	400	0.7[b]	Stevenson and Marshall (1974)
Cuba*	55,000	0.5	Buesa Mas (1964)
Lamotrek* (Pacific)	44	0.45[b]	Stevenson and Marshall (1974)
Bermuda*	1,035	0.4	Bardach and Menzel (1957)
Raroia* (Pacific)	400	0.09	Stevenson and Marshall (1974)

*Catch and fishing effort appear in Fig. 10.
[a]MSY based on catch-effort relation over series of years.
[b]Probably near the MSY because of heavy fishing intensity.
[c]See Fig. 9.

different in each of the ecosystems, but the ranges are so broad that most of the ecosystems overlap considerably. As a consequence, ecosystem type alone is not a precise predictor of the potential yield of a particular fishery, at least at the coarse level of ecosystem classification employed here.

There is a positive association in Fig. 11 between the primary productivities of ecosystems and their fish yields, but the relation is not very tight. Some ecosystems with similar primary productivities have very different fish yields, and other ecosystems with similar fish yields have very different primary productivities. Furthermore, the overall range of fish yields through all ecosystems is much greater than the overall range of primary productivity, indicating that fish yield is not simply responding in proportion to primary productivity regardless of the ecosystem. Primary productivity is not useful to predict fish yields unless the ecosystem is specified.

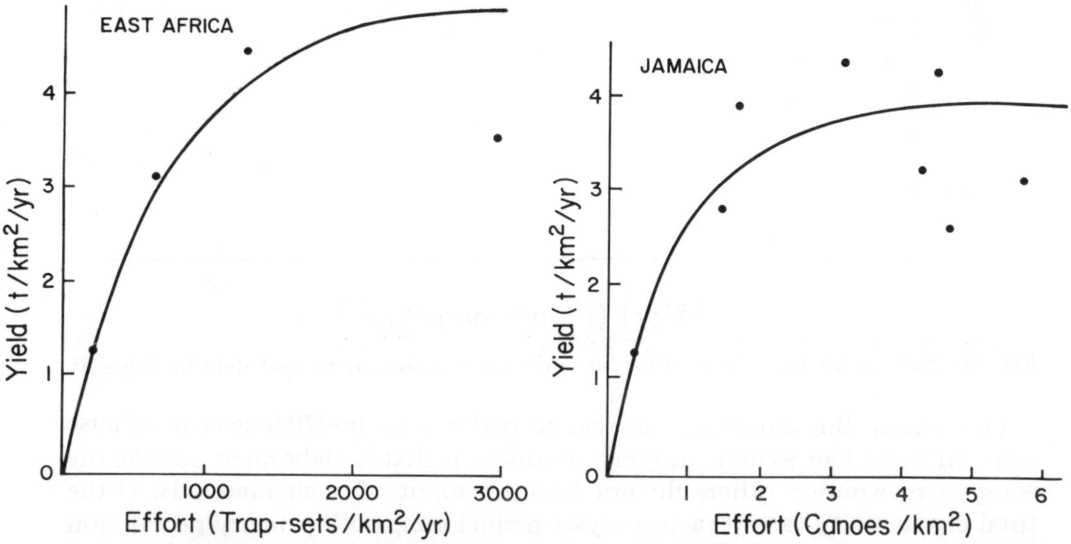

Fig. 9. Estimation of maximum sustainable yields for coral reefs [from Munro 1978 (Jamaica) and Gulland 1979 (East Africa)]. The points in the graphs represent different reef fishing locations in the same geographic area.

Table 10 shows the estimated range of efficiencies with which primary productivity is turned into fish yields in each of the ecosystems. (The real range in efficiencies for each ecosystem is probably more narrow because measurement errors would tend to broaden the range.) Lakes and reservoirs are quite similar. Rivers can have a higher efficiency, but it may be an artifact due to external inputs. The limited information on lagoons and estuaries indicates that the lower end of their range falls within the same range of efficiencies as lakes. The upper end is higher because of intensive aquaculture. Continental shelf fisheries appear to have a slightly higher efficiency than lakes, while upwelling areas have much higher efficiencies due to short food chains. The efficiency of coral reefs, which typically have a multitude of species, can be somewhat lower than the rest of the continental shelf. The efficiency of open oceans is less by an order of magnitude, presumably because of longer food chains.

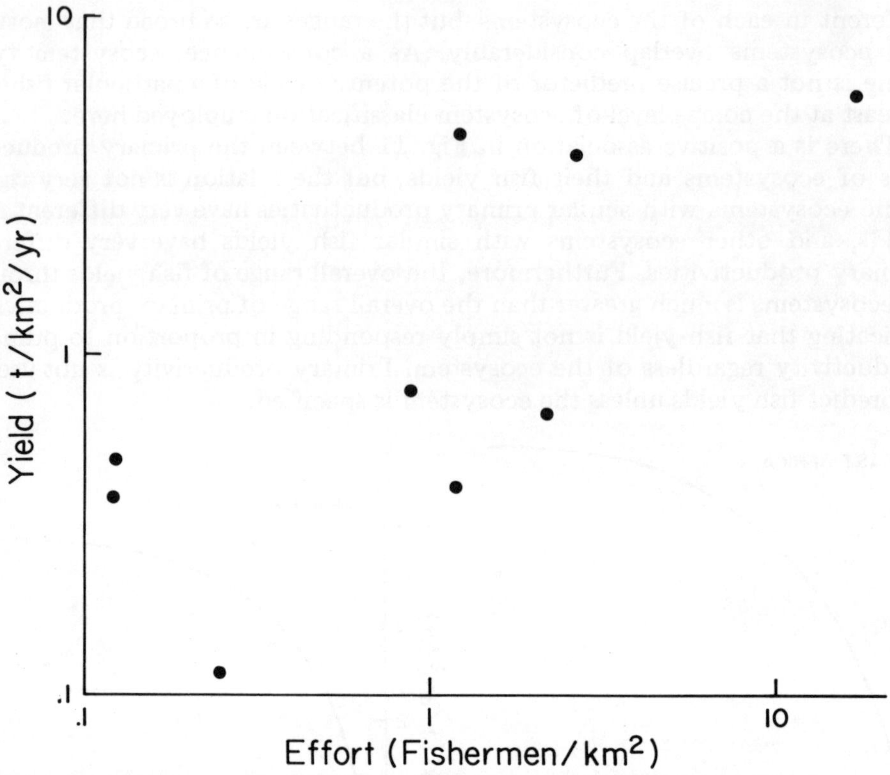

Fig. 10. Fish yields vs. fishing effort in coral reefs (based upon reef data in Table 9).

One reason the ecosystems appear to differ in their efficiencies is because some of them can exploit primary production that is elaborated outside the ecosystem, whereas others do not have an input of such materials. If the total productivity base of each ecosystem—including both primary production and organic matter from outside—were used in place of primary production alone, the relation between the productivity base and potential fish yield might be more universal regardless of the ecosystem. However, even with a better measure of the productivity base, different ecosystems could still differ in the food-chain structure which determines the efficiency of translating the resource base into fish yields.

Ecosystem and primary productivity can together predict potential fish yields better than either one alone. Nonetheless, there is considerable variation in yields that remains unexplained even when both are taken into account (Figs. 3 and 10), though it is possible the predictions could be more precise if the ecosystem classification were more refined. Even though such predictions can be helpful for rough inventories of potential fish yields, it is unlikely they will ever be precise enough for managing particular fisheries in the absence of other information about yields.

Yield Assessment and Management

The potential fish yield from an ecosystem cannot be inferred from catch records without reference to the fishing effort behind those records. The relationship between catch and effort is customarily displayed by means of graphs where different levels of effort are found at different locations within the same fishery (e.g., Fig. 9) or where different levels of effort have occurred at different times in the history of the fishery (e.g., Fig. 6). It is possible from catch-effort graphs to see if the fishing effort has been intense enough for the potential yield of the fishery to be expressed in actual yields. However, the yield from a multispecies fishery is not only a matter of how much fishing, but also what kind of fishing. Potential yields from different ecosystems can only be assessed in this context.

Table 10. Ratio of fish yields to primary productivity.[a]

Ecosystem	Range	Geometric mean
Coastal upwelling	.005 - .013	0.0081
Rivers	.005 - .01	0.0071
Ponds	.001 - .01	0.0032
Lagoons and estuaries	.0008 - .01	0.0028
Continental shelf	.0003 - .003	0.00095
Lakes	.0004 - .0016	0.00080
Reservoirs	.0002 - .002	0.00063
Coral reefs	.0002 - .0008	0.00040
Open ocean	.00001 - .0002	0.000045

[a]The ratios in this table are based on primary productivity estimates in Bunt (1975), Beadle (1974), Conner and Adey (1977), Cushing (1969), Edwards (1978), Gerlotto et al. (1976), Hempel (1973), Kaliyamurthy (1973), Kinsey (1979), Koblentz-Mishke et al. (1970), Likens (1975), Plante-Cuny (1977), and Rodriguez (1963). The ratios were calculated as the carbon yield of fish (assumed to be one-tenth of wet weight) divided by primary productivity (in carbon units).

There are two principal ways that the total harvest from a multispecies fishery can be increased by manipulating the kind of fishing:

— more intricate harvesting;
— restructuring the food web.

We have already observed that the highest yields from coral reefs occur where there is intense, intricate fishing. Higher yields can also be achieved by altering food web structure by:

— shortening the food chain between primary production and fish production;
— reducing fish consumption by predators which compete with fishermen.

Examples are the introduction of herbivorous or sediment-feeding fish to a fishery or the implementation of special measures to reduce predation (e.g., cage culture or intense fishing of predators).

The fishery of Lake Victoria (East Africa) illustrates these points. Lake Victoria has a multispecies fishery (12 commercial genera and more than a hundred species) which, like many other inland fisheries, shows signs of overfishing: declining fish sizes and disappearance of major commercial species. The diversity of fishing intensities and fishing gear around the shoreline of Lake Victoria has led to a similar diversity of yields and species composition in the catch. The result is a series of unplanned "experiments", which Marten (1979a, 1979b) has analyzed statistically to summarize the impact upon the stocks of the amount of fishing and the kind of fishing. The main interpretive tool was a curvilinear regression of catch versus effort,

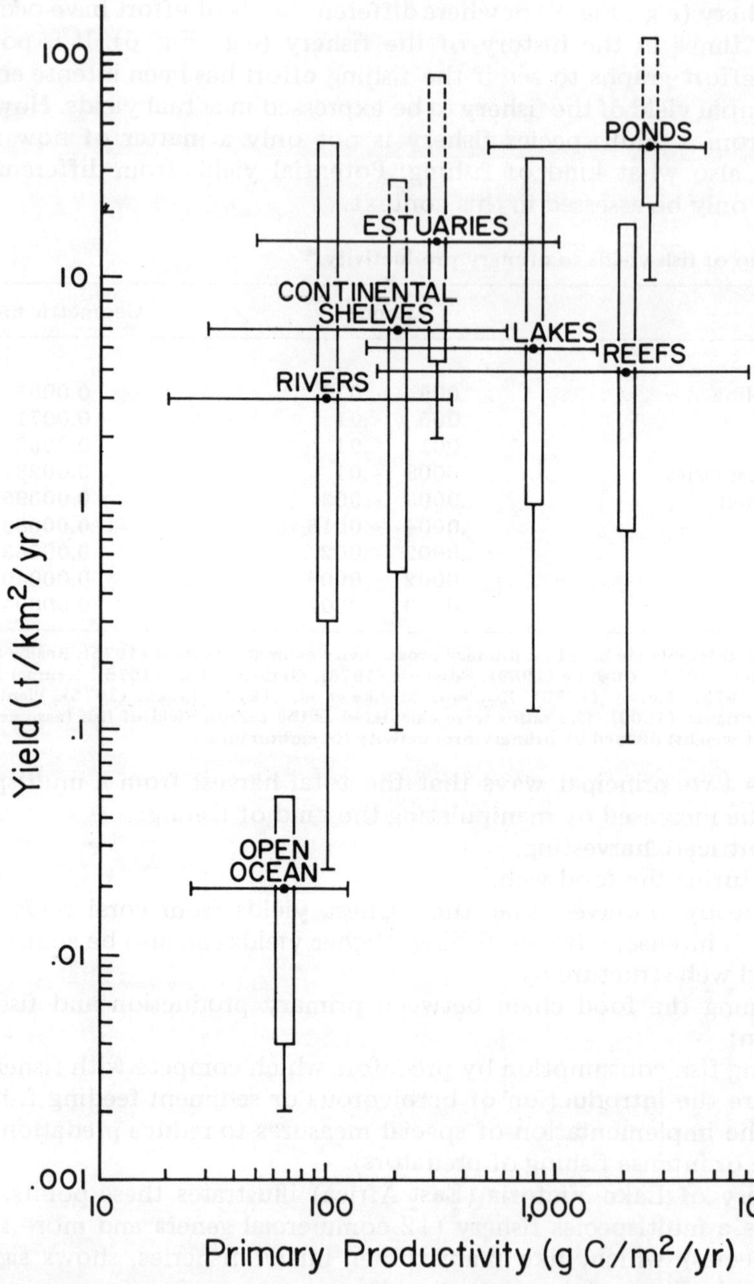

Fig. 11. Ranges of fish yields and primary productivities in various tropical ecosystems. Dots at the intersection of ranges represent modal values. Thickened portions of the bars represent the range of maximum sustainable yields. Dashed projections at the top of the ranges for estuaries and ponds represent elevated yields from aquaculture with fertilization (but not supplemental feeding). The dashed projection for continental shelves represents higher yields which occur in areas of upwelling. Primary productivity estimates are based on the references listed in Table 10.

in which the total catch (summed over all species) occupied one dimension and effort occupied six dimensions corresponding to six categories of fishing gear.

Above a certain fishing effort, the total multispecies catch in Lake Victoria is not affected much by fishing effort *per se*, but it is very much affected by the kind of fishing gear employed. There is no mix of gear which is optimal for all species in the fishery. What is optimal for one species may underexploit or destructively exploit another species. Fishing gear also has indirect ecological effects upon fish species that may not even be captured by that kind of gear, because of predation and competition, and these effects may lead to successional changes in the species composition of the fishery.

The optimal mix of gear for the fishery as a whole is a compromise. In the case of Lake Victoria, the optimal mix emphasizes intensively harvesting species at the end of short food chains (which are part of Lake Victoria's native fish fauna) and fishing down large fish that prey upon these species. There is no indication that even the heaviest fishing leads to lower yields if the optimal mix is employed. This suggests the practical conclusion that maximizing the yield from a multispecies fishery may be as much a matter of developing the infrastructure to encourage the right kind of intensive fishing as of restricting fishing practices which appear harmful.

Most of the management attention in fisheries to date has been devoted to yields, but the management of multispecies fisheries may be equally a matter of insuring a desirable species composition in the fishery. Although heavy fishing may not in itself significantly reduce the total yield from a multispecies fishery, it is quite common for heavy fishing (or the wrong kind of fishing) to change the composition, and therefore economic value, of the fishery. A change in species composition can also influence total yields to some extent because of food chain and predation effects.

A Habitat Perspective

A finer view of aquatic ecosystems than has been customary for fisheries purposes will be necessary for more effective management and yield assessment of multispecies fisheries. Substrate type is one way of distinguishing different habitats within the broader ecosystem, and Wanjala (1978) has shown that different sections of the Lake Victoria shoreline with different substrates (e.g., stony or muddy) are inhabited by different fish communities. The same is true in Hawaii, where different demersal fish communities are found at different depths offshore (Ralston and Polovina, in press); and different inshore substrates, such as lava and sand, each have their characteristic fish communities, potential yields and management needs, even though all are part of the broader shore ecosystem (Hawaii Coastal Zone Fisheries Management Program 1980). Each of these habitats has its own characteristic fisheries succession in response to fishing and each requires specific management decisions.

A habitat approach emphasizes attention to species composition (including a classification of fish communities), how different fish communities are associated with different habitats, and how community composition responds to human activities (including fishing practices) that impinge upon the fishery (Smith et al. 1973; Marten 1981). This perspective does not mean that massive amounts of quantitative data are required in much more detail than before. The data need only be roughly quantitative, but they must be sensitive to species composition. (This approach to species management is analogous in many respects to range management, where attention is given to forage species composition and how this changes under grazing pressure.) Habitats can be mapped (Aecos 1979), and because fishermen are generally precise about where they fish, it is possible at least in theory to maintain catch records on a habitat basis.

A habitat perspective in multispecies fisheries management and assessment may mean that the maintenance of habitat quality will be as important as the regulation of fishing effort. Although the production of a multispecies fishery can be manipulated to some extent by adjusting the harvesting regime, the yield from that fishery can be reduced immensely by habitat destruction. This includes destructive fishing practices such as the use of dynamite, poisons, and seines and trawls that scrape across the bottom and disrupt the production of fish food or spawning of fish. Equally important, however, are non-fishing activities that may cause even more serious destruction of fishery habitat, such as the siltation of coral reefs due to runoff from mining activities or pollution due to sewage, industrial effluents or oil spills. Pollution problems will increase as industrialization increases in the tropics, particularly under impetus from some countries to transfer their most polluting industries to developing countries.

This kind of fisheries management and assessment is complicated. It is not realistic to depend upon ecological theories to predict what will happen in every specific situation. Multispecies fisheries management and yield assessment will have to remain empirical, based on observations of how fish community composition and yields change under different circumstances, taking advantage of "experiments" provided by existing fishing activities in different places with different fishing conditions and different histories. It will have to be pragmatic and adaptive in the sense described by Holling (1978) for adaptive environmental management, relying upon monitoring to anticipate unwanted "surprises", and developing new approaches for keeping options open in dealing with such surprises.

Acknowledgments

We wish to thank J.M. Kapetsky (lagoon fisheries), Robin Welcomme (rivers and reservoirs), and Jerry Wetherall (Pacific tuna) for useful data and references, as well as Henry Regier, Daniel Pauly and Moe Yesaki for miscellaneous material and data.

References

Aecos, Inc. 1979. Hawaii coral reef inventory. Island of Oahu. Rep. to U.S. Army Corps of Engineers.

Alcala, A.C. 1981. Fish yield of coral reefs of Sumilon Island, Central Philippines: implication for coral reef resources management in the Philippines. Nat. Res. Counc. Philipp. Bull. 36(1): 1-7.

Almazan, G. and C. Boyd. 1978. Plankton production and tilapia yield in ponds. Aquaculture 15: 75-77.

Anon. 1979. Trends in total marine fish production in India—1976. Mar. Fish. Inf. Serv. T and E. Ser. 9: 7-22.

Aoyama, T. 1973. The demersal fish stocks and fisheries of the South China Sea. SCS/Dev/73/3. 46 p. FAO, Rome.

Arfin, O. and Z. Arfin. 1976. Fisheries in the floodplain area of south Sumatra, a case study of the Lubuk Lampam in 1973. Paper presented at the the IPFC Symposium on the Development and Utilization of Inland Fishery Resources, 27 Oct.-5 Nov., 1976. Colombo, Sri Lanka. IPFC/76/SYM/10. FAO, Bangkok.

Bardach, J.E. and D.W. Menzel. 1957. Field and laboratory observations on the growth of certain Bermuda reef fisheries. Proc. Gulf Caribb. Fish. Inst. 9: 106-112.

Bardach, J.E., J.H. Ryther and W.O. McLarney. 1972. Aquaculture: the farming and husbandry of freshwater and marine organisms. John Wiley and Sons, New York.

Bazigos, G.P. et al. 1977. The present state of the fishery of the Magdalena River Basin, Columbia. FAO Working Paper No. 2. FI:DP/COL/72/552. FAO, Rome.

Beadle, L.C. 1974. The inland waters of tropical Africa. Longman, New York.

Belloc, G. 1938. L'Etang de Bruglia. Rapp. P.-v. Réun. Cons. Int. Explor. Mer. 11: 433-473.

Bhukuswan, T. and S. Pholprasith. 1976. The fisheries of Ubolratana reservoir in the first ten years of impoundment. Proc. Indo-Pac. Fish. Counc. 17: 195-205.

Buesa Mas, R.J. 1964. Las pesquerías cubanas. Cent. Invest. Pesq. Contrib. 21. 93 p.

Bunt, J.S. 1975. Primary productivity of marine ecosystems, p. 169-184. *In* H. Leith and R.H. Whittaker (eds.) Primary productivity of the biosphere. Springer-Verlag, New York.

Cadwalladr, D.A. and J. Stoneman. 1966. A review of the fisheries of the Uganda waters of Lake Albert. East Afr. Forestry Fish. Res. Org. Suppl. Publ. No. 3.

Calkins, T.P. 1975. Geographical distribution of yellowfin and skipjack tuna catches in the eastern Pacific Ocean and fleet and total catch statistics, 1971-1974. Inter-Amer. Trop. Tuna Comm. Bull. 17.

Carvajal, R.J. 1972. Posibilidades de conversión de parte de la laguna de Piritu. (Estado Anzoategui) en salina artificial. Lagena 29: 7-11.

Chukajorn, T. and O. Pawapootanon. 1976. Annual catch statistics of freshwater fishes taken from seven reservoirs in northeastern Thailand. Proc. Indo-Pac. Fish. Counc. 17: 206-221.

Cobb, J. 1901. Commercial fisheries of the Hawaiian Islands. Rep. U.S. Fish. Comm. 427-433.

Conner, J.L. and W.A. Adey. 1977. The benthic algal standing crop and productivity of a Caribbean algal ridge. Atoll. Res. Bull. 211.

CTFT. 1957. Notes et documents sur la pêche et la pisciculture (Dahomey, Vallée inférieure de l'Ouêmé). Serie D.G. No. 2. Centre Technique Forestier Tropical, Nogent-sur-Marne, France.

CTFT. 1969. Resultats de observations realisées sur l'évolution de la pêche lagunaire de 1966 à 1969. (République du Dahomey). Centre Technique Forestier Tropical, Nogent-sur-Marne, France.

Cushing, D.H. 1969. Upwelling and fish production. FAO Fish. Tech. Pap. No. 84. 40 p.

Dobrovici, N.B. 1971. Rapport au Gouvernement du Niger sur le développement et la rationalisation de la pêche sur le fleuve Niger. FAO/UNDP (Tech. Assist.) Rep. Fish. no. 2913. 47 p. FAO, Rome.

Dubey, G.P. and S.N. Chatterjee. 1976. Case study of Ghandi Sagar Reservoir Madhya Pradesh, India. Proc. Indo-Pac. Fish. Counc. 17: 185-194.

Duerre, C. 1969. Report to the Government of Zambia on fishery development in the Central Barotse flood plain. (Second phase, May 1967-April 1968) FAO/UNDP(TA) Rep. Fish. 2638. 80 p.

Durand, J.R. et al. 1978. Statistiques de pêche en lagune Ebrié (Côte d'Ivoire): 1976 et 1977. Doc. Sci. Centre Rech. Océanogr. Abidjan 67: 114.

Edwards, R.R.R. 1978. Ecology of a coastal lagoon complex in Mexico. Estuarine Coastal Mar. Sci. 6: 75-92.

FAO. 1972. Atlas of the living resources of the seas. FAO, Rome.

FAO. 1976. Yearbook of fishery statistics. Catches and landings, 1975. FAO Yearb. Fish. Stat. 40 p.

FAO. 1979. Comparative studies on freshwater fisheries. FAO/FIR Tech. Rep. 92 p.
Fernando, C.H. 1976. Reservoir fisheries in S.E. Asia: past, present, and future. Paper presented at the Indo-Pacific Fisheries Council Symposium on the development and utilization of inland fisheries resources, 27 Oct.-5 Nov. 1976. Colombo, Sri Lanka. IPFC/76/SYM/46.
Frey, D.G. 1947. The pond fisheries of the Philippines. J. Mar. Res. 4: 247-258.
Frey, D.G. 1969. A limnological reconnaisance of Lake Lanao, Philippines. Ver. Int. Ver. Theor. Angew. Limnol. 17: 1090-1102.
Gamboa, B.R., A.G. Garcia, J.A. Benitey and T. Okuda. 1971. Estudio de las condiciones hidrográficas y químicas en el agua de la laguna de Lacarigua. Bol. Inst. Oceanogr. Univ. Oriente 10: 55-72.
Ganf, G.C. 1975. Photosynthetic production and irradiance-photosynthesis relationships of the phytoplankton from a shallow equatorial lake (Lake George, Uganda). Oecologia (Berl.) 18: 165-183.
Garcia, S. 1975. Los recursos pesqueros regionales de Tuxpan, Veracruz a Tampico, Tamps. y su posible industrilización. Inst. Nac. Pesca (Mexico). Ser. Inf. INP/SI.
Gerlotto, F., S. Hem and R. Briet. 1976. Statisques de pêche en lagune Ebrié, année 1975. Cent. Rech. Océanogr. Abidjan Serv. Stat. 1: 1-35.
Gulland, J.A. 1971. The fish resources of the ocean. Fishing News (Books) Ltd., West Byfleet, England.
Gulland, J.A. 1974. The management of marine fisheries. Scientechnica (Publishers) Ltd., Bristol.
Gulland, J.A. 1979. Report of the FAO/IOD workshop on the fishery resources of the western Indian Ocean south of the equator. Mahe, Seychelles, 23 October-4 November, 1978. IOFC/Dev/79/45. FAO, Rome.
Haussman, U.B. 1971. Abstract food webs in ecology. Math. Biosci. 11: 291-316.
Hawaii Coastal Zone Fisheries Management Program. 1980. Hawaii coastal zone fisheries management study. Hawaii Division of Fish and Games, Honolulu.
Hayne, D.W. and R.C. Ball. 1956. Benthic productivity as influenced by fish predation. Limnol. Oceanogr. 1: 162-175.
Hempel, G. 1973. Productivity of the oceans. J. Fish. Res. Board Can. 30: 2184-2189.
Henderson, H.F. and R.L. Welcomme. 1974. The relationship of yield to morphoedaphic index and numbers of fishermen in African inland fisheries. CIFA Occas. Pap. No. 1. 19 p. FAO, Rome.
Hernandez, R.R.A. and M.G. Calderon. 1974. Inventario preliminar de la flora y fauna acuática de la Bahía de Jiquilisco. Ministerio de Agricultura y Granadería, Dirección General de Recursos Naturales Renovables, Servicio de Recursos Pesqueros, El Salvador.
Hickling, C.F. 1970. Estuarine fish farming. Adv. Mar. Biol. 8: 119-213.
Hill, R.B. 1978. The use of nearshore marine life as a food resource by American Samoans. Pacific Island Studies Program, Univ. Hawaii, 170 p. Mimeo.
Holling, C.S., Editor. 1978. Adaptive environmental assessment and management. John Wiley and Sons, New York.
ICCAT. 1980. Collective volume of scientific papers. Int. Comm. Conserv. Atl. Tunas 10.
INDERENA. 1974. Estadísticas, actividad pesquera en Colombia. Instituto para el Desarollo de los Recursos Renovables Naturales. Oficina de Planeación, Ministerio de Agricultura. 66 p.
Indrasena, H.H. 1970. Limnological and freshwater fisheries development work in Ceylon, p. 45-47. *In* Proceedings of the IBP Section PF (freshwater production) meeting of Inland Water Biologists in S.E. Asia, 5-11 May 1969. Djakarta, UNESCO Field Science Office for S.E. Asia.
Jhingran, V.G. and A.V. Natarajan. 1969. A study of the fisheries and fish populations of the Chilka Lake during the period 1957-1965. J. Inland Fish. Soc. India. 1: 49-126.
Jhingran, V.G. and V. Gopalakrishnan. 1973. Estuarine fisheries resources of India in relation to adjacent seas. J. Mar. Biol. Assoc. India. 15: 323-334.
Juhl, R. and J. Suarez-Caabro. 1972. La pesca en Puerto Rico. Agro. Pesq. 4: 1-52.
Kaliyamurthy, M. 1973. Preliminary studies on the primary production of the Pulicat Lake in relation to its fisheries. Indian J. Anim. Sci. 43: 230-242.
Kinsey, D.W. 1979. Carbon turnover and accumulation by coral reefs. Oceanography

Department, Univ. of Hawaii, Honolulu. Ph.D. thesis.

Klima, E.F. 1977. An overview of the fishery resources of the West Central Atlantic Region. FAO Fish Rep. 200: 231-252.

Koblentz-Mishke, O.J., V.V. Volkovinsky and J.C. Kabanova. 1970. Plankton primary production of the world, P. 183-193. *In* W.S. Wooster (ed.) Scientific exploration of the South Pacific. National Academy of Sciences, Washington, D.C.

Konare, A. 1977. Collecte, traitement et commercialisation du poisson en plaines inondables, p. 32-45. *In* CIFA Working Party on River and Floodplain Fisheries. (Mimeo)

Kudhongania, A.W. and A.J. Cordone. 1974. Batho-spatial distribution patterns and biomass estimates of the major demersal fishes in Lake Victoria. Afr. J. Trop. Hydrobiol. Fish. 3(1): 15-31.

Lasserre, G. 1979. Bilan de la situation des pêches: aux Pangalanes Est. (Zone Tamatave-Andevovanto) au Lac Anony (region Fort-Dauphin). Perspective et Amenagement. Consultant's Report to MAG/76/002.

Le Mare, D.W. 1949. The prawn pond industry of Singapore. Annu. Rep. Fish. Fed. Malaya and Singapore, p. 121-124.

Likens, G.W. 1975. Primary productivity of inland aquatic ecosystems, p. 185-202. *In* H. Leith and R.H. Whittaker (eds.) Primary productivity of the biosphere. Springer-Verlag, New York.

Lin, S.Y. 1968. Milkfish farming in Taiwan. Fish Culture Report 3. Taiwan Fish. Res. Inst., Keelung, Taiwan.

Mann, M.J. and N. Ngomirakiza. 1973. Evaluation of the pelagic resources in the Burundi waters of Lake Tanganyika and the evolution of the fisheries. Afr. J. Trop. Hydrobiol. Fish. Spec. Issue II: 135-141.

Marshall, N. 1980. Fishery yields of coral reefs and adjacent shallow-water environments, p. 103-109. *In* S.B. Saila and P.M. Roedel (eds.) Stock assessment for tropical small scale fisheries. International Center for Marine Resource Development, University of Rhode Island.

Marten, G.G. 1979a. The impact of fishing on the inshore fishery of Lake Victoria (East Africa). J. Fish. Res. Board Can. 36: 891-900.

Marten, G.G. 1979b. Predator removal: its impact on fish yields in Lake Victoria (East Africa). Science 203: 646-647.

Marten, G.G. 1981. Ecological data requirements for managing Hawaii's coastal zone fishery. Coastal Zone Fishery Management Program, Division of Fish and Game, Honolulu, Hawaii.

Matuszek, J.E. 1978. Empirical predictions of fish yields of large north American lakes. Trans. Amer. Fish. Soc. 107: 385-394.

Melack, J.M. 1976. Primary productivity and fish yields in tropical lakes. Trans. Amer. Fish. Soc. 105: 575-585.

Menasveta, D., S. Shindo and S. Chullasam. 1973. Pelagic fishery resources of the South China Sea and prospects for their development. SCS/DEV/&3/6. FAO, Rome.

Mothwani, M.P. 1970. Report to the Government of Nigeria on fishery investigations on the Niger and Benue Rivers in the northern region and development of a programme of riverine fishery management and training. FAO/UNDP(TA) Rep. Fish. 2771. 196 p.

Moulherat, J.L. and M. Vincke. 1968. Etude en vue de development des pêches aux Pongalanes—Est (Zone Tamatave-Andevoranto). CTFT, Madagascar, Tananarive.

Munro, J.L. 1978. Actual and potential fish production from the coralline shelves of the Caribbean Sea. FAO Fish. Rep. 200: 301-321.

Murphy, G.I. 1972. Fisheries in upwelling regions—with special reference to Peruvian waters. Geoforum 11: 63-71.

Natarajan, V.V. 1976. Ecology and the state of fishery development in some of the manmade reservoirs in India. Proc. Indo-Pac. Fish. Counc. 17: 258-267.

Nemoto, T. 1971. La pesca en el lago de macaraibo. Projecto de Investigacion y Desarollo Pesquero MAC-PNUD. FAO Inf. Tec. No. 24. 56 p.

Oglesby, R.T. 1977. Phytoplankton summer standing crop and annual productivity as functions of phosphorus loading and various physical factors. J. Fish. Res. Board Can. 34: 2255-2270.

Okuda, T. 1965. Consideraciones generales sobre las condiciones ambientales de la laguna y el Rio Unare. Bol. Inst. Oceanogr., Univ. Oriente (Venezuela) 4: 136-154.

Pakrasi, B.B., P.R. Das and S.C. Thakurta. 1964. Culture of brackish-water fishes in impoundments in West Bengal. Proc. Indo-Pac. Fish. Counc. 11:

Pauly, D. 1976. The biology, fishery and potential for aquaculture of *Tilapia melanotheron* in a small West African lagoon. Aquaculture 7: 33-49.

Petersen, G.H. and M.A. Curtis. 1980. Differences in energy flow through major components of subarctic, temperate and tropical marine shelf ecosystem. Dana 1: 53-64.

Pillay, T.V.R. 1954. The ecology of a brackish water Bheri with special reference to the fish-cultural practices and the biotic interaction. Proc. Natl. Inst. Sci. India 20: 399-427.

Plante-Cuny, M.R. 1977. Pigments photosynthêtiques et production primaire du microphytobenthos d'une lagune tropicale, la lagune Ebrié (Abidjan, Côte d'Ivoire). Cah. O.R.S.T.O.M. Ser. Océanogr. 15: 3-25.

Poll, M. and H. Renson. 1948. Les poissons, leur milieu et leur pêche au bief supérieur du Hualalsa. Bull. Agric. Congo. Belge 39: 247-248.

Qasim, S.Z. 1979. Primary production in some tropical environments, p. 31-70. *In* M.G. Dunbar (ed.) Marine production mechanisms. Cambridge Univ. Press, New York.

Rabanal, H.R. 1961. Status and progress of *Chanos* fisheries in the Philippines. Occas. Pap. 61/8, Indo-Pac. Fish. Counc.

Ralston, S. and J.S. Polovina. A multispecies analysis of the commercial deep-sea handline fishery in Hawaii. Fish. Bull. 80(3). (In press)

Reed, W. 1969. Report to the Government of Nigeria on fishing technology relating to river and swamp fisheries of Northern Nigeria. FAO/TA Fishery Technologist, Rep. FAO/UNDP(TA) 2711. 90 p.

Reizer, C. 1974. Definition d'une politique d'aménagement des resource halieutiques d'un écosysteme aquatique complexe par l'étude de son environnement abiotique, biotique et anthropique. Le fleuve Sénégal Moyen et Inférieur. Docteur en Sciences de l'Environnement. Dissertation Avlon. Fondation Universitaire Luxembourgeoni, 4 vols. 525 p.

Rodriguez, G. 1963. The intertidal estuarine communities of Lake Macaraibo, Venezuela. Bull. Mar. Sci. Gulf Caribb. 13: 197-218.

Ryder, R.A., S.R. Kern, K.H. Loftus and H.A. Regier. 1974. The morphoedaphic index, a fish yield estimator-review and evaluation. J. Fish. Res. Board Can. 31: 663-688.

Saila, S.B. 1975. Some aspects of fish production and cropping in estuarine systems, p. 473-493. *In* L.E. Cronin (ed.) Estuarine research: chemistry, biology and the estuarine system. Vol. 1. Academic Press, New York.

Sarnita, A.S. 1976. Some aspects of fisheries and their development in man-made lakes in Indonesia with special reference to Lake Jatiluhur, West Java. Proc. Indo-Pac. Fish. Counc. 17: 272-291.

Schuster, W. 1952. Fish culture in the brackish-water ponds of Java. Indo-Pac. Fish. Counc. Spec. 1: 1-143.

Smith, S.V., K.E. Chave and D.T.O. Kam. 1973. Atlas of Kaneohe Bay: a reef ecosystem under stress. University of Hawaii Sea Grant Program, Honolulu.

SCS. 1973. Pelagic fishery resources of the South China Sea and prospects for their development. SCS/DEV/73/6. FAO, UNDP, Rome. 68 p.

SCS. 1976a. Report of the workshop on the fishery resources of the Malacca Strait—Part II. Jakarta, 29 March-2 April 1976, SCS/GEN/76/6. 85 p.

SCS. 1976b. Report of the BFAR/SCSP workshop on the fishery resources of the Visayan and Sibuyan Sea areas, 18-22 October 1976. Iloilo, Philippines, SCS/GEN/76/7. 26 p. South China Sea Fisheries Development and Coordinating Programme, Manila.

SCS. 1978. Report of the workshop on the demersal resources of the Sunda Shelf, Part II. November 7-11, 1977. Penang, Malaysia. SCS/GEN/77/13. 120 p. South China Sea Fisheries Development and Coordinating Programme, Manila.

SCS. 1979. Report of the workshop on demersal and pelagic fish resources of the Java Sea, 5-9 December 1978. Semarang, Indonesia. SCS/GEN/79/20. 60 p. South China Sea Fisheries Development and Coordinating Programme, Manila.

Sreenivasan, A. 1978. Fish production and fish population changes in some South Indian reservoirs. Indian J. Fish 23(1/2): 134-152.

Stevenson, D.K. and N. Marshall. 1974. Generalizations on the fisheries potential of coral reefs and adjacent shallow-water environments, p. 147-156. *In* Proceedings of the Second International Coral Reef Symposium, Vol. I. Great Barrier Reef Committee, Brisbane.

Tampi, P.R.S. 1959. The ecological and fisheries characteristics of a salt water lagoon near Mandapam. J. Mar. Biol. Assoc. India 1: 113-130.

Tang, Y.A. 1967. Improvement of milkfish culture in the Philippines. Curr. Aff. Bull. Indo-Pac. Fish. Counc. 4a.

Toews, D.R. and J.S. Griffith. 1979. Empirical estimates of potential fish yield for the Lake Bangwelu System, Zambia, Central Africa. Trans. Amer. Fish. Soc. 108: 241-252.

Turner, J.L. 1977a. Some effects of demersal trawling in Lake Malawi (Lake Nyasa) from 1968-1974. J. Fish Biol. 10: 261-271.

Turner, J.L. 1977b. Changes in the size structure of Cichlid populations in Lake Malawi resulting from bottom trawling. J. Fish. Res. Board Can. 34: 232-238.

Wanjala, B. 1978. The association and distribution of fish species in Nyanza Gulf, Lake Victoria, East Africa, 1975. Oregon State University, Corvallis. M.S. thesis.

Wass, R.C. The shoreline fishery of American Samoa—past and present. UNESCO seminar on Marine and Coastal Processes in the Pacific Motupore Island Research Center, Papua, New Guinea, 14-17 July 1980. (In press)

Welcomme, R.L. 1972a. The inland waters of Africa. CIFA Tech. Pap. 1. 117 p.

Welcomme, R. 1972b. Rapport au Gouvernement du Dahomey sur l'évolution de la pêche intérieure, son état actuel et ses possibilités. FAO/UNDP(TA) Rep. Fish. 2938. 97 p.

Welcomme, R.L. 1976. Some general and theoretical considerations on the fish yield of African rivers. J. Fish Biol. 8: 351-364.

Welcomme, R.L. 1979. Fisheries ecology of floodplain rivers. Longman, New York.

Wheeler, J.F.G. and F.D. Ommanney. 1953. Report on the Mauritius-Seychelles fisheries survey, 1948-49. Colonial Off. Fish. Publ., H.M.S.O., London.

Yesaki, M. Primary production, ecosystem diversity, and fish yield. (Unpublished manuscript).

Zambia. 1965. Fisheries research bulletin, 1963-64. Lusaka, Game and Fish. Dep. Fish. Res. Bull., Zambia, 1963-4.

Zambia. 1971. Fisheries statistics (Natural Waters) 1970. Central Statistics Office, Lusaka.

Zambia. 1974. Annual report for 1974, Department of Fisheries, Lusaka, Zambia.

Discussion of the Paper by Drs. Marten and Polovina

Dr. Gulland suggested that a graph of total catch/km^2/year versus the number of fishermen/km^2, as in Fig. 1, may be of more use than the authors assumed. If a fishery approaches its maximum yield at the same density of fishermen, regardless of how large or small the MSY may be, then the cluster of level points in Fig. 1 would suggest that yields approximate the MSY whenever there are two or more fishermen/km^2. This question deserves further study. However, Dr. Marten was still of the opinion that lakes with a very low level of production could reach their maximum yield at a significantly lower number of fishermen.

Dr. Gulland asked what area was being used for river fisheries, and Dr. Marten replied that it was the area covered by the river at flood level. Dr. Gulland felt that it could make a big difference if the flood plain is included because it can add considerable production to the river.

Dr. Larkin commented that the level of exploitation of a lake depends very much on its location in relationship with the people who exploit it, citing the lakes in the north of Canada which are virtually unexploited because there are so few people around them. Dr. Sainsbury added that it is also a matter of what people are willing to do with the fishery. He wondered what yields would be like in the open ocean if people were prepared to eat myctophids or in coral reefs if they would eat pomacentrids. He has MSY estimates for the Australian northwest shelf that range from 4,000 t/year for an Australian-style fishery based on harvesting only large fish, to 30,000 t, if one is prepared to fish everything that is there.

Mr. Jones was impressed with the high productivity of shallow lakes and noted the importance of shallow, inshore areas as fish nursery areas. The total productivity of areas like the North Sea may be higher than is recognized because the higher inshore productivity is usually not taken into account. Dr. Murphy questioned the general importance of inshore areas as nurseries, citing the northwest shelf of Australia as an example where there is no evidence that inshore areas are nurseries. Dr. Marten commented that the inshore areas of Lake Victoria are nurseries for tilapia and that abuse of those inshore areas with seines appears to have contributed to the decline of tilapia. Dr. Pauly noted that graphs of fish size against depth invariably show the larger fish to be at greater depths. Dr. Gulland noted that it is difficult to know how much of the fish production has actually occurred inshore or offshore because many fish migrate from one area to the other.

Mr. Simpson pointed out that coral reefs that are directly offshore may have different yields from the barrier reef type. He also remarked that the high bird populations in many coastal lagoons may be taking a substantial portion of the fish production.

Dr. Sale suggested that, in addition to different food chains being responsible for the wide range of yields in the same ecosystem, different taxonomic groups and associated differences in the physiology of the fish may also be important. Some kinds of fish may be more efficient at converting their food to usable fish flesh than others. Mr. Jones cited the ability of *Sarotherodon niloticus* and one of the zooplankters in Lake George to digest the abundant blue-green algae that are indigestible for most animals. Dr. Marten speculated that the variation in digestibility of the organisms along the food chain could also influence the overall efficiency of transforming primary production to fish yields.

Following the observations on Lake Victoria that decline in total catch at high fishing efforts were a consequence of destructive fishing practices rather than an increase in fishing effort *per se*, there was a discussion of whether the total catch versus effort curve drops or remains high with increasing effort.

Dr. Pauly noted that the stocks of small fish species in the Gulf of Thailand have collapsed under fishing pressure, apparently due to the additional burden of heavy predation from larger fish, as has been the case with small fish like *Haplochromis* in Lake Victoria. However, Dr. Murphy felt that trawling the Gulf of Thailand should put as much fishing pressure on the small species as the large species and that a manipulation of gear-specific effort is only applicable to artisanal fisheries. Dr. Sainsbury expressed the need for better catch statistics because it is difficult to know with the present information just how hard the small species in the Gulf of Thailand have been fished.

The Management of Tropical Multispecies Fisheries

J. A. GULLAND

Marine Resources Service
Fishery Resources and Environmental Division
Fisheries Department
FAO of the United Nations
Via delle Terme di Caracalla
00100 Rome, Italy

GULLAND, J.A. 1982. The management of tropical multispecies fisheries, p. 287-298. *In* Pauly, D. and G.I. Murphy (eds.) Theory and management of tropical fisheries. ICLARM Conference Proceedings 9, 360 p. International Center for Living Aquatic Resources Management, Manila, Philippines and Division of Fisheries Research, Commonwealth Scientific and Industrial Research Organisation, Cronulla, Australia.

Abstract

A brief review is presented of some of the measures taken to manage tropical multispecies fisheries in different parts of the world. An examination is also made of other government decisions, e.g., on investment, that can affect the amount of fishing and the way in which these resources are used.

The contrast is drawn between the pattern of management that is supposed to occur in advanced countries–careful scientific research, preparation of scientific advice, decision on the desired measures, and implementation and enforcement–with what happens in practice in developing countries. It is suggested that the traditional pattern is not even working too well in developed countries (e.g., in the North Sea). The paper examines how the whole process of taking management decisions (in the wide sense) might be modified to take account of the realities of life in developing countries (particularly the smaller ones with limited research capabilities) and of the complexities and uncertainties of multispecies fisheries.

Introduction

Fishery management is often thought of only in terms of the pattern that has developed in Northern Europe and North America. A fishery, usually on a single species, is troubled by falling catch rates (and possibly also falling total catch); biological research shows that this is due to too much fishing, and further research determines what pattern of fishing would be "optimum", in some sense or other; and in due course controls (catch quotas, size limits, etc.) are applied that will move the fishery towards this optimum pattern.

Little management in this sense has been done in tropical waters for compelling reasons–the available mathematical models of the dynamics of fish populations are not immediately applicable to tropical situations; the basic information to apply any model is generally sparse; there are few scientists to carry out the necessary studies, and often there is not the administrative structure to implement and enforce the detailed restrictive measures typical of fishery management in the narrow sense.

Finally it could be argued until recently that no tropical fisheries were being "overfished" and needed management. This is certainly no longer true. It may be noted here that the same reservations can also be made about the situation in many developed countries, and there are few fisheries which are in fact being successfully managed at the present time according to the idealized pattern described here.

Nevertheless some management decisions are being made. If, as is now being generally done "fishery management" is taken to include most decisions (and all governmental decisions) affecting the amount and pattern of fishing, the number of these decisions in all countries is in fact large and growing. These decisions are often taken with little reference to any biological knowledge of the stocks. For example it is quite common for the treasury to agree to tax-free imports of engines or fuels for fishing boats in order to assist small coastal communities, but only rarely are these decisions preceded by biological analysis showing whether or not the resources can support the likely increase in fishing effort. The results have often not been happy.

The purpose of this paper is to examine ways in which biological research can lead to better management decisions (in the wide sense) being taken in respect of tropical multispecies fisheries. To do this, particular attention will be given to determining the types of biological advice needed, and the institutional and other mechanisms for making that advice available, and for ensuring it is acted upon. Attention will be given to the purely biological problems only to the extent of examining how sufficient biological knowledge can be made available to provide adequate advice. Some of the frustrations in studying multispecies fisheries for the purposes of giving management advice could be avoided by recognizing that complete knowledge of the biological system is unnecessary, as well as being almost impossible to obtain.

Types of Management Decisions

Three broad types of management decisions can be distinguished—(i) decisions on management in the most narrow sense, i.e., on the introduction of regulations such as catch quotas, minimum sizes, etc.; (ii) other actions designed more to protect the special interests of a particular group, e.g., banning trawling within a certain distance of the coast to protect the artisanal fishermen and (iii) decisions to encourage growth of the fisheries, e.g., on investment. To these should be added the social controls on fishing that have been traditional in some areas, e.g., the Pacific Islands (Johannes 1977) which kept the pattern of fishing in balance with the resource, at least until outside influences broke the system.

Management actions, in the narrow sense, have been scarce in tropical demersal fisheries. Very little has been done to control directly the total amount of fishing through catch quotas, or limits on the licenses issued, even when, e.g., in the Gulf of Thailand trawl fishery, it has been generally accepted for some time that the fishing effort is too high. Limits have been set on the number of licensed trawlers in parts of Malaysia, but there seems to have been a corresponding increase in the number of unlicensed vessels. The main exceptions to this lack of effective control on effort have been a few shrimp fisheries, particularly those operated by large industrial enterprises. Thus the Indonesian Government has kept the growth of the shrimp fishery in West Irian under control by limiting the licenses issued to the numbers which assessments showed to be reasonable. However, once such measures have been successful in maintaining or increasing the economic return from the fishery, the pressures to increase the number of licenses can be too powerful for a fisheries department (which is often politically weak) to withstand (see Marr, this vol.).

Regulations to control the sizes caught have been more frequent, and many countries have legislation setting the minimum sizes of trawl meshes. Actions to enforce these mesh sizes are much less common, and without this, or an understanding by the fishermen of why the general use of larger meshes would be to their advantage, the actual meshes used are often well below the legal minimum. For example, Jones (1976) reported that around the South China Sea the legal requirements were not being followed, and it is not believed that the situation has greatly improved since the time of his study. This is not just a problem of developing countries. Around the Mediterranean, for example, studies showed that while most countries have regulations that require a minimum mesh of 40 mm, the actual sizes used by most trawlers in the Mediterranean were much smaller. These regulations seem generally not to have been based on any biological assessment of the effects, but rather on a belief that a mesh regulation would be, in principle, a good thing. Much of the failure to implement measures is due to a past lack of communication between administrators, fishermen and scientists, of which the absence of an assessment is one aspect.

Regulations to protect the interests of particular groups of fishermen seem to have been better obeyed, presumably because there has been an identifiable group who believe the rules are in their interest, and who will complain to the authorities if they are not complied with. The commonest of these rules are those prohibiting the use of certain gears or types of vessel–particularly trawlers–within a certain distance of the coast or within a certain depth zone. These measures are probably effective in reducing or eliminating physical interference and direct conflict, but in most cases there is not enough information on the distribution and movements of the various species of fish caught by the different gears to say whether they reduce or eliminate the competition for the same stocks of fish. Decisions to introduce such measures are usually based more on the desire to keep potentially competing fishermen out of each other's way, and so far as possible out of each other's sight, than on a biological study of the effects of these and other alternative measures.

A great variety of government actions are taken to encourage the development of fisheries. These range from very direct involvement, e.g., investment in a government-owned fleet of new vessels, though encouragement of private investment, e.g., by the provision of low-interest loans, technical assistance and advice on the use of more effective gears, to the provision of shore facilities, or improved communication between landing places and the main markets. The effect of any individual action is often difficult to judge and the biggest changes in the fishery may result from minor actions. For example, the growth of the Thai trawl fishery started with a small technical assistance program which introduced single-boat trawling. Once the economic advantages of this type of fishery had been demonstrated, the fishery grew without much direct government involvement.

Overall, though, the effect of development actions of one sort or another on the fisheries is large, and probably much larger than the effects of management actions. The role of stock assessment and biological advice in reaching the decisions has in the past been minimal. Where advice has been sought it has been concerned with immediate questions, such as the catch rates to be expected from a new vessel. Very little information has been paid to the long-term effects, and the possible impact on the stocks of, say, the increased fishing effort to be expected along a hitherto isolated coast after the construction of an all-weather access road. This situation is now changing. In particular the regional development banks are becoming increasingly interested in the potential value of large-scale investment in fishery development and recognize the importance of having good resource assessments before starting on such investments.

Advice Required

Each type of management decision requires a different type of advice. The following section examines the advice needed for each type of decision and gives special attention to where single-species analyses are not adequate. The first type—management regulations in the narrow sense—appears in principle to require the most detailed advice. If there were adequate administrative machinery to apply complicated regulations, e.g., separate catch quotas for each species in each of a number of different areas, then correspondingly complex advice would be required. However, few tropical countries could enforce such complicated measures. It may indeed be questioned whether such measures could be enforced even in the most advanced countries. Simpler and more easily enforced measures must be used for which less complex advice may be sufficient.

The measures that can actually be applied will depend on the type of fishery. In small-scale fisheries it is extremely difficult to prevent anyone fishing who wants to fish from doing so, especially when there are few other jobs. Measures to control the total effort by small-scale fisheries are probably politically unacceptable, and would in any case be very difficult to enforce (virtually impossible for catch quotas); the possibilities of control on the growth of these fisheries to ensure that as far as feasible it matches the capacity of the available resources, are discussed in the following section.

In practice regulations on small-scale fisheries have to be limited to controls on the patterns of fishing–the types of gear that can be used, closed areas or seasons, or on the sizes or species that can be sold (though the last may be even more difficult to enforce than the others). Adequate advice on most of these can be provided without complex biological studies. For example the prohibited gears will be largely damaging and non-selective methods like dynamiting or poisoning. Any closed areas or seasons are likely to be chosen to protect particularly vulnerable stages of particular species, e.g., spawning runs up rivers, which could be wiped out by nets across the entrance. The need for measures of this type can be determined by straightforward single-species analyses.

Management of the larger-scale commercial fisheries presents more opportunities. Though a variety of gears are used for an even greater variety of species, the main current needs for management appear to lie in the trawl fisheries for shrimp and bottom fish. These catch a great mixture of species, with shrimp seldom accounting for more than about a quarter of the total weight caught, and usually much less. The economic importance of shrimp is much greater. In some fisheries (e.g., U. S. Gulf of Mexico, West Irian, Gulf of Carpentaria) virtually all fish are discarded, and landings are 100% shrimp. In others (e.g., the small trawlers fishing out of Cochin, India) the fishery is directed specifically at shrimp, though incidental catches of fish are landed, while others are mainly directed at fish, with shrimp making up enough of the catch to make the difference between a profitable or an unprofitable fishery. Many of these are known or believed to need management. The measures considered can, as usual, be divided into two classes—control of the total amount of fishing, and control of the pattern of fishing, i.e., the sizes and species caught.

The latter type of control is typified by the imposition of a minimum mesh size. Other approaches, e.g., by stopping fishing when or where small fish or shrimp are particularly abundant have similar effects and raise similar biological problems. The immediate extension of single-species mesh assessment to the multispecies trawl fisheries of Southeast Asia has been discussed by Jones (1976). By considering the shape of the fish and parameter ratios M/K and F/K he shows that the optimum mesh size for any species can be closely estimated as a function of the maximum size. This avoids the tedious work of carrying out selectivity experiments and yield per recruit analyses for a large number of species. The optimum mesh size for the fishery may then be estimated as an average, weighted according to the importance (weight and value) of each species in the fisheries, of the optimum mesh size for individual species—though, as Jones points out, the weighting must be done carefully because the curve of yield as a function of mesh size is strongly asymmetric, and in general, yield decreases faster above the optimum size than below.

This procedure is really the summation of many single-species assessments rather than a true multispecies assessment. The latter should take account of the possible interaction between species. Increasing the mesh size used will change the balance of fishery pressure on different species. The abundance of the larger species should increase in absolute terms, but they will still be exposed to fishing for much of their life, whereas some smaller species, with maximum sizes around the selection size for the larger mesh will hardly be fished at all. The fact that these species may become underutilized should be taken into account in simple procedure for assessing the optimum mesh size; what is not taken into account is any effects that the relative increase of these species might have on the larger species. Among adult fish this increase will presumably provide more food for the larger predators, and therefore be to their advantage, and to the advantage of the fishermen. However, there may be competition in the earlier stages between species with different maximum sizes. Particularly if they are related it can be imagined that the characteristics (food requirements, types of predators) of larval or small juvenile fish might be so similar that the density-dependent effects that determine, for example, the stock/recruitment relation apply more to groups of species taken together rather than to individual species. This would give a selective advantage to the smaller species, and hence in the long run might lead to a depletion of the preferred species, and a loss to the fishery. Checking on this hypothesis is not easy, but an examination of the biology of the younger stages of selected species might confirm or otherwise the existence of substantial overlap in needs.

The other possibility is that the increase in absolute numbers of the larger fish with the bigger mesh could have an effect on the smaller fish on which they feed, particularly the recruits or pre-recruits of the more valuable species. For example, Pauly (1982) has suggested that recruitment of shrimp in the Gulf of Thailand is inversely related to the abundance of fish. This is an extension of the general stock/recruitment problem, to which there seems no easy answer.

Control of the Amount of Fishing

In temperate waters, in single-species fisheries, the scientific rationale for these controls may be based either on yield-per-recruit models of the Ricker/ Beverton and Holt type, or by production models relating catch, effort and catch-per-unit-effort. Even with the limited number of species in the North Sea, for example, the extension of the first type to take account of species interactions is proving so difficult that the management decisions (e.g., on the 1981 quota for cod) are still based on single-species analysis. Given the poorer database and the much greater number of species there is little hope of the direct use of the analytical methods as the main basis for providing advice in tropical demersal fisheries.

The difficulties of applying analytical models to a great range of species are really of two kinds–operational and conceptual. Operationally the work of obtaining the estimates for the necessary parameters (F, M, K, etc.) and making the calculations has proved time-consuming for the handful of species in the North Sea (even now the estimates of M for most of the North Sea are not good), and repeating the same work for, say, the couple of hundred species in the Gulf of Thailand would be impracticable. The estimation aspect has been tackled by Pauly and others who, following and expanding on the work of Beverton and Holt (1959), have noted that there are clear patterns in the natural parameters so that usable estimates can be obtained from a few easily made observations (e.g., see Pauly, this vol.). This approach gives very quickly the form of the yield-per-recruit curve, leaving as the major estimation problem the location of the present fishery on the curve, i.e., estimating the current fishing mortality.

The conceptual problem is that this still leaves the assessment of the fishery as a whole as the sum of single yield-per-recruit curves, with no account of species interactions. This is clearly important, as shown by the differences in trends in abundance between species in the Gulf of Thailand, but incorporation of this interaction in the analytical models in a quantitative rather than a purely descriptive manner has not proved easy. In the

North Sea the effects of predation by one commercial species by another has been taken into account by the so-called "legion" or "phalanx" extension to the cohort analysis (Pope 1980) but it is difficult to see how this can be readily extended, even in a simplified form, to a large number of species. In any case some of the more important types of interaction in the North Sea seem to be those affecting recruitment—for example the outburst of high recruitment in most gadoids in the late nineteen-sixties—which are even less easy to take into account. This is not to say that analytic models are not useful in providing advice—at the very least yield-per-recruit calculations for a few selected species can give support (or otherwise) to results obtained in other ways—but it does seem likely that the main source of quantitative management advice (e.g., what would be the effect of reducing the fleet size by 10%) will have to come from production models.

The use of production models in multispecies has been met head on in the Gulf of Thailand and elsewhere, by examining the relations between total catch (usually as total weight of all species, but also as weight of marketable species, or total value), total effort, and catch-per-effort. In the Thai fishery the catch-per-effort has been obtained from research vessel data, which avoids many of the problems of possible changes in species preference involved in using commercial catch and effort data. The statistical problems of the production of a spurious correlation can also be avoided in this case by plotting catch as a function of catch-per-effort (as an index of stock biomass).

This procedure is attractive because it is simple and gives results that can be clearly understood by the policymaker, especially if the fishing effort is expressed in units appropriate to possible management policy, e.g., the number of trawlers of a standard size, that are operating. The method can show at once that, for example, a further increase in the number of trawlers will reduce the value of the catch, and that the greatest value would be obtained with a fleet say 80% of the present size. The question is whether the method is reliable. Might the variety of species involved and their interactions invalidate the conclusions so that increased effort actually increase the catch, or conversely that the optimum fleet size is even less than 80% of the present? It has been argued that provided the fish stocks adjust quickly to changes in the amount of fishing, the data automatically include the net results of interspecific reactions. Certainly the decline in the catch-per-effort of all species together in the Gulf of Thailand is the aggregate of very different rates of declines between species, and these differences presumably reflect interspecific effects, as well as differences in the direct response of individual species to the impact of fishing. The awkward assumption is that the system adjusts quickly to changes in fishing effort. The data set for the Gulf of Thailand is longer than for any other tropical area, but even so covers less than 20 years. Though the average life span is less than in temperate areas, some of the larger species live several years, so that the whole data series is only a few generations. Given that the intensity of fishing has been changing continuously over this period, there is clearly a probability that the stock at any one time would be in a state different from the equilibrium state corresponding to a sustained fishing effort at the current level, and a possibility that this difference could make a practical difference to the conclusion and the biological advice offered, e.g., on the optimum number of trawlers. This possibility might be tackled by examining the effects on the conclusions of considering the level of effort corresponding to the current stock as being the average effort over a varying number of years up to and including the current year. Alternatively the changes in the relative abundance of various species can be examined in order to formulate hypotheses of how these changes are induced by fishing, and therefore what sort of delays might be involved before the effects of fishing have their full effect. For example, if one major change is that species A increases because of better recruitment due to the reduction in the numbers of large individuals of species B, then there would be a delay equal to the period for fishing to have its full effect on B, plus the time for the improved recruitment to grow into the exploited stock of species A.

The other problems with the most simple-minded application of a production model is that it says nothing about the possible gains from changing the ratios of fishing mortality between different species. Pope (1979) has shown that while the data from a fishery on two or more stocks with a fixed ratio of F's on the various species will fit a production model perfectly, the greatest combined yield obtained under this restriction may be much less than that obtained by allowing the ratio of the F's to vary. Intuitively it is obvious that the production of a multispecies fishery could be improved by controlling the fishery on each species individually, perhaps by deliberately 'overfishing' undesirable species, or underexploiting those low-valued species that are the food of more valuable species. The position giving the greatest combined yield is, however, of purely theoretical interest unless there is a practical possibility of making the required adjustment to the ratios of fishing mortalities. While it is certain that in no fishery is it possible to achieve all possible ratios (two species may be so mixed on the fishing grounds that they must inevitably undergo approximately equal fishing mortalities), it is almost equally certain that in any fishery it is possible to achieve some changes in the ratio. In principle this might be done by setting catch quotas for each species (or group of species) but the experience of this system, and the associated rules on incidental catches, in developed countries does not suggest that this would work well in tropical developing countries with many more species and, generally, weaker administrative structures. A system of graduated levies or grants, such that fishing on one species would be discouraged by imposing a tax or levy on each tonne landed, while other species for which heavier fishing was desirable would attract a grant or premium has theoretical advantages, and might be easier to enforce. However, it has not been applied in practice so far. For the present the best chance of adjusting the balance of effort between species in tropical demersal species would seem to be controls on the area or season of fishing or the type of gear or vessel used.

For any given fishery the number of alternatives is strictly limited. That is, the assessment problem, which in the simple approach deals with only one degree of freedom (total amount of fishing) and for a full multispecies assessment might have 100 degrees of freedom (the F for each of 100 species) may be reduced to perhaps three or four (total effort, timing of any closed season, location of any closed areas, and perhaps specification of type of gear). This is, at least in principle, a much more feasible task.

The normal stages involved in assessment and management might be largely reversed. Instead of making assessments of the effects of changing biological parameters (fishing mortality, size at first capture, etc.) and then considering what specific measures would have the necessary effects, the first stage might be to identify, in general terms, possible measures, and then make the assessment. Two types of identification would be needed. First, the biological data would have to be examined to see what groupings by areas/season/gear/vessel do give significant differences in species (or size) composition in the catches - unless there are differences, then controls will have no effect. Second, the operation of the fisheries and of the administration of fisheries should be examined to see what controls would be practical, in terms of such things as economic disruption to the fishermen, effectiveness of enforcement, etc. Only then need the actual assessments be made, when it can be known what combinations of fishing mortality on different species should be considered.

Protection of Special Interests

The commonest example of this type of management action are the measures taken to protect the interests of small-scale inshore fishermen against possible damage by larger enterprises. Possible interactions between different fisheries after the fish are landed are not considered here, though it must be recognized that if markets are limited, the development of an industrial fishery can reduce the prices received by the artisanal fishery - though conversely the industrial fishery may open new markets and improve prices all round. Equally, within the context of the present meeting, the problems of direct physical interference, e.g., the destruction of fixed gear by trawlers, will not be examined, though this is a major justification for restricting the activity of large vessels near the coast. The concern here is the possible impact of fishing by one group of fishermen (particularly larger industrial vessels) on the stocks of fish exploited by another group (particularly the smaller-scale fishermen). Though some indirect effects are possible, through one fishery reducing the stocks of species which interact with species that are exploited by another fishery, these effects are likely to be smaller than the direct effects, when both fisheries harvest the same species. Also the indirect effects are at least as likely to be favorable (reduction of predators or competitors) as harmful, and the main concern over possible effects is when they are, or are believed to be harmful. Thus the necessary scientific advice can be based largely on single species assessments.

Quantitatively the necessary assessments can be very simple. The policy-maker needs to know for example that introducing an offshore trawl fleet of 50 vessels will cause a significant reduction in the catches of an existing inshore fishery. However, he is not greatly concerned with the precise value of their reduction, i.e., he needs to know whether it is 50% or 5%, not whether it is 50% or 40%. The requirements for qualitative studies may be greater. While the assessments for each species can be quite simple and crude, careful attention has to be given to the exact species being assessed. Unless the two fisheries are operating on the same area (and possibly not even then), it cannot be assumed that they are exploiting the same species. In many families important to tropical demersal fisheries there are a number of similar species-some mainly found inshore, some mainly offshore, and some move between inshore and offshore. Unless the exact species composition is checked it can easily be assumed that because the two fisheries catch, say, many croakers, they must be exploiting the same stock. Also important is some knowledge of migration and dispersion. Little is known about the movements of most species of tropical demersal fish, and they may vary from being resident and localized to making movements over hundreds of miles. Without some knowledge, which might come from tagging, or from examining detailed seasonal information on distribution and size composition, it could be believed that developing an offshore fishery would have no effect on a coastal fishery 50 miles away, when in fact there was a rapid interchange of fish between the areas.

Development Planning

Action by governments to encourage the growth and development of their national fishery-obtaining loans from international development banks for new vessels, providing low-interest loans to fishermen, improving shore facilities such as cold stores, etc.—are probably much more important than the decisions concerning regulations of fishing effort, mesh size, etc. in determining the pattern of fishing, and its match with the optimum pattern as determined by national objectives. In the past many of these decisions were made with little reference to scientific evidence, other than an examination of current catch rates. There is now a growing recognition among many of the responsible organizations (regional development banks, etc.) that decisions to initiate or to expand a fishery should be preceded by analyses of the likely effects of the development on the abundance of the stocks and hence on the catch rates in the fishery concerned, and in any other fishery on the same or related stocks.

The simplest advice sought has been on the productive capacity of the stock-often expressed as a wish to know the "MSY" from the stock. This can be a dangerous over-simplification, especially when the scientific advice is interpreted as meaning that if the estimate of MSY is, say, 25,000 tonnes, and provided the projected total catch is less than 25,000 tonnes, then all will be well. Apart from the risk that the MSY might be overestimated, this attitude ignores all the changes in the stock, and in catch rates that occur as any fishery expands, and which become very important well before the MSY is reached. To be useful, and effective in preventing wrong development decisions, the biological advice must be comparable to that provided in planning management decisions. It also should be presented as early as possible in the process of planning development—all too often biologists are brought in too late, and are placed in the difficult position of objecting to an investment which most often has already been agreed upon.

Provision of development advice of this kind is both easier and more difficult than providing management advice. It is easier in the sense that less detail and reliability is required. Often it will be sufficient to know that a proposed investment will not raise the fishing effort to a level at which catch rates will drop unduly so that the investment can go ahead, or on the other hand that there are sufficient doubts about whether the resources can support additional investment, so that alternative uses for those investment funds can be sought. It is much more difficult in the sense that assessments are required of the effects of levels of fishing that have not so far been experienced, whereas management advice is usually concerned with predicting the effects of reducing effort to some level that has been experienced in the past. The latter difference becomes more significant in a multispecies fishery since the opportunities to depart from a simple extrapolation are greater. Where several fisheries, on different mixes of species occur, the expansion of one fishery may cause interaction between species (and hence between fisheries) which had previously been negligible, and undetected, to become significant and in need to be taken into account when deciding on development.

References

Beverton, R.J.H. and S.J. Holt. 1959. A review of the lifespans and mortality rates of fish in nature and their relation to growth and other physiological characteristics, p. 142-180. *In* G.E. W. Wolstenholme and M. O'Connor (eds.) CIBA colloquia on ageing. Vol. 5. The lifespan of animals. London, Churchill.

Johannes, R.E. 1977. Traditional law of the sea in Micronesia. Micronesica (J. Coll. Guam) 13: 121-127.

Jones, R. 1976. Mesh regulation in the demersal fisheries of the South China Sea area. SCS/76/WP/34. 75 p. South China Sea Fisheries Development and Coordinating Programme, Manila.

Pauly, D. 1982. A method for the estimation of the stock-recruitment relationships of shrimps. Trans. Amer. Fish. Soc. 111: 13-20.

Pope, J. 1979. Stock assessment in multi-species fisheries, with special reference to the trawl fishery in the Gulf of Thailand. SCS/DEV/79/19. 106 p. South China Sea Fisheries Development and Coordinating Programme, Manila.

Pope, J. 1980. Phalanx analysis: an extension of Jones' length cohort analysis to multispecies cohort analysis. ICES CM 1980/G:19 (mimeo) 18 p.

Discussion of Dr. Gulland's Paper

Several participants noted that, in a given fishery, the value of the catch might not change in the same way as the total weight. In some fisheries, particularly in developed countries the aim of management is often to preserve some particularly desirable species or species mix. In most developing countries all species are acceptable on the market, though there are big differences in the prices received for different species. High-value species, including shrimp, may be worth ten or twenty times the trash fish (including small individuals of desirable species), which is most often used only for duck food or fish meal.

The problem of the incidental catches of fish in fisheries directed primarily at shrimp was raised as a special and urgent problem of multispecies management. The situation could be eased if it were possible to separate the exploitation of fish and shrimp. In many fisheries the high value of the shrimp catch effectively subsidizes the overexploitation of fish. However, in others it is only the presence of shrimp in the catch that allows for an economically viable exploitation of the fish.

It was pointed out that while changes in species composition can be expected as a result of exploitation, these need not always be in an undesirable direction. On the northwest shelf of Australia for example, it seems that the Taiwanese fishermen expected, and planned for, a shift in species composition towards the more valuable shrimp and squid.

It was agreed that the pattern of fishing can be as important as its total amount. Thus, for example, while there is generally little evidence from multispecies stocks of any decrease in total catch, even under extremely high levels of fishing, a drop could easily occur if fishing concentrated on very small fish. In Lake Victoria catches have fallen apparently because of intense seining of the breeding areas of tilapias. In this connection there was some discussion of the generally accepted rule that for most species of fish the smaller individuals are found inshore, and the larger ones in deeper water. While this rule is generally true, the distribution of juveniles of some of the tropical demersal species is unknown. On the northwest shelf of Australia, for example, a careful sampling of the inshore areas did not yield any juveniles of the main commercial species.

The question of the stability of unexploited stock was raised. It was stressed that there is no reason to expect that, in the absence of any interference by man, the stock should remain constant in total abundance or species composition. Some fluctuation may be expected and these could make interpretation of changes after the beginning of fishing difficult, especially since there is an obvious tendency for a fishery on a given stock or species to develop when it is particularly abundant.

The use of catch quotas as a method of controlling a fishery was raised. It was pointed out that though quotas are widely used in temperate waters, they have several disadvantages when applied to tropical areas. Few countries are in a position to enforce them; calculation of the appropriate quota for a short lived species whose abundance might be rapidly changing is difficult, and more especially single-species quotas in a multispecies fishery could lead to a highly undesirable situation; for example a quota on shrimp could lead to severe overexploitation of many fish species.

While there is some pessimism about the immediate possibilities of implementing and enforcing any restrictive management measure in most tropical areas, there are still many other opportunities for influencing the way in which the fisheries in these areas develop, especially if action is taken before a large overcapacity has grown up.

The Realities of Fishery Management in the Southeast Asian Region

JOHN C. MARR
P.O. Box 772 Rancho California
Temecula, California 92390
U.S.A.

MARR, J.C. 1982. The realities of fishery management in the Southeast Asian region, p. 299-307. *In* Pauly, D. and G.I. Murphy (eds.) Theory and management of tropical fisheries. ICLARM Conference Proceedings 9, 360 p. International Center for Living Aquatic Resources Management, Manila, Philippines and Division of Fisheries Research, Commonwealth Scientific and Industrial Research Organisation, Cronulla, Australia.

Abstract

Fishery management, particularly in Southeast Asia, is concerned primarily with people-problems rather than resource-problems. It cannot be successful unless viewed in the context of rural development. Conventional constraints to management include lack of theory, lack of data, lack of trained personnel, lack of institutional infrastructure, lack of physical infrastructure and gear conflicts. Unconventional constraints include large numbers of artisanal fishermen distributed over exceedingly long coastlines, responsibilities diffused throughout government, intra-government conflicts, international competition for the same resources and a close identity with nature leading to a fatalistic view about the course of events. Requirements for a successful fishery management system are listed. In spite of constraints and requirements, there are nevertheless reasons for optimism including the establishment of EEZs and examples, of which three are cited, of the ability and willingness of responsible individuals in government to identify critical problems and to take difficult decisions/actions with respect to such problems.

Introduction

Fishery science, and fishery management based upon the findings of fishery science, have been conceived and have evolved mainly in the context of economically developed countries, in which fishery development has been carried out almost entirely by the private sector. As such, the approaches have been resource-oriented, usually concerned with the preservation of the resource and the production of the maximum biological yield from it. They have not, at least until comparatively recently, been fishery-oriented and certainly have not been people-oriented.

However, about 25 years ago fisheries were discovered by economists and then, with some time lag to be sure, the concepts of economics as they relate to fisheries were discovered by fishery scientists (then primarily biologists) and to some extent by fishery managers. The ensuing evolution away from a strictly resource-orientation has made it possible for Crutchfield (1980) to recently state that: "The basic problem of rational utilization of marine fisheries is not scientific ignorance. The real weakness lies in our institutional mechanisms for getting something done, and for making the regulated fishing industry itself a part of the analytical and decision-making process . . . ". Also, this evolution made it possible for one of the major economically developed fishing countries to promulgate a comprehensive law relating to the fishery resources of the 200-mile extended economic zone (EEZ) established by that law (U.S.A. 1976). The law provides a mechanism for producing fishery management plans concerned with "optimum yield" and states as one of its purposes "to take into account the social and economic needs of the States".

While in a broad sense fishery management is concerned with people-problems, this is true in a more particular sense in Southeast Asia where about 50% of the fishery production arises from the small-scale fisheries carried out by some 3.5 million artisanal fishermen. With an average family size of six, this means that some 21 million people are directly dependent for their livelihood upon the income produced by such fishermen (Smith 1979). In this context, the objective of fishery management is clearly to improve the standard of living of the fishermen and their families. As Smith et al. (1980) have pointed out, small-scale " . . . fisheries must be viewed as part of the larger, particularly rural, economy." Unfortunately, the artisanal fishing communities are rarely effectively included in rural development schemes. But, as Panayotou (1979), and others, have pointed out, "the solution to the problems of resource management are to be found outside of the resource sectors." Thus, fishery management cannot be expected to be successful in improving the quality of life for fishermen unless it is imbedded in the broader matrix of rural development. What follows should be viewed in this context.

The extent to which effective fishery management can be included in rural development in Southeast Asian countries depends in part upon government perception of the need for technical advice on fishery management. Obviously, this perception will vary according to the part of, and level of, government at which the question is raised. The technocrats–fishery biologists and fishery economists—undoubtedly perceive a great need for such advice; planners, who are concerned primarily with "growth", may feel less need; and those in government who make political decisions and must be responsive to public reaction, may recognize even less need since short-term considerations most frequently prevail at this level. Clearly, this suggests an important constraint to effective management.

Conventional Constraints

What may be considered conventional constraints to fishery management include lack of theory, lack of data, lack of trained personnel, lack of institutional infrastructure, lack of physical infrastructure and gear conflicts. These constraints have long been recognized and require only brief mention here.

Lack of theory relates to the fact that the modern theories of population dynamics (which must be taken into account in the management decision-making process) are largely based upon single-species fisheries of high latitudes in which the species are usually long-lived, whereas the fisheries of Southeast Asia are multispecies fisheries based upon a large number of species which are usually short-lived. However, in recent years it has been recognized by Andersen and Ursin (1977) and others that single-species are part of a complex multispecies ecosystem; the present volume is evidence of increasing interest in the "multispecies problem."

Generally speaking, in the Southeast Asian countries there do not exist long series of catch, effort and age data for the species which are the objects of fisheries such as exist for many higher latitude species (see Simpson, this vol.). This lack tends to be a two-sided problem; there are no data to examine in the light of existing theory and there are no data to analyze for the development of theory relevant to multispecies fisheries. However, Pauly (1979) has suggested that there may, in fact, be more sources of data than are generally appreciated and, further, that there may be appropriate methodology for at least their preliminary analysis.

The lack of adequate numbers of trained personnel is frequently commented upon and the provision of training facilities urged. This is a somewhat simplistic approach, to say the least. In most Southeast Asian countries there are, in fact, many individuals with training in fishery science who are not effectively involved in fisheries owing to inadequate government pay scales. Such individuals may leave government fishery service altogether or may remain nominally active while actually devoting most of their time to more profitable activities.

In many cases the institutional infrastructure necessary to supply the essential scientific information, identify the broad problems of fishery management in an appropriate context, take the necessary policy decisions and put management plans effectively in place are simply not present. However, with increased interest in fisheries, increased awareness of resource limitations and the virtually universal establishment of 200-mile EEZs, there is some reason for optimism in this regard (see below).

While the establishment of EEZs makes fishery management feasible by establishing national resource ownership, and thus making it possible to deal with the problem of unlimited access, it has at the same time created enforcement problems of unanticipated magnitude. Countries all over the world, including those in Southeast Asia, simply do not have the physical infrastructure with which to effectively enforce management regimes.

Gear conflicts are almost protean components of fisheries. Usually, but not always, they involve two or more different kinds of gear taking not only the same species but the same species at essentially the same life-history stage. However, the major gear conflicts in Southeast Asia often involve the same species at somewhat different life-history stages. These conflicts involve, on the one hand, the "inshore" artisanal fishermen using beach seines, traps, gill-nets, etc. and, on the other hand, the "offshore" commercial fishermen most commonly using trawls. This conflict is a major problem, since the artisanal fisheries involve so many people. Draconian measures may be required to deal effectively with this problem (Sardjono 1980), but it is most commonly attacked by measures requiring the trawlers to stay outside of waters of a certain depth or a specified distance from shore (Khoo 1980; Pathansali and Jothy 1974). The effectiveness of the latter measures as far as the resources are concerned is questionable, since the species caught commonly occur on the inshore grounds as juveniles and on the offshore grounds as adults. Such measures may be effective in eliminating or at least reducing physical confrontation between inshore and offshore fishermen (Goh 1976), but even this often fails.

Unconventional Constraints

What may be characterized as unconventional constraints include some which are geographic/demographic, institutional, international and cultural.

Geographic/demographic constraints are especially important in Southeast Asia, particularly in the archipelagic States of Indonesia and the Philippines. Exceptionally long coastlines with very large numbers of artisanal fishermen distributed along them pose management problems of overwhelming magnitude.

Among the institutional constraints, two in particular should be noted. First, the responsibility for the various components of fishery management may be widely spread throughout government, so that it may be difficult or impossible to put management plans into effect (Marr et al. 1976). Second, fishery departments and other departments frequently do not communicate very well with each other at best, and may actually be in conflict, at worst. One responsible official has told me that in his country the planning department will not accept any resource information other than that which would indicate the possibility of substantial increases in yield. Another official in another country stated that the foreign ministry frequently made decisions on fishery matters without consulting the fishery department.

In Southeast Asia an almost universal international constraint has been the planned expansion of fisheries by a number of countries based upon the same resources in the South China Sea (Marr 1976). Even with the establishment of EEZs, as the result of which there is no remaining high seas area in the South China Sea, competition for these resources will remain an important problem because some migratory species will occur in the EEZs of two or more countries and because the EEZ boundaries will not be respected by some fishermen. An official of yet another country has ventured the view that some governments, including his own, would not be willing to enter into international management systems because of their unspoken intent that their nationals fish in the EEZs of other countries.

Finally, there are cultural constraints to fishery management. One of these is possible loss of face. The last individual mentioned above also said that it would be difficult for his government to participate in an international management body, since this would have the consequence of requiring a public statement or admission that the requisite data for management had not been collected. A more widespread constraint is the sense of identity with nature. This leads to what might be termed a "fatalistic" view about what happens including what happens to fishery resources, rather than to the more western view that the course of events can be altered by human actions.

Management Requirements and Future Outlook

A successful fishery management system in Southeast Asia would have a number of requirements, most of which appear to be so obvious as to be truisms. Yet such a system is not yet in place, in part because some of the elements are frequently overlooked and in part because it is easier by several orders of magnitude to specify such a system than it is to put it into operation. In any case the requirements for a successful fishery management system include the following:

— Explicit establishment of specific management goals. Frequently goals are selected tacitly rather than explicitly and frequently conflicting or opposing goals are selected, i.e., to maximize biological yield and to maximize employment in the capture sector.
— The establishment of appropriate institutional and physical infrastructure.
— The recognition that the problems of coastal small-scale fishing communities are part of the broader problems of rural development.
— Realistic planning taking into account the finite, but renewable, nature of the resources.
— Management on a logical basis that takes into account the nature of multispecies fisheries and the general lack of conventional data.
— Management on a basis that is culturally acceptable, as well as physically possible.

As a result of the many biological and socioeconomic constraints to effective fishery management in Southeast Asia and of the requirements for an effective management system, it would be easy to be pessimistic about the future course of events. But, on the contrary, I am optimistic. Before citing examples of my reasons for optimism, I would make the point that no matter how great our interest, concern and involvement, we who are not Southeast Asians should not expect Southeast Asians to adopt management measures that would be appropriate in our own countries. I believe that the Southeast Asian countries will find unique solutions to their unique problems.

The reasons for optimism include, first of all, the declarations of the EEZs which establish national resource ownership and therefore the possibility of effective management. Secondly, there are examples, of which three will be mentioned, of the ability and willingness of those in government to identify critical problems and to take difficult decisions and actions with respect to such problems.

The establishment of the 200-mile EEZs found Thailand with a large distant-water trawling fleet which supplied in the order of 60% (660,000 t) of the Thai "domestic" landings by virtue of catches made in waters now included in the EEZs of other countries. The effects of the loss of such a large proportion of the landings obviously could have substantial political effects beyond the immediate losses to producers and consumers. Government joined with the fishing industry (in this case, the owners of the trawlers) in successful attempts to develop cooperative ventures between the Thai companies and entities (government or private) in neighboring and even some more distant countries, thus avoiding at least some of the loss of protein in the Thai market, economic disaster for vessel owners and increased fishing pressure in Thai waters of the Gulf of Thailand, already seriously overfished. In addition, the government, with bilateral and development bank assistance, is attempting to stimulate the further development of aquaculture as a source of additional protein, employment, investment opportunity and export products.

In Indonesia the conflict between the coastal artisanal fishermen and the trawlers around densely populated Java and the southeastern end of Sumatra has been an important, and at times violent, problem since the introduction of trawling. Attempts were made to solve the problem, or at least reduce it, by reserving an inshore zone for the sole use of the artisanal fishermen. This regime was difficult to enforce and was not effective. Therefore, trawling around Java has been prohibited completely as of September 1980 and the southeastern end of Sumatra as of 1 January 1981 (Sardjono 1980), reserving the entire resource for the use of the artisanal fishermen.

A common approach to small-scale fishery development (an aspect of management) has been to attempt to increase total catches and catch rates by the mechanization of boats and introduction of synthetic nets. This approach in its most simplistic form ignored possible limitations to growth imposed by the magnitude of the resource, i.e., made the tacit assumption that the resource was infinite. This has been true in the Philippines as it has been elsewhere. However, there is now increasing recognition in the Philippines of the fact that many of the coastal resources are already overfished and that additional/alternative sources of income are necessary if the quality of life of the artisanal fishermen is to be improved. As Smith et al. (1980) have pointed out, a shift in attitude from "development" to "management" is taking place.

References

Andersen, K.P. and E. Ursin. 1977. A multispecies extension to the Beverton and Holt theory of fishing, with accounts of phosphorus circulation and primary production. Medd. Dan. Fisk. Havunders. (New Ser.) 7: 319-435.

Crutchfield, J.A. 1980. Fisheries management: the end of the beginning. Aust. Fish. 39(4): 42.

Goh, C.T. 1976. The fishing conflict in Penang and Perak: personal memoir. Kajian Ekon. Malays. 13(1/2): 17-25.

Khoo, K.H. 1980. Implementation of regulations for domestic fishermen, p. 49-59. *In* F.T. Christy, Jr. (ed.) Law of the Sea: problems of conflict and management of fisheries in Southeast Asia. Proceedings of the ICLARM/ISEAS Workshop on the Law of the Sea. ICLARM Conference Proceedings No. 2, 68 p. International Center for Living Aquatic Resources Management, Manila, and Institute for Southeast Asian Studies, Singapore.

Marr, J.C. 1976. Fishery and resource management in Southeast Asia. RFF/PISFA Paper no. 7, 62 p. Resources for the Future, Washington, D.C.

Marr, J.C., G. Campleman and W.R. Murdoch. 1976. Kingdom of Thailand: an analysis of the present, and recommendations for future, fishery development and management policies, programmes and institutional arrangements. SCS/76/WP/45, 182 p. South China Sea Fisheries Development and Coordinating Programme, Manila.

Panayotou, T. 1979. Natural resource management for economic development: lessons from Southeast Asian experience with special reference to Thailand. Staff Paper no. 34, 61 p. Dep. Agric. Econ., Kasetsart University, Bangkok.

Pathansali, D. and A.A. Jothy. 1974. Fish stock management policies—the situation in Malaysia. IPFC/74/SYM21, 6 p. FAO, Bangkok.

Pauly, D. 1979. Theory and management of tropical multispecies stocks: a review, with emphasis on the Southeast Asian demersal fisheries. ICLARM Studies and Reviews No. 1, 35 p. International Center for Living Aquatic Resources Management, Manila.

Sardjono. I. 1980. Trawlers banned in Indonesia. ICLARM Newsletter 3(4): 3.
Smith, I.R. 1979. A research framework for traditional fisheries. ICLARM Studies and Reviews No. 2, 40 p. International Center for Living Aquatic Resources Management, Manila.
Smith, I.R., M.Y. Puzon and C.N. Vidal-Libunao. 1980. Philippine municipal fisheries: a review of resources, technology and socioeconomics. ICLARM Studies and Reviews No. 4, 87 p. International Center for Living Aquatic Resources Management, Manila and Fishery Industry Development Council, Manila.
U.S.A. 1976. House of Representatives, 94th Congress, 2d session. Fishery conservation and management act of 1976: report of the Committee of Conference on H.R. 200, to provide for the conservation and management of fisheries, and for other purposes. U.S. Government Printing Office, Washington, D.C.

Discussion of Mr. Marr's Paper

The initial reaction by the group to Mr. Marr's paper was a pessimistic one. One discussant observed he was struck by the author's ability to maintain his optimism in light of the list of conventional and unconventional constraints. The question was asked whether it is too late to make the required changes or whether our efforts are too small to make any significant impact. Considering the rate of increase in fishing effort in the region over the last decade, the slight possible improvement in constraints to good management appears small indeed. It was suggested that the situation could be described as "not enough, early enough" rather than "too little, too late". The complex of social, political, economic, ethical and bureaucratic constraints is of such dimensions as to completely overshadow the biological considerations and to discourage research on biological aspects.

Several discussants took a more optimistic viewpoint, however, noting that recognition of the need for change and the need for management are beginning to be recognized and that some examples of effective management steps had been identified in the paper. Regarding the "fatalistic" philosophy in Asia, the observation was made that the paper makes it sound as though a one-way flow of philosophical understanding was called for. Rather, an exchange of points of view or a two-way flow seems more appropriate. The people in the developed countries who have these notions of highly sophisticated management are going to have to get used to some of the "fatalistic" attitudes in the East. On the other hand, over the last decade or two there has been a dramatic dissemination of the technological attitudes of the West by people in the East resulting in marked changes in knowledge and attitudes.

A discussant took exception to the term "fatalistic" used by Marr. Its frequent use with reference to Asian peoples implies they won't change when in fact they are changing and will have to continue to change. He stated that fisheries management changes are only a part of the changes required by society as a whole. The biological problems are not as large as the social problems and good fisheries management requires much more than good biological data. It is less a question of knowing enough than one of being able to use what we already know. Others took exception to the assumption that we will be able to change many of the social factors influencing fisheries. Another participant noted the existing biological problems were very big ones and should not be minimized.

Exception was taken to the feeling that we should be pessimistic if fishery management measures cannot be implemented during the next few months or years. The North Sea example was cited where 30 years ago it was fairly clear that fishing mortality on cod should be reduced by 50% and that mesh size should be increased from 50 mm to 90 mm. Mortality rates were gradually reduced over a period of 20 years and mesh sizes have been changed from 50 to 70 mm (and in some cases are approaching 90 mm) over 30 years. We can be optimistic if our time scale is appropriate, a reasonable time scale being of decades rather than of months. A comparison was made of the situation in Southeast Asia in 1950 and today. Important progress has been made during this period including training of scientists and the building of research capabilities in several countries.

Progress has been made on all the constraints. The more serious constraints are the political and social will on the part of fisheries management and fishermen. However, even there a lot of progress has been made. In Thailand, for example, fishermen will not anymore quarrel with the fact that too many trawlers exist; they only quarrel with the idea of giving up their own fishing rights. Thinking need only progress slightly beyond this to permit beneficial action.

An additional voice of optimism was sounded by a participant who noted considerable advances in a number of aspects. Administrators, Fishery Director Generals and others have a much better approach to and understanding of the subject now than they did 5 years ago. The general understanding of the scientists is also steadily improving. There is a recognition in many countries that rural problems, including fisheries, have to be faced in a multifaceted way. There is progress but significant improvements will take a long time.

A question was raised concerning the statement, "in this context the objective of fishery management is clearly to improve the standard of living of the fishermen and their families". Is this the objective of fishery management? An example was cited in which fishermen were being trained for other trades in a "fisheries" high school since the only means of achieving a long-term improvement of the standard of living of fishermen is to reduce their total number.

Discussion of economic versus total yield or social objectives led to no resolution of this point. Even an optimum species mix cannot be defined in a biological sense at present and decisions will have to be made on the basis of economics and local preferences.

Mention was made of the assumption reiterated by Marr that artisanal fishermen catch the young of fish harvested by offshore trawlers. It was stated that this may not always be the case and instances were cited where the opposite is true.

A question was raised concerning whether or not joint fishing ventures are going to help the overall fishery problems. Examples were given of recent joint ventures and international sharing of resources that are not working well primarily because of dishonest practices. It was suggested that only limited benefits can be expected from international cooperation in fishing and that the real problems of overfishing will not be solved by these activities.

Discussion returned to the point that some positive action has been taken in Indonesia with the trawler ban there. The effectiveness of enforcement of this ban was questioned and other examples of enforcement problems were cited. The general impression however, was that Indonesia was likely to enforce this action.

As a follow-on to the trawling ban discussion, the point was made that we don't know from a biological standpoint what the impact of reduced trawling will be. Biologists cannot say with certainty that the ban is justified because it is not known whether overfished tropical stocks will return to their original state when fishing is reduced.* At what point in the exploitation of multispecies stocks are irreversible changes made? It is important that biologists be able to answer this question, but at present they cannot. Biologists are continually confronted with surprising events even in thoroughly studied situations, such as the North Sea. Close monitoring of the Indonesian fisheries will be of great importance.

The history of fishery management around the world has been one of waiting until things have gone seriously wrong, then realizing the need for management. Experience in fisheries has been that management is easy only if you are putting the brakes on something that is moving into deep trouble. Generally, the Southeast Asian countries feel they are not yet in serious trouble.

It was noted that one type of activity is never legitimate, that is activity that results in fish-habitat destruction. Examples are use of dynamite, crushing of corals or destruction of mangroves. Participants noted, however, that the value of mangroves to fisheries may have been overestimated in the past and that overstating the case for mangrove preservation could be damaging to fisheries interests in the long run.

Attention was called to the difference between making management decisions and giving scientific advice on biological aspects of fishery management. Marr's paper paints a realistic picture of how management decisions are made, and even though we may not like the process, that is how things happen. The scientist's business is giving scientific advice, not making management decisions. In the scientific arena, there is room for optimism because we are making progress. As long as the advice we provide is improving we are doing our part of the job. A supplement to this statement is that a big responsibility of scientists is to educate administrators.

A final point made was that many of the trained fishery scientists in developing countries are not effectively involved in fisheries for economic, bureaucratic or other reasons. A common problem is that their incomes from government jobs are inadequate and they are forced to expend their energies on non-fisheries employment.

---

*Editorial note: since the workshop was held, a brief paper was published which suggests that stocks of the type discussed here may recover remarkably fast when fishing pressure is reduced. See: Saeger, J. 1981. Do trawling bans work in tropical waters? ICLARM Newsletter 4(1): 3-4.

Directions for Future Research in Tropical Multispecies Fisheries

P. A. LARKIN
(Workshop Chairman)
Institute of Animal Resource Ecology
University of British Columbia
Vancouver, Canada V6T 1W5

LARKIN, P.A. 1982. Directions for future research in tropical multispecies fisheries, p. 309-328. *In* Pauly, D. and G.I. Murphy (eds.) Theory and management of tropical fisheries. ICLARM Conference Proceedings 9, 360 p. International Center for Living Aquatic Resources Management, Manila, Philippines and Division of Fisheries Research, Commonwealth Scientific and Industrial Research Organisation, Cronulla, Australia.

Abstract

This paper summarizes the deliberations and conclusions of participants at the ICLARM/CSIRO workshop on "The Theory and Management of Tropical Multispecies Stocks," held on 12-21 January 1981 at Cronulla, Australia. The subjects reviewed include models for research and assessment and their data requirements; understanding of multispecies fisheries; and characteristics of contemporary fisheries management in Southeast Asia. Recommendations for future research are given.

Introduction

After presentation and discussion of the prepared papers, the participants formed three working groups concerned with the status of:

—Models for research and assessment and their data requirements.
—Understanding of multispecies fisheries systems.
—Characteristics of contemporary fisheries management in Southeast Asia.

The reports of these three working groups are presented below. With the completion of the reports, participants then formed three new working groups to prepare summary comments and recommendations. Summaries of these deliberations follow the reports.

Status Reports

MODELS FOR RESEARCH AND ASSESSMENT AND THEIR DATA REQUIREMENTS

Three classes of models are potentially useful for research and assessment of tropical multispecies fisheries, and are differentiated by their treatment of the multispecies nature of marine ecosystems. The first class comprises those models containing no explicit recognition of interspecies interaction, and includes single-species models, the total biomass Schaefer model, and various statistical models. In the second class are those models that allow for such interactions, but in a non-mechanistic manner. These models are typically framed as sets of differential equations, and have been termed "metaphorical" or "phenomenological" models. The third class comprises those models that attempt a mechanistic description of processes and interactions in the ecosystem.

Single-species models

The tasks of modelling and management of multispecies fisheries can be greatly simplified if the multispecies assemblage can be treated as a set of non-interacting single species. A major advantage of this approach is the existence of an established body of theory and methods for dealing with single species. It is recognized that both the data and existing biological theory indicate that interspecies interactions occur; but it is not clear that interactions exert sufficient influence on and control of the system that ignoring them (as in the single-species approach) would lead to erroneous conclusions. On the other hand, single-species assessments may be substantially in error if species interactions are substantial in reality.

The data (and the theory) will admit several interpretations. For example, the fact that some species in the Gulf of Thailand apparently increased in abundance with exploitation is, on the face of it, inexplicable by single-species modelling. However, the fact that only research vessel data and not total catch are known, and uncertainty that the system was in an unexploited equilibrium at the beginning of the data series, make it impossible to reject the single-species approach. The group also noted that environmental changes may further complicate interpretation.

In view of the pragmatic advantages of single-species models, the extent to which the single-species approach is valid or invalid warrants further attention. This requires an experimental approach, rather than re-analysis of existing data, with data being gathered specifically to test single-species assumptions. An example is an experiment in which a single species in an assemblage is reduced severely in abundance, and the abundance of all species then monitored.

The data requirements of single-species approaches depend on the level of complexity of the models. Data needed could range from simple series of catch and effort data to estimates of each of the biological parameters of growth, mortality and catchability. In this context, the group felt that the methods described by Pauly (this vol.) could be of great practical use in the Southeast Asian region. However, it was agreed that their use requires special caution, and that further work, including sensitivity analyses, on these methods was needed.

Total biomass Schaefer model

The total biomass Schaefer model, in which all species are lumped together and essentially then treated as a single species, has often been applied in the tropics (Pauly 1979). The model is ostensibly empirical, with no theoretical basis. However, the existence of interactions between species is not explicitly excluded in a lumped biomass model, and "conservation of mass" arguments might indicate some basis for such a model, provided *all* the biomass in the system was included. Thus, the total biomass may not change radically with exploitation, but the species mix may, possibly with disastrous effects on the viability of an established fishery for a particular species or group of species.

The total biomass model has data requirements of only total biomass and effort. Simpson (this vol.) indicates that even these data, as presently available, are somewhat unreliable. In testing the models, it is doubtful whether data currently exist that would allow examination of whether the biomass trajectory, as total effort decreases from an historical maximum, is consistent with the trajectory followed when effort earlier increased to the maximum. In such a test, it is essential that the system be viewed as dynamic rather than at equilibrium. Again, an experimental approach offers the best possibility for testing the models. For example, the effects of removing a catch of a certain biomass should be the same regardless of the species composition of the catch.

Statistical models

Strictly correlative statistical models in which, for example, the catch next year is estimated from a time series of historical catches, may lump biological, environmental, and economic data. Several examples can be cited in which the estimation of stock sizes and catches may be best accomplished purely through correlations of catches with important environmental influences (e.g., monsoons). Changes in the relative prices of fish species and other marketing variables also could have a substantial effect on next season's catches. The purely statistical approach, possibly more than any other, depends on the availability of reliable data.

Metaphorical models are theoretically superior to those discussed above, in that they contain an explicit account of interactions, although it is still preferable to model such interactions mechanistically. There is an urgent need for metaphorical models that, wherever possible, are constructed in such a way that the parameters are biologically meaningful, even if they are not readily measurable. Otherwise, modellers will be unable to draw upon available biological expertise. These models are of a nature that their parameters can normally be estimated only by fitting them to a time series of relative abundance and catch data. Such estimation methods also appear to be the only means of testing the adequacy of the models. For example, they could be fitted to one half of the data series, and then the other half could be used for testing.

The conclusions reached by Kirkwood (this vol.) that these models provide few generally-applicable management paradigms, seem sound. Nevertheless, these models are at least theoretically superior to the models in the first class. Models incorporating current understanding of biological mechanisms are to be preferred.

Discussions on the third group of models were set in a background in which it was acknowledged that existing theory and data are inadequate for their practical implementation and use for stock assessments. The level of complexity and detail of every biological process should be specifically included, but models should make best use not only of the knowledge but also of the data that are, or are likely to be, available.

Another aspect that should indicate the appropriate level of detail is the available "levers" and objectives of the modelling exercise. These levers include those management actions that are available and feasible for altering the state of the system. In turn, these levers partly dictate the objectives of the modelling and the answers the model is being developed to provide. Careful consideration of these points should clearly indicate the appropriate level of complexity of the model, as well as the possible need to include other than biotic variables (such as those in the environment), economics or social considerations.

Appropriate geographic and time scales must be identified as part of the development of a useful model. Development and description of a model should incorporate suitably comprehensive sensitivity analyses, along the lines described by Majkowski (this vol.). An essential feature of the presentation of results from the model is the compression of the results into a form which the end users could readily assimilate.

The second and third class of models can also be distinguished by their major data requirements. For the second group a long, reliable, time series of catch and effort and species composition statistics is required. By contrast, a majority of the data requirements for the third class could be met by an intensive short-term research effort. The apparent lack of reliability of official statistics in the Southeast Asian region is a matter for concern, and improvement of the quality of these data is essential. Because the underlying mechanisms in the third class are being modelled directly, rather than implicitly as in the second class, the possibility of establishing reliable parameter estimates in a relatively short time frame represents an additional advantage of the third class over its competitors.

Despite our inability to construct models of the third class at present, it is a practically feasible task; a possible approach is the workshop technique currently used successfully at the University of British Columbia and other centers. This approach requires identification and analysis of data from a well-defined area, followed by a workshop with participants including biologists, modellers, administrators, fishermen, economists, etc.

Understanding of Multispecies Fisheries Systems

Differences between ecosystems

There is a range of fish communities which support tropical fisheries. These communities and fisheries differ substantially in a number of respects, such that quite different management procedures may need to be applied to each. For example, tropical lakes, ponds and some coastal bays and estuaries support small-scale fisheries. Techniques of harvesting are easily manipulated to modify the nature of the catch, and manipulation of such systems is feasible as a technique for research investigations upon them. In addition, the fishermen or the social group involved can be expected to show a considerable sense of responsibility for the fish community they are harvesting. In some cases, there may be private ownership of the harvested community.

By contrast, the fish stocks which support extensive, tropical trawl fisheries extend over large areas, and trawling is a method which operates on a broad spatial scale. Manipulation of a stock for research purposes is also very difficult. The fishermen, and often the nations to which they belong, have a lesser sense of responsibility for the stock they are harvesting. Coral reef and inshore small-scale fisheries may operate on an intermediate scale.

Functions of stock assessment

Stock assessment in the broad sense is carried out to provide management advice. It seeks to identify the stock present, to estimate potential yields, and to recommend strategies for approaching these yields. As such, it has both short- and long-term goals: providing immediate advice as to whether a stock is close to full exploitation, and proposing strategies for manipulating the fishery so as to achieve desired catch characteristics. Policies designed to achieve a particular species mix in the catch, or designed to maximize yield of a particular component, are examples of these long-term goals.

There will be economic and other constraints on the kinds of fishing policies that are adopted; the decisions must, and will, be made, whether or not information on stock assessment is available. Accurate stock assessment requires reliable data concerning the stock, but depends upon current understandng of the nature and functioning of biological communities on the one hand, and yield responses of fisheries on the other. In the working group's view, understanding of both these matters is limited.

Nature and functioning of biological communities

The sizes of species populations can be determined through competitive interactions with other coexisting species, through the effects of various environmental factors, such as weather. Sizes of species populations may be closely regulated around an equilibrium condition, or less tightly regulated. Despite the recent widespread acceptance among ecologists of the view that communities are usually equilibrial and competitively structured, data supporting this view are not strong. New evidence is being obtained in a wide variety of communities in favor of a non-equilibrial view, and of regulation by processes other than competition (see Connell 1978; Wiens 1976; Sainsbury, this vol. and Sale, this vol.).

While there are many examples in the literature which show that fish species use subtly different food and other resources, examples sufficiently detailed to demonstrate that these differences in resource use are a response to interspecific competition, are few. Among studies of fish communities, only the work of Werner and Hall (1976, 1977) on three species of centrarchid fish in ponds, is really convincing (Sale 1979). It is likely that ecologists have tended to exaggerate the role of competition in structuring communities.

Somewhat better evidence exists to demonstrate the importance of predatory responses in determining fish community structure (see Jones, this vol.). Planktivorous fish have a pronounced effect on the composition of zooplankton communities in lakes (Brooks and Dodson 1965; Hall et al. 1976). Herbivorous and benthic-feeding carnivorous fish on reefs play an important role in determining the nature of algal and benthic invertebrate communities (Sale 1980, this vol.). Evidence from temperate lakes and pond culture indicates that actions of predatory fish can determine the size and structure of prey-fish communities in these simple systems (Swingle 1950). Marten (1979a) interprets some of the changes in characteristics of the catch from Lake Victoria as due to changes in the effect of predators on prey species caused by reduction in predator abundance through fishing.

The variability in recruitment common in fish populations (see Murphy, this vol.) may indicate an important role of environmental factors other than competition and predation in determining species abundance.

Evidence that the structure of virgin multispecies stocks remains constant through time appears not to have been collected. While it seems likely that in many of these stocks a persistent association of species exists, with some of them continually dominating the assemblage, even this is not established. Such persistence is not necessarily evidence of a stable equilibrium in the assemblage, nor that species composition and relative abundance are resistant to, and able to recover from, disruption (see discussion of Sale's paper, this vol.). Small-scale studies, both spatially and temporally, of coral reef assemblages suggest that these are non-equilibrial, though persistent systems (Sale, this vol.). Available data from other systems are from fished stocks.

To summarize, it would be unwise to approach stock assessment in tropical fisheries under the impression that there are well-defined ecological principles governing the composition and relative abundance of species in communities. Interspecific predatory or competitive interactions may be important in determining structure of tropical fish communities or they may not. The species composition of these communities may or may not be stable in the virgin state. Finally, there is as yet no ecological evidence indicating that tropical systems are organized in a fundamentally different way than are temperate communities, although this possibility must be borne in mind as more data become available.

Yield responses of multispecies fisheries

Catch statistics appear to demonstrate two kinds of changes as fishing effort is continued or increased:

Contrary to expectations from studies of single species, there are examples of heavily exploited multispecies stocks in which yield has continued to increase, or has not necessarily declined in association with changes in species composition. These examples include Lake Victoria, Africa (Marten 1979b) and Lake Michigan, U.S.A. (Smith 1968; Lawrie 1978); and offshore fisheries in the Gulf of Thailand (Pauly 1979), Sunda Shelf (Buzeta 1978), Malacca Strait (SCS 1976), Gulf of Guinea (Domain 1979), western Atlantic (Brown et al. 1976) and the North Sea (Hempel 1978). In reef fisheries in Jamaica (Munro 1977) and east Africa (FAO 1979), data sampled from various sites subject to different fishing effort possibly indicate a similar pattern. (See Figs. 6 and 9 in Marten and Polovina, this vol.) We stress this as a strictly empirical observation, that the failure of the total yield to decline may depend on the nature of the fishing effort used, and that a maximum total catch may still fluctuate in value as its composition changes.

As effort is sustained or increased, changes occur in the relative abundance of species in the catch. These changes, which are usually interpreted as induced by the fishery, may make the catch of substantially different value to the fishermen. Concern is expressed when valuable species drop out and are replaced by less valuable ones. However, some changes may lead to enhanced value of the catch. For example, in some tropical trawl fisheries, squid have replaced fish of lesser value.

While reasonable estimates of the maximum total yield can be derived from catch and effort data once a reasonable level of effort has been reached, causes of the changes in composition can at present only be guessed. For example, in the Gulf of Thailand trawl fishery, leiognathids were a dominant and ecologically important component of initial catches. As effort increased, they became rare and were replaced by flatfish and trash fish of lesser value. Pauly (1979) argues that the fishery-reduced stocks of leiognathids had competitively prevented the flatfish from becoming abundant in the virgin stock. Released from competition, and with less predation on their eggs and larvae, the flatfish became commoner as the fishery continued.

However, there are at least two other plausible explanations of the observations:

— Leiognathids require some bottom structure which was destroyed by the trawling process. This habitat change simultaneously made the area less suitable for leiognathids and more suitable for flatfish.

— Leiognathids were in the process (for reasons unknown) of becoming rarer in the Gulf of Thailand, and flatfish were independently (or, as a result) becoming commoner during the early years of the fishery. The catch data documented a trend that was unrelated to the fishery.

These explanations can be more generally stated as:

— A true fishery-induced change in composition of the stock because of interactions among species.

— A change in abundance of one or more species due to modification of habitat by the fishery method used.

— A change in composition of the stock independent of the fishery.

All of these explanations are unlikely to be correct in any one case, but they are not mutually exclusive and may apply to a given example. Importantly, available catch data do not allow us to discriminate among them. We are not yet in the position of knowing whether a multispecies fishery will always directly induce a change in composition of the stock, or how to manipulate a fishery to effect changes in desired directions. Pond-culture data (Swingle 1950) indicate that such manipulations are possible, at least in some simple, closed systems. Theory advises that high, though not necessarily more valuable, yields should be obtained by fishing down to lower trophic levels. Whether this is true for complex, tropical marine communities is largely a matter of conjecture.

Concluding comments

The lack of clearly-established general principles of community organization suggests that short-term management advice at present can be best derived from empirical analyses of catch and catch-per-effort data. Some faith can be placed in the belief that the total catch will not decline if effort increases, although catch per unit effort will decline and the species mix caught may well change markedly.

Rather than put additional effort into complex and time-consuming methods of stock assessment, effort should be channelled into research towards understanding the mechanisms which determine the species composition of multispecies stocks subjected to varying levels of fishing. Where possible, this research might take advantage of the fishery as a way of manipulating the system under study. This should not be done in an *ad hoc* way. A useful experiment requires accurate information on the nature of the fishing effort applied to a site, and the use of a sampling method independent of the fishery, which can be applied to control sites as well as to sites fished in various planned ways.

This research effort should be based on the assumption that differing types of fishery may depend on stocks that are organized by qualitatively different rules. It should not assume that tropical multispecies systems are similar to one another and fundamentally different from temperate systems.

Characteristics of Contemporary Fisheries Management in Southeast Asia

The problem

The Director of Fisheries in a typical tropical country faces several issues—increasing fish production, reducing conflicts, minimizing over-capacity and economic waste, and reducing threats to existing fish catches from overfishing or activities other than fishing—for all of which he needs scientific advice.

Tropical countries generally need more fish to improve the food supply of their existing populations, and to meet the demands of increasing populations. In many of these countries, the shortage of meat or consumer preferences mean that fish play a particularly important role in the diet. Under the new Law of the Sea, most nations also have to deal with requests for access to underutilized resources in their Exclusive Economic Zone by other countries. In the tropics, this particularly concerns tuna, which are not considered in this volume. There are several other fisheries where scientific advice is needed to deal with requests for access. These include access by Taiwanese vessels to the northern Australian shelf or by Thai trawlers to waters off Burma and Bangladesh.

Clashes between traditional artisanal fisheries and fishermen using more modern gear (especially trawls) have occurred in many countries, and have involved physical violence in India, Indonesia and elsewhere. Some countries have tackled this problem by banning the use of the more powerful gear within certain distances of the coast (or within certain depths). This has not always been effective and Indonesia has found it necessary to introduce a complete ban on trawling in the waters of Java and Sumatra.

Over-capacity results in falling catches of individual fishermen, falling or disappearing profits, and excessive costs of fuel, etc. There is also higher employment than is strictly necessary, but this is often held to be, if anything, a positive feature. This may be true in the short run, so long as there is no alternative employment, but it is probably not true in the long run, and in any case results in reduced income to the individual fisherman. Over-capacity, with its attendant social and economic problems, is a growing problem in many parts of Southeast Asia and other tropical regions. It is particularly serious in the Gulf of Thailand and in several shrimp fisheries; it is also occurring in a number of small-scale fisheries due to sheer numbers of fishermen, even though only simple gears are being used.

There is little evidence that total catches have fallen in tropical fisheries due to overfishing. Though catches of individual species have dropped, these often have been made up by increases of other species. However, there are considerable uncertainties about the trends in total catch from the Gulf of Thailand (omitting catches by Thai vessels outside the Gulf).

Other users of the sea and of the coastal areas threaten fisheries in several tropical areas. Mangroves provide nursery areas for a variety of valuable species, including several shrimps. In most countries with extensive areas of mangroves these are being reduced by cutting for charcoal, reclamation for aquaculture and other uses. In several countries, noticeably the Philippines, coral reefs are being rapidly destroyed by dynamite and other destructive fishing methods, and by removal of coral for building and other uses. Pollution, in the narrow sense, is at present not so urgent a problem, but industrial and domestic wastes threaten some mollusc fisheries, and these are also threatened by dredging for tin and zinc, especially in Malaysia.

Advice required

In tackling these problems, the Director of Fisheries needs scientific advice. The questions he needs to have answered are discussed in the paper by Gulland (this vol.), and are summarized below, but it was stressed that in practice some of these questions are not asked and decisions are taken without appropriate scientific advice. Some of the failures in fisheries management and in the planning and direction of fishery research, are due to failure on the part of the administrator to understand what information could and should be provided by the scientist; and by the scientists, to appreciate what is needed by the administrator.

To plan the expansion of catches, the first priority is to identify the resources, and then to give rough estimates of how much can be caught, and what effort (e.g., numbers of vessels of different types) would be required. Also important is to know the likely impact of such development on existing fisheries, including fisheries on different species which might be affected because of interspecific reactions. It often may not be obvious which these fisheries are, and some caution may be advisable pending research.

Problems of over-capacity and overfishing require similar advice, but in more detail. Qualitative statements (e.g., it should be possible to at least double the yield from a given fishery) need to be replaced by more quantitative advice, such as the presentation of the curve of yield as a function of fishing effort, with the estimated position of the present fishery on that curve. The advice needs to be matched to those management actions that are likely to be practicable, so that the immediate and long-term effects can be predicted.

Other problems, such as pollution and land-reclamation, also require quantitative advice, giving estimates of expected losses to fisheries. Since government action on these matters usually involves striking a balance between different interests (e.g., shrimp fishermen and those cutting mangroves for charcoal), fishery interests must be provided with persuasive and clearly argued estimates of the amount they stand to lose.

The ability to provide advice

The ability of scientists in tropical areas to answer the types of questions described in the previous section has to be matched against the quality required in the answers. Ideally, any answers should be based on a comprehensive understanding of how the natural system works, and how it will be affected by fishing. We are still a long way from having that understanding for any tropical fishery.

In practice, much lower standards may be adequate. The Director of Fisheries only needs advice in sufficient detail, and with sufficient precision to help him come to policy decisions, though the advice must be reliable, and must be presented in such a way that it can be readily understood and accepted by non-scientists. At this level it is possible now to give advice for several tropical fisheries, and advice for many more could be provided without having to carry out fundamentally new research. The problems in any given fishery usually relate more to the absence of necessary data, the lack of people trained to analyze these data in a suitable way, and poor communication between the Director of Fisheries and those who should be giving him biological advice.

The most obvious shortcoming of information in most fisheries is in the basic statistics. Even the total catch is not well known for some fisheries; for example, the trends in total catch of demersal fish from the Gulf of Thailand (as opposed to the catch by Thai trawlers, including fish from waters outside the Gulf) are far from clear. Effort data are scarce and are often of little use because the figures are not matched to the catches taken by that effort. Even quite crude effort data (e.g., on numbers of vessels) can be useful in monitoring a fishery provided the data refer to a fairly uniform group of vessels and give both the catch and the effort used in taking that catch. It is important that the scientist using these data has a good understanding of how the fishery operates and is aware of any changes in fishery practice that might affect the relation of catch per unit efort to the abundance of the stock. High precision is much less important. Figures that could be known to be within ± 20% of the actual catch would be usually fully acceptable and a great improvement on the situation in many fisheries. This might be better achieved by sampling schemes than by attempting a complete census.

Biological information, in relation to current fishery problems, seems less important. One aspect, which is growing in importance with the establishment of national zones of fishery jurisdiction, is the better identification of stock structure, and of those stocks that are shared between different countries. This is suspected to be most important for the more active pelagic fish, e.g., *Scomberomorus*, and tagging of these fish might be a fruitful activity. In addition, studies are needed of the degree to which movements of a few species, (e.g., the large active predators) can cause significant interaction between multispecies fisheries operating in different areas, even when there is no interchange for most of the species.

As regards the analysis of data, it was felt that despite their obvious theoretical limitations, simple methods, such as relating catch per unit effort to effort, are adequate for dealing with most of the urgent fisheries problems. For example, from the simple analysis of the commercial catches and the survey data in the Gulf of Thailand, it is certain that there are far too many trawlers operating and, until effective action is taken to reduce the numbers, it is not important to be able to determine, from more sophisticated analysis, whether the optimum number should be, say, 40% or 60% of the present number.

A greater problem is a lack of scientists in many countries who can do even these simple analyses and present the results to the policymakers in a convincing way. In part, this arises from structural problems within the countries.

The scientist in a government fishery research institute often has a lower status and is more poorly paid than scientists in universities or other people in government service. The institutes therefore tend to have a relatively rapid turnover of relatively inexperienced staff and lack scientific leadership. Their staffs also tend to be poorly trained, particularly in the specialized work of providing advice to managers. While training must be a long-term program, reaching back to the universities, some immediate improvement could be achieved.

Workshops, preferably at the regional level, to which scientists working on particular fisheries could bring their data for review and analysis, would be useful. By bringing in one or two outside experts who are familiar with new ideas and up-to-date methods, such workshops could be effective in training and in producing new assessments. Several such workshops have been conducted in Southeast Asia, notably by the FAO/UNDP South China Sea Programme, but they need to be continued regularly. To ensure that those working on the fisheries, rather than more senior persons, attend the workshops will probably necessitate external support for some time.

The proceedings of such workshops could go some way towards increasing the amount of published information on tropical fisheries. At present, too much of the data being collected remains unpublished and many scientists are not aware of the necessity of getting them published, or of the procedures involved in bringing information into a publishable form. It was suggested that assistance in technical editing and scientifically reviewing the work could be most helpful.

Improving publications, including better use of outside journals, is only one aspect of the wider problem of improving communication. Within the scientific community there is the need for better awareness of what data are available, and improved access to unpublished data, other than confidential data (e.g., relating to the operation of an individual commercial fishing company). A particularly urgent need is for improved communication between the scientists and decision-makers inside and outside the government fishery department. One suggestion was for workshops or seminars in which computer modelling groups, such as at the University of British Columbia, could be used to make senior administrators familiar with the issues involved in resource management and development. It was felt that it might not be appropriate yet (for example, because of the risks of uncritical acceptance of computer-based output of untried models) to use these techniques to provide advice on specific fisheries, but they could be extremely valuable as part of a seminar on management principles.

Summary Comments and Recommendations

This Workshop began with the expectation that there might emerge from the deliberations some recipes, or at least some rules of thumb, for guiding the management of multispecies fisheries in tropical waters. Some hopes were also expressed by the convenors that a variety of computer programs for multispecies models might be run comparatively, and critically assessed for their utility to managers. What actually unfolded was much less sophisticated, and far more suited to current circumstances.

The first sobering thought was that any basis for management, any "model" in the sense of modelers, must be tested with data if it is to be applied with confidence. For Southeast Asia and perhaps for many other parts of the world, there is simply not enough reliable information at hand to test the utility of complex multispecies models.

Statistics of catch with approximations of species composition, accompanied by related statistics of effort, are a bare minimum for even the crudest kinds of multispecies assessments. Beyond that minimum, it is useful to know the vital statistics of the populations of the major species, their age composition and growth rates and, ideally, to have an appreciation of the factors that determine their recruitment. But such information is not at hand, nor is there an immediate prospect that it will be.

One might hope that ecological theory could be brought to the rescue, but while it is a rich source of ideas for multispecies research, it is of only general and qualitative value as a guide to management. The central question of whether biotic or abiotic factors regulate populations and community structure is as controversial today as it has been for the past 40 years. It is not even certain from ecological theory whether populations released from the pressures of fishing return to their unharvested states by predictable paths.

It is a matter of necessity to approach the question of multispecies management with less rigor and more pragmatism. Fortunately, there are good prospects that single-species assessments coupled with crude appreciations of species compositions can provide a valuable basis for management decisions for the time being. For any species, a substantial decline in average length and/or a change in maturity with age or length associated with a decreasing catch per unit effort are sure signs of the effects of exploitation. When total catch declines despite an increase in effort, there is evidence that the maximum sustainable yield has been exceeded. When these warning signs persist and are followed by sharp changes in species composition, it is apparent that harvesting is at levels which demand close attention by management.

The second sobering thought was that fisheries resources will not be managed on a biological basis unless biologists can explain what should be done and be heard by administrators. Explaining the intricacies of multispecies interactions involves venturing beyond current levels of understanding and should probably not be attempted. Preferable is to concentrate on the pattern of declining yields of preferred species and the relatively simple evaluation of the extent to which there is evidence of their overfishing. Even such straightforward presentations may not be appreciated, not because there is a lack of comprehension on the part of administrators, but rather because social and economic considerations concerning the fishermen and the fishing industry may seem to be overriding, at least in the short term.

In the long term, the case for sound biological management is compelling, for the welfare of the resource is fundamental to its usefulness. It is accordingly wise to press on with data acquisition and research activities that will lead to new levels of understanding as a basis for more effective management in the future. At the same time, it is important to build the human resource base that is the essential *modus operandi* of management. Until there is an adequate cadre of resident fisheries scientists in Southeast Asia and other tropical areas, adequately equipped with the necessary facilities for scientific research and embedded in sympathetic administrations, it is not to be expected that management of fisheries will progress much beyond present levels.

The third, and final sobering thought, echoes a theme mentioned in the introduction–fisheries constitute only one element in a much broader social and economic context. The increases in the past two decades in the intensity of fisheries in Southeast Asia are part of sweeping changes associated with the transition embraced by the word "developing."

It cannot be denied that, in many respects, development is associated with better conditions for humanity, but it is equally undeniable that accelerated uses of natural resources pose profound questions of conflict for which there are no easy answers. Is a bay better used as a receptacle for industrial pollutants or as a rearing area for juvenile fishes? There are also profound questions of social adjustment, exacerbated by the pace of development in relation to the human life span. Who is to tell a 50-year old fisherman that he should become a factory worker?

While these considerations were not to the fore in the discussions, they were recognized as the basis for many of the problems that currently beset Southeast Asian fisheries. To some extent, perhaps to a large extent, the rational use of fisheries resources may depend on the success attached to the implementation of comprehensive plans for national development in which fisheries play a role that is consistent with their capacities. Meanwhile, the fisheries will continue to be prosecuted and the opportunities to observe their responses should be exploited to gain the understanding necessary for the future.

There are some bright spots in the overall pattern of events of the last two decades that should not be forgotten. In consequence of the many international and national initiatives related to fisheries, there is now an increasingly longer and more comprehensive series of statistics, a substantial body of resident expertise and a growing scientific literature concerning tropical fisheries. From these beginnings, it should be possible to build a lasting base for better biological management in the future. The recommendations of the discussion groups were developed in this frame of mind. The recommendations are in three groups: (1) data needs and requirements; (2) research priorities; (3) improving management.

Data Needs and Requirements

Many of the routine techniques of stock assessment assume the existence of a database that is commonly available from systematic recording and sampling of commercial catches. Completing this database for Southeast Asian fisheries has a high priority. The requirements are:

- Reliable catch by species and associated effort data.
- Length composition by species or, if appropriate, by groups of species. Where discarding of part of the catch at sea is known to occur, it will be necessary to sample discards for length composition as well as by species to enable conversion of length composition of landings to length composition of catch.

- Indices of abundance calculated from records of catch and effort, and expressed in units of catch per unit of standardized fishing effort. Research vessels or selected commercial vessels may be used for this purpose.
- Age composition of selected species as a basis for using standard techniques of assessment and for calibrating length-structured models.
- Related to these data requirements is the problem of obtaining satisfactory species identification. With the large number of species, special efforts are needed to provide field workers with easily used taxonomic aids.

It is emphasized that these data should be published, or kept in a readily accessible form. Data should be analyzed as they become available and backlogs of data currently at hand should be worked up.

To facilitate the attainment of these requirements it is recommended that:

— National fisheries agencies in tropical developing countries should give high priority to improving the coverage and reliability of statistical data concerning fisheries.

— International agencies should issue (or reissue in some cases) collections of literature on data gathering, particularly with regard to multispecies tropical fisheries.

— National and international agencies should accelerate programs for assisting species identification. Well-based, easy to use, taxonomic handbooks should be made readily available. Color photographs are particularly useful for such handbooks.

— International agencies should sponsor regional workshops for data analysis by fisheries scientists, and follow up with workshops involving both administrators and fisheries scientists. These workshops should stress the use of modern computing devices to facilitate data processing.

Research Priorities

Research activities should cover a spectrum that includes both short-term projects aimed at resolving current problems and long-term projects designed to build understanding. In large measure, these two kinds of activities are complementary and inseparable, and both should be components of a balanced and integrated fisheries research program.

In the short term, there is a high priority for development of stock assessment methodology through:

— Length-structured models. Because of the difficulties in routinely aging large numbers of fish in tropical waters, it is often not practical to use the age-structured models developed for use with temperate species. Research should be directed to the continued development, testing and dissemination of length-structured models for use in all aspects of stock assessment. A special program of age determination should also be carried out to compare results obtained from the two approaches as a means of validating the length-structured models.

— Relationships between yield, effort and species composition. Empirical investigations are required to relate yield, species composition and fishing effort directly. This might be done by comparing catches, species compositions and fishing effort at a number of different places that experience different levels of fishing effort and different mixes of fishing gears. Another approach would be to fish, either experimentally or commercially, in a particular study area. A specific geographic area could be identified and a planning group, composed of scientists, fishermen and, if appropriate, administrators, be set up to monitor the results of such a study as a basis for future fisheries policy.

— Resource identification. An important objective is to identify each geographical area within which stocks interact more strongly with each other than with stocks elsewhere. The intention is to define areas that can be treated as management units. The types of studies needed include morphological studies, tagging, electrophoretic measurements, identification of nursery areas and hydrographic studies. Particular attention should be given to the identification of migratory, as distinct from non-migratory, stocks and their impact on the structure of local communities.

— Selectivity and catchability of gears. In some areas trawling is impractical and fishing is by such gears as pots, traps or gill-nets, for which there may be no indices of catchability and selectivity. Estimation of the catchability and selectivity of such gears, as well as further work on trawl gears, would be an important research project.

— Experimental management. The most efficient way of answering many management questions is to undertake comparative studies of the consequences of different types of regulation. Ideally, these experiments should be done on a scale that is applicable directly to management. Advantage should also be taken of any "natural experiments", i.e., monitoring the effect of any major change in regulation or intensity of a fishery. In particular, all fisheries development projects should have built in "before and after" studies to enable assessment of their impacts.

For the long term, there are several potential research projects that could lead to better understanding. Many advances in theory and methodology will be made by workers outside the field of fisheries. Accordingly, the list below emphasizes research that is most immediately relevant to tropical fisheries and assumes that it will take place in a world-wide and general scientific context.

Research aimed at the better understanding of species interactions

A long-term, multidisciplinary research program is required to improve understanding of the ways in which species interact. A principal objective is to understand the ecological roles of the various species. This should provide a basis for understanding the processes involved when there are changes in species composition. The ultimate objective is to make useful predictions of the likely effects of exploitation on species composition.

This program should include:

— Stomach analyses and anatomical comparisons (e.g., of mouth shape, form of teeth, length of gut) aimed at defining the trophic positions of the various life stages of each species.

— Experimental manipulation of self-contained species assemblages in areas where it is possible to monitor any changes that might occur.

— Observations of feeding behavior, and predator avoidance behavior under natural conditions wherever this may prove practical.
— Studies of recruitment with particular reference to larval ecology and interrelations among species in nursery areas.

Research aimed at development of theory

Contemporary theory of multispecies fisheries is in an early but rapid stage of development. There is an urgent need for relatively simple models that capture the essential properties of multispecies systems, and that can be tested for their adequacy as bases for management. The development of models should go hand in hand with collection of ecological information and experimental management. Included should be:
— Evaluation of the validity of the existing multispecies models in situations where there is a suitable database.
— Sensitivity analyses of existing models both for single and multispecies assessments.
— Development of an ensemble of simple multispecies models for possible application to tropical fisheries.
— Development of models based on ecological understanding that enable prediction of changes in species composition and yield under conditions for which there is not yet sufficient experience to make predictions on an empirical basis. Such models might include a deductive exploration of species successions or total yields to be expected from different food-web structures and harvesting regimes.
— Investigation and modelling of the recruitment of one species as a function of the abundance of that species, the abundance of other species in the same ecosystem and of environmental factors. Both spatial and seasonal variation on appropriate scales should be considered. Reef systems, because they are replicated, may have some advantages for such studies, especially since there are possibilities for imposing different regimes of harvesting on a controllable scale. The northern shelf of Australia also offers considerable potential for such a study.

Research projects and reviews aimed at some specific problems

— Impact of fishing on species composition. Present ecological information on multispecies fisheries is not sufficient to guide the manipulation of fisheries to attain or retain a desirable species composition. The purpose of this empirical study would be to provide guidelines for anticipating deleterious changes which might occur in the catch composition of specific fisheries and suggest means for avoiding those changes. The study might also suggest means for inducing desirable changes in species composition and could be useful for predicting changes in total yields that occur because of changes in species composition. The study would catalog changes in species compositions which have occurred under different environmental and fishing conditions from records of catch for fisheries in different locations. Catch statistics of local governments would be supplemented by experimental fishing to give unbiased estimates of species composition. To the extent that the study involves spatial variation, it would be important to have information on stocks that are geographically independent in terms of fishing impacts, but are still part of the same system in terms of ecology. The study should be accompanied by a survey to identify kinds of changes in catch composition that have sufficient economic or social impact to merit attention.

It would be necessary for participating agencies to structure their catch statistics to fit the needs of the study. This should be done by workshops to standardize catch and effort statistics, and to ensure identification to the new level required for assessments of species composition, particularly so that fish with different food habits are distinguished in records. Operationally, this is a major step and would require adequate investigation before the workshops.

Other materials that would be useful for workshops would include a fully developed case study from a fishery that has experienced a change in species composition, to communicate the need for the study and to clarify the kinds of stock changes the study is intended to identify. A microcomputer model might be used to demonstrate the kinds of processes involved and make the contents of the study more graphic to those who participate. Once the study was underway, it would be important to have rapid central processing of information and feedback to participants to maintain motivation.

— Conflicts between different sectors. The purpose would be to provide information that bears upon conflicts among users of fish stocks, such as can occur between domestic or foreign trawlers and small-scale fishermen in coastal fisheries. Initial data would include the species and life stages that are taken by existing commercial (trawl) small-scale fisheries to determine the extent of competition for the same stocks. The consequences of new trawling restrictions, as are occurring in Indonesia, should also be monitored.

— The impact of habitat degradation upon fisheries. A synthesis of existing information is needed on the impact of human activities that affect fisheries by damaging habitats. In some cases, fishing activities are responsible (e.g., fishing with dynamite and disturbance of bottoms by trawling), but in many other cases (e.g., use of reefs for construction material, deforestation, mining, release of industrial effluents and other pollutants, destruction of mangrove forests) non-fishing activities act detrimentally. The information assembled would be useful in evaluating tradeoffs for those making decisions on resource use or investments that influence fisheries.

— Fisheries intensification and its implications. A review of the kinds of fisheries intensification that increase total catch and those that do not would be useful to those forming development plans or considering fisheries investments. In some circumstances, increasing effort or efficiency (e.g., mechanization in a fully exploited small-scale fishery) should not be expected to increase total yield. Other kinds of development may intensify exploitation of underexploited species or areas and increase total yield. Still other developments (e.g., artificial reefs) may increase biological production, thereby increasing yield. In a related vein, reservoirs often lack elements of a fish community necessary to fully channel biological production into fish yields. Serious social and economic dislocations can occur when fish yields from a reservoir decline after fishermen have become dependent on an initial peak in fish production. The introduction of herbivorous or sediment-feeding fish, such as tilapia or carp, can often lead to significant increases in fish production. A manual concerned with introducing new species to reservoirs could be helpful.

Improving Management

It was repeatedly brought out in discussions that at present in Southeast Asia there is virtually no biological basis to the management of fisheries. The necessary information is often lacking. Where it is available, it may not have been analyzed. Where it has been analyzed and presented to administrators, it may not have led to regulations. Where regulations have been imposed, they may not have been enforced. In short, there is much that can be done to improve the management of Southeast Asian fisheries.

Perhaps the most serious obstacle to effective management stems from the lack of communication between scientists, administrators, decision-makers and fishermen. Fishermen may not be convinced that it would be in their own long-term interest to cooperate in measures to manage the resource. Administrators or decision-makers are rarely influenced by the fact that too much fishing effort is an inefficient use of resources. They may be prepared to take action only if they can be convinced that a particular resource is about to collapse. Scientists do not always appreciate that socio-economic factors may have to take precedence in the short term over the long-term objective of attaining a maximum biological yield.

It is recommended, therefore, that "dialogs" be held to promote better understanding among scientists, decision-makers and fishermen. The particular need for improved communication may be different in different regions. Examples are:

— Between different government departments in situations where the division of responsibility leads to conflicting objectives.

— Between fishermen and scientists in situations where a resource is threatened, and where this threat is not obvious to the fishermen themselves. Sometimes it may be possible to communicate directly with fishermen's representatives. In the case of artisanal fisheries, however, it may be necessary to set up an educational program with talks and films to communicate a particular message.

— Between governments in situations where there is a shared resource.

To a very large extent, management of fisheries in Southeast Asia has been more concerned with reconciling conflicts between different socio-economic groups than with attaining long-term goals, such as maximum biological yields.

However, where it is recognized that there is too much fishing and an inefficient use of resources, it should be made clear by scientists to decision-makers that:

— Legislation prohibiting or proscribing certain courses of action is not likely to be effective if there is no effective means of enforcement.

— All measures which subsidize the fishery are likely to be counter-productive in the long term since they cause fishing effort to increase, and further reduce the net return from the fishery. In extreme cases they may lead to stock collapse.
— In extreme cases, for example where the artisanal fishery supports millions of people, the only effective measure in the long term, will be to encourage the development of activities that will attract fishermen into non-fishing activities.

In the past two decades there have been notable advances in knowledge of Southeast Asian fisheries. As well, through the efforts of national governments and international agencies, there have been conspicuous improvements in the capacities in the region for fisheries management. Given a slower pace of development, it might well have transpired that Southeast Asian fisheries would have been managed against a sufficient background of scientific information to have been an example to other parts of the world. In the event, however, fisheries have expanded at such a rate that biological studies have lagged far behind what is required for informed management. In consequence, the fisheries of the region today are evidently exploited at very high levels; there is little prospect of respite, and every expectation that current yields of many preferred species will not be sustained.

In these circumstances, it is important that both national and international agencies should maintain their present levels of fisheries scientific activity and augment them, especially along some of the lines that were suggested in these discussions.

References

Brooks, J.L. and S.I. Dodson. 1965. Predation, body size, and composition of plankton. Science 150: 28-35.

Brown, B.E., J.A. Brennan, M.D. Grosslein, E.G. Heyerdahl and R.C. Hennemuth. 1976. The effect of fishing on the marine fish biomass in the northwest Atlantic from the Gulf of Maine to Cape Hatteras. ICNAF (Int. Comm. Northwest Atl. Fish.) Res. Bull. 12: 49-68.

Buzeta, R.B., Editor. 1978. Report of the workshop on the demersal resources of the Sunda Shelf, 31 October-4 November 1977, Penang, Malaysia. Part I. SCS/GEN/77/12. 51 p. South China Sea Fisheries Development and Coordinating Programme, Manila.

Connell, J.H. 1978. Diversity in tropical rain forests and coral reefs. Science 199: 1302-1310.

Domain, F. 1979. The demersal resources (fish), p. 77-119. *In* J.P. Troadee and S. Garcia (eds.) The fish resources of the Gulf of Guinea from Angola to Mauritania. FAO Fish. Tech. Pap. No. 186.

FAO. 1979. Report of the FAO/IOP workshop on the fishery resources of the western Indian Ocean south of the equator, 23 October-4 November 1978, Mahé, Seychelles. IOFC/DEV/79/45. 102 p. Development Report, Indian Ocean Programme, FAO, Rome.

Hall, D.J., S.T. Threlkeld, C.W. Burns and P.H. Crowley. 1976. The size-efficiency hypothesis and the size structure of zooplankton communities. Annu. Rev. Ecol. Syst. 7: 177-208.

Hempel, G., Editor. 1978. North sea fish stocks—recent changes and their causes. A symposium held in Aarhus, 9-12 July 1975. Rapp. P.-V. Reun. Cons. Inst. Explor. Mer. 172. 449 p.

Lawrie, A.H. 1978. The fish community of Lake Superior. J. Great Lakes Res. Int. Assoc. Great Lakes Res. 4: 513-549.

Marten, G.G. 1979a. Impact of fishing on the inshore fishery of Lake Victoria (East Africa). J. Fish. Res. Board Can. 36(8): 891-900.

Marten, G.G. 1979b. Predator removal: effect on fisheries yields in Lake Victoria (East Africa). Science 203: 646-648.

Munro, J.L. 1977. Actual and potential fish production from the coralline shelves of the Caribbean Sea. FAO Fish. Rep. 200: 301-321.

Pauly, D. 1979. Theory and management of tropical multispecies stocks: a review, with emphasis on the Southeast Asian demersal fisheries. ICLARM Studies and Reviews 1, 35 p. International Center for Living Aquatic Resources Management, Manila.

SCS. 1976. Report of the workshop on the fishery resources of the Malacca Strait, 29 March-2 April 1976, Jakarta. Part II. SCS/GEN/76/6. 113 p. South China Sea Fisheries Development and Coordinating Programme, Manila.

Sale, P.F. 1979. Habitat partitioning and competition in fish communities, p. 323-331. *In* H. Clepper (ed.) Predator-prey systems in fisheries management. Sport Fishing Institute, Washington, D.C.

Sale, P.F. 1980. The ecology of fishes on coral reefs. Oceanogr. Mar. Biol. Annu. Rev. 18: 367-421.

Smith, S.H. 1968. Species succession and fishery exploitation in the Great Lakes. J. Fish. Res. Board Can. 25(4): 667-693.

Swingle, H.S. 1950. Relationships and dynamics of balanced and unbalanced fish populations. Ala. Polytech. Inst. Agric. Exp. Stn. Bull. 274. 74 p.

Werner, E.E. and D.J. Hall. 1976. Niche shifts in sunfishes: experimental evidence and significance. Science 191: 404-406.

Werner, E.E. and D.J. Hall. 1977. Competition and habitat shift in two sunfishes (Centrarchidae). Ecology 58(4): 869-876.

Wiens, J.A. 1976. Population responses to patchy environments. Annu. Rev. Ecol. Syst. 7: 81-120.

Geographic Index

Name Index

Included here are exclusively the name of *persons*, i.e., of authors, persons mentioned in acknowledgements and discussants; name of persons that have become concepts (e.g., *Lotka-Volterra* equations, *Schaefer* Model) are not included here and must be accessed through the text.

Miller, J.M. 243
Mills, E.L. 204

Taxonomic Index

Included here are all references to groups of plants and animals with a clear taxonomic status. Where possible the main entries are given under the scientific names, to which the common names have been cross-indexed.

List of Participants

Mr. William Gazey
Institute of Animal Resource Ecology
University of British Columbia
2075 Westbrook Mall
Vancouver, B.C. V6T 1W5
Canada

*Mr. D. Gilbert***
Fisheries Research Division
P.O. Box 297
Wellington
New Zealand

Dr. John A. Gulland
Chief, Marine Resources Services
Fisheries Department (FIRI)
FAO of the United Nations
Via delle Terme di Caracalla
00100 Rome
Italy

*Dr. Norman G. Hall***
West Australia Marine Research Laboratory
P.O. Box 20
North Beach
Western Australia 6020
Australia

Dr. Maurice K. James
Lecturer in Systems Engineering
Department of Civil and Systems Engineering
James Cook University of North Queensland
Townsville, Queensland 4811
Australia

Mr. Rodney Jones
Department of Agriculture and Fisheries for Scotland
Marine Laboratory
P.O. Box 101
Victoria Road, Aberdeen AB9 8DB
Great Britain

Dr. Geoffrey Kirkwood
CSIRO Marine Laboratories
Division of Fisheries Research
P.O. Box 21
Cronulla, N.S.W. 2230
Australia

Professor Peter A. Larkin
(Workshop Chairman)
Institute of Animal Resource Ecology
University of British Columbia
2075 Westbrook Mall
Vancouver, BC V6T 1W5
Canada

Dr. Jacek Majkowski
CSIRO Marine Laboratories
Division of Fisheries Research
P.O. Box 21
Cronulla, N.S.W. 2230
Australia

*Mr. John C. Marr**
P.O. Box 772
Rancho California
Temecula, California 92390
U.S.A.

Dr. Gerald C. Marten
Research Associate
Environment and Policy Institute
East-West Center
1777 East-West Road
Honolulu
Hawaii 96848
U.S.A.

Dr. John L. Munro[a]
Chief Biologist
Fishery Research Branch
Fishery Division, DPI
P.O. Box 2417
Konedobu
Papua New Guinea

Dr. Garth I. Murphy
(Workshop Convener)
Senior Principal Research Scientist
CSIRO Marine Laboratories
Division of Fisheries Research
P.O. Box 21
Cronulla, N.S.W. 2230
Australia

*Dr. Richard A. Neal***
Deputy Director-General
ICLARM, MCC P.O. Box 1501
Makati, Metro Manila
Philippines

Dr. Daniel Pauly
(Workshop Technical Secretary)
Associate Scientist
ICLARM, MCC P.O. Box 1501
Makati, Metro Manila
Philippines

*Dr. Jeffrey Polovina**
NMFS Southeast Fisheries Center
Honolulu Laboratory
P.O. Box 3830
Honolulu, Hawaii 96812
U.S.A.

Dr. Keith Sainsbury
Research Scientist
CSIRO Marine Laboratories
Division of Fisheries Research
P.O. Box 21
Cronulla, N.S.W. 2230
Australia

Dr. Peter F. Sale
School of Biological Sciences
The University of Sydney
Sydney, N.S.W. 2006
Australia

Mr. Arthur C. Simpson
106 Fambridge Road
Malden, Essex CM9 6BG
Great Britain

*Professor Kevin P. Stark**
Head, Department of Civil and Systems Engineering
James Cook University of North Queensland
Townsville, Queensland 4811
Australia

---

[a]Present address: ICLARM, MCC P.O. Box 1501, Makati, Metro Manila, Philippines
*Absent authors and co-authors
**Observers